EMPLOYEES AND EMPLOYERS IN SERVICE ORGANIZATIONS

Emerging Challenges and Opportunities

T0330849

21st Century Business Management

EMPLOYEES AND EMPLOYERS IN SERVICE ORGANIZATIONS

Emerging Challenges and Opportunities

Edited by
Arvind K. Birdie, PhD

AAP APPLE
ACADEMIC
PRESS

Apple Academic Press Inc.
3333 Mistwell Crescent
Oakville, ON L6L 0A2 Canada

Apple Academic Press Inc.
9 Spinnaker Way
Waretown, NJ 08758 USA

©2017 by Apple Academic Press, Inc.

First issued in paperback 2021

Exclusive worldwide distribution by CRC Press, a member of Taylor & Francis Group
No claim to original U.S. Government works

ISBN 13: 978-1-77-463694-7 (pbk)
ISBN 13: 978-1-77-188473-0 (hbk)

Library and Archives Canada Cataloguing in Publication

Employees and employers in service organizations : emerging challenges and opportunities / edited by Arvind K. Birdie, PhD.
(21st century business management book series)
Includes bibliographical references and index.
Issued in print and electronic formats.
ISBN 978-1-77188-473-0 (hardcover).--ISBN 978-1-315-36585-5 (PDF)
1. Service industries--Personnel management. 2. Service industries--Employees. I. Birdie, Arvind K., author, editor II. Series: 21st century business management book series

| HD9980.5.E48 2017 | 658.3 | C2017-900748-3 | C2017-900749-1 |

Library of Congress Cataloging-in-Publication Data

Names: Birdie, Arvind K., editor.
Title: Employees and employers in service organizations : emerging challenges and opportunities / editor, Arvind K. Birdie, PhD.
Description: Toronto ; New Jersey : Apple Academic Press, [2017] | Series: 21st century business management | Includes bibliographical references and index.
Identifiers: LCCN 2017002947 (print) | LCCN 2017008012 (ebook) | ISBN 9781771884730 (hardcover : alk. paper) | ISBN 9781315365855 (ebook)
Subjects: LCSH: Service industries--Employees. | Service industries--Personnel management.
Classification: LCC HD9980.5 .E467 2017 (print) | LCC HD9980.5 (ebook) | DDC 658.3--dc23
LC record available at https://lccn.loc.gov/2017002947

Apple Academic Press also publishes its books in a variety of electronic formats. Some content that appears in print may not be available in electronic format. For information about Apple Academic Press products, visit our website at **www.appleacademicpress.com** and the CRC Press website at **www.crcpress.com**

Dedicated to the love and joy of my family—my parents, Gagan, Permendra and Shaurya for being my lifelong lighthouse and my safe harbor

ABOUT THE EDITOR

Arvind K. Birdie, PhD

Arvind K. Birdie, PhD, is Program Director, PGDM and Associate Professor at Vedatya Institute, Gurgaon, India. Dr. Arvind has been consistently recognized for her teaching abilities. For more than fourteen years she has taken post-graduate and graduate courses in management. As an avid reader her strength lies in teaching various interdisciplinary subjects with equal ease. In addition to academic teaching and training, she has organized management development programs for corporate and academicians. She is a regular presenter at various international and national conferences, and she has published papers in refereed journals. Her areas of interest include leadership, work-life balance, virtual work, and positive psychology. Dr. Birdie is the recipient of the Prof. Mrs. Manju Thakur Memorial Award 2016 for Innovative Contributions in Research/Test Construction/Book Publication for her book *Organizational Behavior and Virtual Work: Concepts and Analytical Approaches* (Apple Academic Press, 2016). The award was presented during the 52nd National and 21st International Conference of the Indian Academy of Applied Psychology, held at the Department of Psychology, University of Rajasthan, Jaipur, February 23rd–25th, 2017. She enjoys travelling and spending time with her family in free time.

CONTENTS

ABOUT THE 21ST CENTURY BUSINESS MANAGEMENT BOOK SERIES

Series Editor:
Arvind K. Birdie, PhD
Program Director PGDM, and Associate Professor,
Vedataya Institute, Gurgaon, India
Email: arvindgagan@gmail.com

CURRENT BOOKS IN THE SERIES

Employees and Employers in Service Organizations:
Emerging Challenges and Opportunities
Editor: Arvind K. Birdie, PhD

The Future of Organizations: Workplace Issues and Practices
Editor: Arvind K. Birdie, PhD

Other topics/volumes are planned on these topics:

1. Globalization and Emerging Leadership
2. Positive Psychology and Today's Organizations
3. Changing Consumer Behavior and Organizations
4. The Impact of Technological Advancement on Organizations
5. Emerging Employer and Employee Relations
6. Designing Future Organizations and Emerging Sectors
7. Aging in South Asia and the Impact on Emerging Businesses
8. Issues in Intercultural Management
9. The Role of Spirituality in Management
10. Increasing Workforce Diversity in Organizations
11. Creating Innovation in Organizations
12. Purchasing Power and Happiness in Customers

LIST OF CONTRIBUTORS

Arvind K. Birdie
IIMT School of Management, Vedatya Institute, Gurgaon, India. E-mail: arvindgagan@gmail.com.

Samo Bobek
Faculty of Economics and Business, University of Maribor, Razlagova, Maribor, Slovenia. E-mail: samo.bobek@um.si.

Jelena Horvat
University of Zagreb, Faculty of Organization and Informatics, Varaždin Pavlinska 2, HR 42000 Varaždin, Croatia. E-mail: jelena.horvat@foi.hr.

Lalatendu Kesari Jena
Department of Humanities and Social Sciences, IIT Kharagpur, India. E-mail: lkjena@iitkgp.ac.in.

Sanjeev Kumar Jha
Educational Survey Division, National Council of Educational Research and Training, New Delhi, India. E-mail: krjhasanjeev03@gmail.com.

Navdeep Kaur Kular
Vedatya Institute, Gurgaon, India. E-mail:novikular@rediffmail.com.

Vandana Madhavkumar
GRG School of Management Studies, Coimbatore, Tamil Nadu, India. E-mail: vandana@grgsms.com.

Sasmita Misra
Narsee Monji Institute of Management Studies, India. E-mail: sasmita.misra@nmims.edu.

Atasi Mohanty
Centre for Educational Technology, IIT Kharagpur, India. E-mail: atasim@cet.iitkgp.ernet.in.

Durgamohan Musunuri
Professor & Head, Department of Management Studies, JSS Academy of Technical Education, NOIDA 201301, Uttar Pradesh, India. E-mail: durgamohan27@gmail.com.

Uma Nagarajan
Bhartiyar University, Chennai, India. E-mail: uma.anita@gmail.com.

Ranjan Pattnaik
Department of Humanities and Social Sciences, IIT Kharagpur, India. E-mail: dr.ranjanpattnaik@iitkgp.ac.in.

Rabindra Kumar Pradhan
Department of Humanities and Social Sciences, IIT Kharagpur, India. E-mail: rkpradhan@hss.iitkgp.ac.in.

Urban Sebjan
Faculty of Economics and Business, Raziagova, University of Maribor, Maribor, Slovenia. E-mail: urban.sebjan1@um.si.

Simona Sternad-Zabukovšek
Faculty of Economics and Business, University of Maribor, Razlagova, Maribor, Slovenia. E-mail: simona.sternad@um.si.

Polona Tomnic
Faculty of Economics and Business, Raziagova, University of Maribor, Maribor, Slovenia. E-mail: polona.tominc@um.si.

Parul Wasan
Manav Rachna International University, Faridabad, Haryana, India. E-mail: parul.wasan@yahoo. co.in.

Girijesh Kumar Yadav
National Institute of Occupational Health, Indian Council of Medical Research, Ministry of Health & Family Welfare, Govt. of India. E-mail: girijeshkrydv@gmail.com.

LIST OF ABBREVIATIONS

APA	American Psychological Association
ATMs	automated teller machines
AVE	average variance extracted
BI	behavioral intention
BIS	business information systems
BRICS	Brazil, Russia, India, China, South Africa
CBRC	China Banking Regulatory Commission
CCMDs	career decision-making difficulties
CDM	career decision making
CEPR	Centre for Economic and Policy Research
CR	composite reliability
DU	deemed to be universities
ERP	enterprise resource planning
EU	European Union
EU-OSHA	European Union for Occupational safety and Health
GDP	gross domestic product
GoF	goodness-of-fit
GSS	General Social Survey
HEIs	higher educational institutions
HES	Higher Education System
IBOS	International Business Owners Survey
ICT	information and communication technology
IS	information systems
IT	information technology
KM	knowledge management
LES	leadership effectiveness scale
MIS	management information systems
NCR	National Capital Region
NIOSH	National Institute for Occupational Safety and Health
OPC	organizational-process characteristics
PCIL	personal characteristics and information literacy
PLS	partial least squares
POB	positive organizational behaviour

PsyCap	psychological capital
PU	perceived usefulness
PwC	Price Waterhouse Cooper
SCT	system and technology characteristics
SEM	structural equation modeling
SSA	Sarva Shiksha Abhiyan
STC	system and technological characteristics
TAM	technological acceptance model
TPB	theory of planned behaviour
TRA	theory of reasoned action
VUCA	volatile, uncertain, complex, and ambiguous
WLB	work–life balance

PREFACE

Organizations are changing at a fast pace in reaction to globalization and technological advancements, and new structures are emerging organizations. These changes are bringing changes in behavior in organizations. These days employees believe in less loyalty toward organizations in comparison to baby boomers. Generation Y believes in working smarter, not harder.

With aging in South Asia, where the major portion of the population is young, young entrepreneurs are creating wonders in the business world, with employees choosing boundaryless and protean careers. New concepts of flat structure for employers are becoming in demand for today's businesses to function in an ever changing environment digital world without boundaries.

Service industry in the past has metamorphosed into a revolution not only in the United States but is evident in all developed and developing countries also. At the dawn of the 21st century, all highly industrialized countries have become "service economies," at least when measured in terms of share of the workforce employed in service industries.

More and more companies are service oriented, and with globalization and technological advancement, expectations and skills required by the organizations are changing at a fast pace from multitasking, working 24×7, and there are emerging challenges and opportunities for organizational behaviorists, such as work–life balance, globalization, redefining leadership, contingent work force, stress management, telecommuting, work force diversity, ergonomics, life satisfaction/subjective well-being, etc.

The above-discussed concepts are new for employers and employees, and very few qualitative and quantitative research studies with practical implications for policymakers and employers have been performed. This book explores the upcoming challenges/opportunities for employees, and employers particularly in the orientation of service organizations in the 21st century. This book provides insights for implication and examines old theories of organizations in light of changing concepts as well as emerging concepts. The book explores the themes of work–life balance, spirituality, and emerging positive psychology concepts, such as psychological capital, knowledge management, mindfulness. These are a big change from the workforce generation of so-called Y Generation and its motivation, etc. The approach in the chapters is by reviewing and conducting empirical studies, round tables, and focus discussions with employees and employers of the service industry by the authors.

I would like to acknowledge with gratitude the support I received from the contributing authors. Their rich insights and shared perspectives have enabled us to bring a lot of unique concepts of interdisciplinary ways to understand today's business world. We are indebted to organizations and many industry practitioners who provided support and discussed their ideas with the authors. I would like to acknowledge the support received from my parent organization, Vedatya. I owe special thanks to my mentor Dr. Madhu Jain. My heartfelt thanks to Dr. Vinnie Jauhari, Director, IIMT School of Management for her inspiration and support.

I would like to express my deepest gratitude to my parents, in-laws, my husband—Gagan, and our wonderful son, Shauryavir, for having supported me all along. I would also like to thank my brother, Permendra Singh, for his encouragement and always being there for me. Without their patience, support, and motivation, this work would never have been completed. My family has always been a great anchor and always had a great role in all my accomplishments. Their being with me is the greatest joy, which enables me to make a meaningful contribution.

I deeply value the strong support I received from the extraordinary team at Apple Academic Press. A deep gratitude and my heartfelt thanks to Ashish Kumar, President, AAP for believing in me to initiate the series and constantly encouraging me to accomplish this work. I deeply value his guidance. My special and heartfelt thanks to Sandra Sickels, Vice President, Editorial and Marketing for support in terms of using her sensitivity and good eye to design cover page and flyers for the series. She was immensely helpful throughout the project from starting till final come out. I also express special thanks to Rakesh Kumar, for valuable support with index, referencing, suggested readings and production work, and Sheetal Kumar from editorial team.

To all these truly excellent people, and to many others, too, I offer my personal warm regards.

PART I
Generation Y

CHAPTER 1

PROTEAN CAREER AND BOUNDARYLESS CAREERS: THE CAREER ORIENTATION OF GENERATION Y

ARVIND K. BIRDIE[1] and VANDANA MADHAVKUMAR[2*]

[1]Department of Post Graduate Studies & MBA, IIMT School of Management, Vedatya Institute, Gurgaon, Haryana, India

[2]GRG School of Management Studies, Coimbatore, Tamil Nadu, India

*Corresponding author. E-mail: vandana@grgsms.com

CONTENTS

ABSTRACT

For today's organizations, right people are key differentiators that give them competitive advantage. Due to globalization, career contexts and work environments have changed. Workforce diversity has increased due to technological advancements and the use of outsourcing, part time, and temporary employees. In conjunction with environmental changes, individuals are also altering their career attitudes and behaviors. This comes as a response to the increase in lifespan (and thus working lives), changes in family structures (such as dual-career couples), and the growing number of individuals seeking to fulfill needs for personal learning, development, and growth. One of the most important challenges for HR managers is to acquire and retain the Gen Y employees (born between late 1970s and mid-1990s) who are likely to comprise the bulk of employees.

In times of frequently changing work environments with decreased job stability, it is often more possible for an employee to pursue his career with not one single organization during his or her lifetime but with multiple over time. The decline of the traditional organizational career requires new ways of viewing careers. Over the last decade, two new perspectives on careers have emerged and become popular in the organizational literature: the protean career and the boundaryless career. This chapter reviews this new concept of individual career development practiced by the Gen Y employees whose expectations and preferences are different from their predecessors.

1.1 INTRODUCTION

the path with a heart

—*Herb Shepard*

Traditionally, career is often described as an individual's work-related journey, often associated with the employing organization. Careers were seen as linear and occurring within stable hierarchical organizational structures (Levinson, 1978). This form of career, also referred to as the "organizational career," had limited job mobility as employees stayed with the same organization throughout their professional life. Individual careers in organizations were seen as predictable and secure; an individual entered an organization and strove to rise through the ranks in an attempt to reach higher positions with clearly defined boundaries. Individuals who performed satisfactorily could be expected to be promoted. An employee's loyalty was rewarded by an organization in the form of job security and lifetime employment. The implicit psychological contract, which refers to a set of mutual expectations between the individual and the employer (Hall & Mirvis, 1996), stipulated that as long as employees provided satisfactory work they could rely on long-term employment with the prospect of promotion in the organization (Baruch, 2006; Hansen, 1997). Career was the responsibility of the organization and employees were committed towards their organization (Sullivan & Baruch, 2009).

In contrast, contemporary careers appear to be more unpredictable, non-linear and vulnerable. Today organizations operate in a complex business environment characterized by stiff competition, disruptive technology advances, globalization and outsourcing (Cooper & Burke, 2002; Kuchinke & Park, 2012). Organizations are therefore no longer in a position to guarantee job security and stable career growth to employees. Employees do not remain with a single employer for their lifetime as the idea of lifetime employment with a single organization no longer holds. Where psychological contract between individual and organization used to be long-term and relational, it has become short term and transactional. The new psychological contract requires individuals to engage in continuous learning and to modify their work-related self-perceptions and identities (Baruch & Hall, 2004). In the contemporary world of work, organizations can expect employees to be loyal only as long as the employee's short-term expectations are met. In turn, individuals can expect organizations to be loyal only as long as their skills and performance fulfill the organization's current needs (Hall & Mirvis, 1996).

One consequence of the changed psychological contract is that employees' job insecurity has increased (Baruch, 2006). During the 20th century, employees might have expected the organization to plan and control their careers, but in an era of increased uncertainty they need to take greater responsibility for their own career development. Hansen (1997) emphasized that because individuals "can no longer rely on their work for security and stability, [they] will become self-directed persons who develop their own careers, gain respect for others and value difference. They will learn to expect change" (Hansen, 1997, p. 247). Employees will need to become increasingly adaptable and multiskilled, which implies that continuous professional development and learning how to learn will take on greater importance in individuals' careers. It is in this context that

Sullivan and Baruch (2009, p. 1543) define career as "an individual's work-related and other relevant experiences, both inside and outside of organizations that form a unique pattern over the individual's life span." The definition identifies mobility between jobs, employers, occupations, and industries (Sullivan & Baruch, 2009). Thus, a new approach to career emerged that transfers the responsibility and risk of managing careers from organizations to individuals. The traditional employee–employer contract has been replaced by transactional relationship (Frenandez & Enache, 2008; Rousseau, 1989) and less loyalty from both sides (Hall, 2002).

Considering these developments, a perspective on careers by Hall (1976a,b) and Arthur and Rousseau (1996) is characterized by a shift of responsibilities for career progress from organizations to individuals and by emphasizing on the individual's freedom and growth as a core value (cf. Hall, 1976a,b, 2002, 2004). The two key concepts that emerged within this career perspective were the protean career attitude (Hall, 1976a,b, 2004) and the boundaryless career attitude (Arthur, 1994; Arthur & Rousseau, 1996). A protean career attitude has been characterized as involving a broader perspective, a developmental progression, and viewing a career as a calling and a way to self-fulfillment (Hall, 2002). A person with a boundaryless career attitude is characterized by high mobility and prefers to navigate physically and/or psychologically across many organizations (Sullivan & Arthur, 2006).

The "protean career" argues that being self-driven and moving in consonance with one's own values is necessary for continuous personal transformation and to achieve psychological success (Hall, 1976a,b; Hall & Chandler, 2005, Sargent & Domberger, 2007). In such terms, protean career actors constantly reevaluate their relationship with the organization (Hall, 2002). This stance versus the organization is also reflected in the "boundaryless career" which emphasizes freedom and agency, and is in short, anything, but the traditional career designed by the organization (Arthur & Rousseau, 1996; Cadin et al., 2000; Sullivan & Arthur, 2006). Boundaryless careers can trespass physical boundaries (e.g., organizational or national boundaries) and psychological boundaries (e.g., defying organizational norms for personal reasons) (Peiperl & Jonsen, 2007; Sullivan & Arthur, 2006). Individuals combining the characteristics of both protean and boundaryless careers are suggested to be in the best position to adapt to changing contexts and achieve psychological success (Briscoe & Hall, 2006). Being the alternative to organizational careers, the boundaryless and protean career concepts refer to a wide range of possible careers.

1.1.1 PROTEAN CAREER

Individuals in order to survive in the current business environment termed as VUCA world (volatile, uncertain, complex, and ambiguous) need to become more self-reliant in managing their careers. This means knowing what they want from their careers, developing the skills/knowledge/network that is necessary to achieve their goals and being able to "change with change." These changes in career management gave rise to the term "protean career," first proposed by Hall (1976a,b) that describes a career approach that is not dependent on the organization but proactively managed by the individual. The origin of which comes from *Proteus*, the Greek sea-god who could change forms as the situation demanded. A more formal definition is provided below:

> *The protean career is driven by the person, not the organization, based on individually defined goals, encompassing the whole life space, and being driven by psychological success (rather than) objective measures of success such as pay, rank or power.*
>
> —*Briscoe*

The protean career therefore shifts the focus of career management to the individual while the organization's role is to provide employees with opportunities for growth and development. A protean career attitude has been characterized as involving a broader perspective and viewing career as a calling and a way to self-fulfillment (Hall, 2002). As Hall (1996c, p. 10) notes, protean career is "a contract with oneself, rather than with the organization," when individuals take responsibility for transforming their career path, in line with their personal aspirations (Grimland et al., 2011). Individuals with protean career orientation take care of their career management and demonstrate greater mobility, a more whole-life perspective and a developmental progression (Briscoe et al., 2006; Hall, 1976a,b, 2002). Hall (2002) alleges that the modern career growth involves work challenges, relationships, and lifelong learning—all of which are required for continued career success. Protean career approach is thus based on continuous learning. The goal of the protean career is subjective career success (Briscoe & Hall, 2006; Grimland et al., 2011; Hall, 1976a,b, 2002). The aim of the employee in the protean career is therefore to develop the skills and competencies that ensure employability in a changing work context.

Protean individuals value individual freedom and growth and define career success in terms of psychological criteria, such as the degree of job satisfaction, self-actualization, personal accomplishment, and a feeling of fulfillment through the pursuit of meaningful work (Hall & Chandler, 2005; Hall & Mirvis, 1996). This description contrasts with a more traditional view where career success is defined in terms of external criteria such as promotions, salary, and occupational status. Briscoe and Hall (2006) further argue that "the Protean Career Orientation does not imply particular behavior, such as job mobility, but rather it is a mindset about the career—more specifically an attitude toward the career that reflects freedom, self-direction, and making choices based on one's personal values" (Briscoe & Hall, 2006, p. 6). Therefore, as rightly put by Briscoe and Hall (2006), a protean career orientation reflects the extent to which an individual adopts such a perspective to their career. According to Hall and Chandler (2005), the extreme form of this protean career perspective would occur when the person's attitude toward his/her career reveals a sense of calling or awareness of purpose in his/her work.

In addition, Briscoe and Hall (2002) describe individuals with protean career as being

(1) values-driven, that is, making career decisions based on their own values as against the organization's values. The person's internal values guide them in managing their career and attain career success; and being

(2) self-directed, playing an independent role in managing their career and being proactive in terms of performance and learning.

(3) Further, individuals with protean career orientation will be proactive and independent and will not depend on external standards (Briscoe et al., 2006).

Therefore, there is a change in the way individuals approach career with professionals becoming more self-reliant, flexible, and mobile.

In operationalizing the protean career attitude, Briscoe et al. (2006) distinguished two key components, namely, (1) an attitude of self-directedness in terms of managing one's own career and (2) a value-driven attitude where the individual's own values, rather than the values of the organization, drive the career. The first component, namely Self-Directed Career Management is tied to the meta-competency of adaptability and is seen in individuals' ability to adapt to changing conditions and to take responsibility for their own career development. The second component, namely Values Driven, is tied to the meta-competency of self-awareness and is seen when individuals' personal values guide their careers and become the yardstick against which they measure their career success (Hall & Chandler, 2005).

Thus, individuals who hold protean career attitudes are intent upon using their own values (vs. organizational values for example) to guide their career ("values-driven") and take an independent role in managing their vocational behavior ("self-directed"). An individual who did not hold protean attitudes would be more likely to "borrow" external standards, as opposed to internally developed ones, and be more likely to seek external direction and assistance in behavioral career management as opposed to being more proactive and independent.

1.1.2   BOUNDARYLESS CAREER

Another term that emerged along with the concept of protean career is "boundaryless" career (Arthur, 1994; Arthur & Rousseau, 1996), which involves a sequence of job opportunities that go beyond single employment settings. A person with a boundaryless career mindset "navigates the changing work landscape by enacting a career characterized by different levels of physical and psychological movement" (Sullivan & Arthur, 2006, p. 9). It involves the breaking down of traditional boundaries (e.g., job boundaries of specialist functions and skills), organizational careers which progress independently of well-trodden career paths and the social boundaries separating work and family roles. A boundaryless career attitude is characterized by high mobility and a preference for navigating physically and/or psychologically across many organizations (Sullivan & Arthur, 2006).

The term boundaryless career was developed to provide a new perspective on old traditional career theories (Arthur, 2008). Boundaryless careers are seen as the interlinking boundaries of organizations and occupations, with other parts of people's lives. Boundaryless careers involve career opportunities that go beyond the boundary of a single employer (Arnold, 2011). Some of the trademarks of a boundaryless career include: transferrable skills, knowledge and abilities across multiple firms, personal identification with meaningful work, on the job learning and the development of multiple networks and peer learning relationships, and individual responsibility for career management.

A boundaryless career attitude too has two components—organizational mobility preference and a boundaryless mindset. Boundaryless career attitude is psychological and varies in the attitude that individuals hold toward initiating and pursuing work-related relationships across organizational boundaries. It does not necessarily imply physical or employment mobility. A person with a decidedly high "boundaryless" attitude toward working relationships across organizational boundaries may also be comfortable, even enthusiastic about creating and sustaining active relationships beyond organizational boundaries. However, a second important boundaryless career attitude is the inclination toward physically crossing organizational boundaries in employment mobility. Someone high in such an organizational mobility attitude would be comfortable with, or even prefer a career that played out across several employers.

Arthur and Rousseau (1996, p. 6) provide six different meanings: they involve movement across the boundaries of several employers, draw validation and marketability from outside one's present employer, are sustained by external networks and information, break traditional organizational assumptions about hierarchy and career advancement, involve rejecting existing career opportunities for personal or family reasons, and are based on the interpretations of the career actor who may perceive a boundaryless future regardless of structural constraints. These diverse career forms have one over reaching characteristic in common they all represent "the opposite of organizational careers—careers conceived to unfold in a single employment setting" (Arthur & Rousseau, 1996, p. 5).

1.1.3 THE PROTEAN AND BOUNDARYLESS MODELS OF CAREER

While most protean individuals exhibit more mobility and a learning orientation, mobility and learning have been identified as correlates of a protean career, but not necessary components of it. Therefore, an overlap between protean and boundaryless is to be expected as protean and boundaryless career attitudes are independent yet related constructs. That is, a person could display protean attitudes and make independent, inner-directed choices, yet not prefer cross boundary collaboration. In comparison, a person could embrace a boundaryless mindset, yet rely on one organization to develop and foster his or her career. As such myriad possibilities exist, these concepts are expected to impact one another to certain degrees in an individual's work experience; however, they may impact behavior in very different ways also. Most of the scholars studying these careers have studied them together.

A protean career has been characterized as (Hall, 1996c) involving greater mobility, a more whole-life perspective, and a developmental progression. Whether these latter dimensions relate to the protean career remains to be seen. In more recent renditions of the protean career model, Briscoe and Hall (2002) have characterized it as involving both a value-driven attitude and a self-directed attitude toward career management. Further, Briscoe et al. (2006) have developed 14-item scale to measure protean career orientation and 8-item scale to assess boundaryless mindset.

1.2   GEN Y AND PROTEAN CAREER ORIENTATION

The latest entrants into the workforce—Generation Y or Millennials are individuals born between 1980 and 1994, falling within the age range of 22–36 years, have attracted a lot of attention from scholars and recruiters. This is because their values, approach, expectations, attitude, and aspirations toward work and career are very distinct from their predecessors. They are set to outnumber the other generations in not so distant future. As they are very different from the other generations, managing, attracting, and retaining them is a challenge for organizations.

Devotion to one company is fast becoming an idea of the past with Gen Y (Erickson, 2008; Meier et al., 2010). It has also been reported that Gen Y members associate themselves less with the organization that employs them and more with the type of work which they perform (Lloyd, 2007). Gen Y employees change jobs frequently so are job hoppers and have no issues in changing employers (Hall, 2002). Therefore, the critical components needed to retain them are challenging work, job training, career advancement, and work environment (Smith, 2008; Terjesen's & Frey, 2008; Yahoo! Hot Jobs/Robert Half International, 2008). They no longer depend on organizations to manage their career and take responsibility of managing their own career. To advance in their careers, they may decide to leave organizations and occupations and make both upward and lateral career moves to gain more skills and experience. Gen Y employees dislike being stuck at one level for a long time and prefer learning and growing.

They would like to grow quickly so they prefer a job that recognizes performance and not tenure (Meier et al., 2010). It has also been shown that Gen Y expects all these traits in a job and will also do whatever it takes to find such a job. They have no problem moving on somewhere that will offer them what they want. Cruz (2007) explains that Millennials are inclined to change organizations, if they perceive better opportunities offering greater levels of appreciation. Sargent and Domberger (2007) and Dries et al. (2008) in their studies posit that the younger generation will be protean in their career orientation. Therefore, it may be proposed that Millennials or the Gen Y employees will demonstrate protean career approach. Though according to Reitman and Schneer (2003), MBA graduates enjoy both self-managed and promised (conventional) career path.

Previous literature available has explored career orientation among Gen Y members and finds the younger generation to be more protean in their career orientation. In a study, Hess and Jepsen (2009) state that the membership of a particular generational group and career stage exert some influence over how employees perceive their protean career obligations and how employees respond to different levels of Protean Career fulfillment. Sargent and Domberger (2007) examine both undergraduates with work experience and individuals in the early stage of their career. They investigate the development of the protean career orientation in the early career stages of students shaped by their personal values, as well as how the early career experiences influence protean orientation. They find protean career is identifiable in the cohort or adults early in their careers.

Hall (2004) in a study traces the link between the protean concept and the context of growing organizational restructuring, decentralization, and globalization and concludes with a suggestion for examining situations where people are pursuing their "path with a heart" with the intensity of a calling. Enache et al. (2008) in their study explore the relationship between boundaryless and protean career attitudes and psychological career success, within today's complex and ever changing organizational context and reveal that the relationship between value-driven inclination of protean career orientation and psychological career success is moderated by the individual's perceived value fit with his or her employing organization. Dries et al. (2008) examine whether four different generations (Silent Generation, Baby Boomers, Generation X, and Generation Y) hold different beliefs about career. They find the majority of participants still having "traditional" careers, although contrary to the belief Generation Y exhibit larger incongruity between career preferences and actual career situation as they show "old-fashioned" belief or preference about career. Dries et al. (2008) offers a possible explanation to the result that Generation Y generally is just dreaming about their future career as they have not yet having been confronted with career reality today, hence the incongruities.

Thus, Gen Y in times of frequently changing work environments with decreased job stability, it is often no longer possible for an employee to pursue a career within one single organization during his or her lifetime with prescheduled linear upward moves over time. More and more frequently, employees have to take responsibility for their own careers Due to the unpredictability of modern day working environment context, maintaining a boundaryless career is very useful in Western countries. As explained above, this new concept developed as an outcome of wider economic and organizational changes (Littler et al., 2003). Supporters of this concept argue that if organizations are becoming increasingly flexible and fluid careers should also become more flexible and fluid (Inkson & Thom, 2010). Having a boundaryless career maintains that an individual should be the agent of their own career, thus an organization is less responsible for its employees' career outcomes (Inkson & Thom, 2010). Some researchers (e.g., Arnold, 2011) go as far as to proclaim that in order to survive and be successful in modern day work environments, it is necessary to have a boundaryless mindset. Protean career has been studied principally with reference to the mid-career stage with the exception of Hall and Mirvis (1995) who examined older workers, and Briscoe et al. (2006) who studied undergraduates. It is therefore concluded that Gen Y will be protean in their career orientation and show boundaryless mindset.

1.3   INDIVIDUALS WITH PROTEAN CAREER ORIENTATION

Individuals who hold protean career attitudes are intent upon using their own values (vs. organizational values for example) to guide their career ("values-driven") and take an independent role in managing their vocational behavior ("self-directed"). An individual who did not hold protean attitudes would be more likely to "borrow" external standards, as opposed to internally developed ones, and be more likely to seek external direction and assistance in behavioral career management as opposed to being more proactive and independent. While most protean individuals might in fact exhibit more mobility and a learning orientation, we posit that mobility and

learning may be correlates of a protean career, but not necessary components of it. A person with a boundaryless career mindset "navigates the
changing work landscape by enacting a career characterized by different
levels of physical and psychological movement" (Sullivan & Arthur,
2006, p. 9). While a boundaryless career attitude is recognized which
is primarily *psychological*, Arthur and Rosseau's emphasis (1996) upon
careers which unfold beyond a single employment setting has frequently
been interpreted as involving interim, *physical* employment mobility. As
such, a second important boundaryless career attitude is the inclination
toward physically crossing organizational boundaries in employment
mobility. Someone high in such an organizational mobility attitude would
be comfortable with, or even prefer a career that played out across several
employers.

1.4   GENDER AND PROTEAN CAREER ORIENTATION

Studies have also explored the role of gender in protean career orientation. There have been mixed findings regarding new career patterns being
more prevalent among women. McDonald (2005) in their study find that
the trend toward protean careers is evident and is more pronounced for
women than for men contrary to some other studies where men were more
career-oriented than women (Markus & Kitayama, 1991; Ng et al., 2008).
Gender differences in protean career orientation have received considerable theoretical attention. Sullivan and Baruch (2009) predict important
gender as well as generational differences in career attitudes and behaviors and called for further research in the areas. Quite a few researches
have established that protean career is more of a characteristic for women
(e.g., Briscoe et al., 2006; Eby, 2001; Valcour & Tolbert, 2003). Kim and
Jyung (2011) found that individual characteristic variables such as gender,
educational background, and voluntary department transfers have high
predictive value in explaining protean career attitudes, while organizational factors appear to have little influence.

While some studies report gender differences (Ng et al., 2008), the
others report no differences (e.g., Agarwala, 2008; Briscoe et al., 2006;
Forrier et al., 2009; Valgaeren, 2008; Vigoda-Gadot & Grimland, 2008;
Volmer & Spurk, 2010). Hall (2004) in a study finds a person's career
orientation is unrelated to gender. Segers et al. (2008) report no gender
differences in self-directedness, but find that women score higher on the

value-driven dimension than men. Cabrera (2009) studies protean career orientation among women who return to work after a gap and find that majority of them follow a protean career orientation to balance their personal lives. Cabrera reports that in line with a protean orientation, the women were self-directed in managing their careers, rejecting the traditional corporate careers and also exhibit a protean orientation in that their decisions were driven by personal values. Mostly all of the women who change their career orientation did so in order to balance their work and nonwork lives. As Cabrera (2008) examined women in their mid-career stages, the difficulty of fulfilling both family responsibilities and work demands led these women to follow protean career which allowed them to achieve subjective career success albeit less monetary rewards.

1.5   IMPLICATIONS FOR ORGANIZATIONS

Changes in economic, social, and technological spheres are affecting organizational flexibility and responsiveness in meeting global market requirements. Gen Y will soon overtake the other generations in the workplace; hence, organizations have to determine the best manner in which to train and develop younger, more mobile workers (Kniveton, 2004) and also manage an aging workforce. Individual's values, attitudes, and motivation influence the individual perceptions of career success (Agarwala, 2008; Derr, 1986) which is subjective (Hall, 1996a,b, 2002).

Organizations can benefit from the deeper understanding that career stories provide. Accommodating employees goals, aspirations, and motivations can be an effective way of retaining key employees. At the same time, this career stories can facilitate mutual understanding and future collaboration with flexible and mobile employees.

Due to the unpredictability of modern day working environment context, maintaining a boundaryless career is very useful in Western countries. As explained above, this new concept developed as an outcome of wider economic and organizational changes (Littler et al., 2003). Supporters of this concept argue that if organizations are becoming increasingly flexible and fluid careers should also become more flexible and fluid (Inkson & Thom, 2010). Having a boundaryless career maintains that an individual should be the agent of their own career; thus, an organization is less responsible for its employees' career outcomes (Inkson & Thom, 2010). Some researchers (e.g., Arnold, 2011) go as far as to proclaim that in order to survive and be successful in modern day work environments, it is necessary to have a boundaryless mind-set.

Gen Y employees also desire a good compensation and reward system based on performance. But HR managers should ensure that there is consistency between the image of the company as a good employer and the reality. Else the Gen Y will not hesitate to move out. These potential employees also value growth opportunities. It is therefore important that employers focus on the target segment by understanding their preferences and expectations and work on developing those attributes. It will also help organizations to promote these initiatives and building a brand image as a good place to work.

1.6   IMPLICATIONS FOR INDIVIDUALS

Pursuing a protean career means, therefore, the development of a new psychological contract. Whereas in the past, the contract was with the organization, in the protean career, the contract is with the self. The protean career is therefore a process that the person and not the organization is managing with the criterion for success being internal (psychological) and not external. Protean career managed by self-according to their personal values is characterized by freedom and growth that are also subjective (intrinsic) and not driven by organizational rewards (Briscoe & Hall, 2006; Hall, 1996a,b, 2002) or objective (extrinsic or materialistic). For young management graduates, Job attribute preferences (both intrinsic and extrinsic) are based on their values, attitudes, and perception of career success when pursuing employment with organizations of their choice. King (2003) in the British context and Sturges et al. (2005) concluded that graduates were yet to develop attitudes of self-direction. Briscoe et al. (2006) and Sargent and Domberger (2007) studied protean career orientation in undergraduates. Sargent and Domberger (2007) found that protean career is identifiable with individuals in their earlier career stage and is particularly salient to the current generation of graduates. Protean career as described by Hall (2002) has goals which are psychological rather than material success (Sargent & Domberger, 2007). Individual's values, attitudes and motivation influence the individual perceptions of career success (Agarwala, 2008; Derr, 1986) that is subjective (Hall, 1996a,b, 2002). Time spent away from work force (such as in the case of maternity leave) is now being used by individuals to increase their education or gain valuable skills in order to build their resume and ease their reentry into the workforce . Individuals have become increasingly driven by their own desires and are taking more responsibility for their own career development and employability, rather than depending on organizational career management (Hall, 2004).

Individuals are traveling career paths that are discontinuous and go beyond the boundaries of a single firm. This is mostly due to the downsizing of organizations in order to become more flexible in response to environmental factors, such as rapid technological advancements and increased global competition. Many employees who were once secure in their jobs have found themselves unsecure situations (Arthur & Rousseau, 1996).

Hall and Mirvis (1995) very well summarized future career contract by

THE PROTEAN CAREER OF THE 21ST CENTURY

- The goal: psychological success
- The career is managed by the person, not the organization
- The career is a lifelong series of identity changes and continuous learning
- "Career Age" counts not chronological age
- The organization provides
  1. Work challenges and
  2. Relationships

- Development is not necessarily
1. Formal training
2. Retraining
3. Upward mobility

  - Profile for success:
  1. From Know-How…To Learn How
  2. From Job Security…To Employability
  3. From Organizational Careers…To Protean Careers
  4. From Work Self…To Whole Self

Source: "New Styles of Management: Career builders or Career Blockers," presented at the 1995 Work and Family conference, "The New Employee Employer Contract: A Work Family Perspective," sponsored by the Conference Board and the Families and Work institute, New York, NY, April, 26, 1995.

1.7 CONCLUSION

Generation Y individuals desire growth opportunities, challenging work, assign importance to compensation and benefits. The career of the 21st century will be protean, a career that is driven by the person, not the organization, and that will be reinvented by the person from time to time, as the individual and the environment change. In the 21st century model career, growth will be a process of continuous learning and identity change and will not be measured by chronological age and life stages. The ultimate goal of the career for the future employee will be psychological success, the feeling of pride and personal accomplishment that comes from achieving one's most important goals in life, be they achievement, family happiness, inner peace, or something else. Focus of Gen Y will be on psychological success, job security will be replaced by goal of employability. This career structure was never possible before Gen Y predecessors. As Hall (1996a,b) sums it:

> *The career is dead—long live the career.*

ACKNOWLEDGMENTS

Our sincere thanks to Michael Arthur, C. Brooklyn Derr, Rosina Gasteiger, Kerr Ink-son, Sherry Sullivan, and Lea Waters, and also Hall for providing insights through their researches and models for building this concept.

KEYWORDS

- **workforce diversity**
- **career orientation**
- **protean careers**
- **boundaryless careers**
- **Generation Y**

REFERENCES

Agarwala, T. Factors Influencing Career Choice of Management Students in India. *Career Dev. Int.* **2008**, *13*, 362–376.

Arnold, J. *Career Concepts in the 21st Century*. ©British Psychological Society, 2011.

Arthur, M. B. The Boundaryless Career: A New Perspective for Organizational Inquiry. *J. Organ. Behav.* **1994**, *15*, 295–306.

Arthur, M. B.; Rousseau, D. M., Eds. *The Boundaryless Career: A New Employment Principle for a New Organizational Era*. Oxford University Press: New York, 1996.

Baruch, Y. Career Development in Organizations and Beyond: Balancing Traditional and Contemporary Viewpoints. *Hum. Resour. Manage. Rev.* **2006**, *16*, 125–138.

Baruch, Y.; Hall, D. T. The Academic Career: A Model for Future Careers in Other Sectors? *J. Vocat. Behav.* **2004**, *64*, 241–262.

Briscoe, J. P., Hall, D. T.; Frautschy DeMuth, R. L. Protean and Boundaryless Careers: An Empirical Exploration. *J. Vocat. Behav.* **2006**, *69*, 30–47.

Briscoe, J. P.; Hall, D. T.: The Interplay of Boundaryless and Protean Careers: Combinations and Implications. *J. Vocat. Behav.* **2006**, *69*, 4–18.

Briscoe, J. P.; Hall, D. T. The Protean Orientation: Creating the Adaptable Workforce Necessary for Flexibility and Speed. Paper Presented at the Annual Meeting of the Academy of Management, Denver, 2002.

Cabrera, E.F. Protean Organizations: Reshaping Work and Careers to Retain Female Talent,' *Car. Dev. Int.* **2009**, *14*(2), 186–201.

Cadin, L.; Bailly-Bender, A. F.; de Saint-Giniez, V. Exploring Boundaryless Careers in the French Context. In: *Career Frontiers: New Conceptions of Working Lives*; Peiperl, M., Arthur, M., Goffee, R., Morris, T., Eds.; Oxford University Press: Oxford, 2000.

Cooper, C. L.; Burke, R. J., Eds. *The New World of Work: Challenges and Opportunities*. Blackwell: Oxford, UK, 2002.

Cruz, C. S. Gen Y: How Boomer Babies are Changing the Workplace. *Hawaii Business*, May 2007. Retrieved from http://www.hawaiibusiness.com/gen-y/.

Derr, B. C. *Managing the New Careerists: The Diverse Career Success Orientation of Today's Workers*. Jossey-Bass: San Francisco, CA, 1986.

Dries, N.; Pepermans, R.; Kerpel, E. De. Exploring Four Generations' Beliefs about Career Is "Satisfied" the New "Successful"? *J. Manage. Psychol.* **2008**, *23*(8), 907–928. DOI:10.1108/02683940810904394.

Eby, L. The Boundaryless Career Experiences of Mobile Spouses in Dual-Earner Marriages. *Group Organ. Manage.* **2001**, *26* (3), 343–368.

Enache, M.; Simo, P.; Maria, J.; Vicenç, S. Examining the Impact of Protean and Boundaryless Career Attitudes Upon Psychological Career Success. *Work Organization and Human Resources Management*; 2008, pp 1939–1948.

Erickson, T. J. *Plugged In: The Generation Y Guide to Thriving at Work*. Harvard Business Press: Boston, MA, 2008.

Forrier, A.; Sels, L.; Stynen, D. Career Mobility at the Intersection between Agent and Structure: A Conceptual Model. *J. Occup. Organ. Psychol.* **2009**, *82*(4), 739–759.

Frenandez, V.; Enache, M. Exploring the Relationship between Protean and Boundaryless Career Attitudes and Affective Commitment through the Lens of a Fuzzy Set QCA Methodology. *Intang. Cap.* **2008**, *4*(1), 31–66.

Grimland, S.; Vigoda-Gadot, E.; Baruch, Y. Career Attitudes and Success of Managers: The Impact of Chance Event, Protean, and Traditional Careers. *Int. J. Hum. Resour. Manage.* **2011,** *23*(6), 1074–1094. DOI:10.1080/09585192.2011.560884.

Hall, D. T. *Careers in Organizations.* Scott, Foresman: Glenview, IL, 1976a.

Hall, D. T. *Careers in Organizations*: Goodyear: Pacific Palisades, CA, 1976b.

Hall, D. T. Protean Careers of the 21st Century. *Acad. Manage. Execut.* **1996c,** *10*, 8–16.

Hall, D. T. Long Live the Career. In: *The Career is Dead—Long Live the Career*; Hall, D. T., Ed.; 1996a; pp 1–12.

Hall, D. T. Protean Careers of the 21st Century. *Acad. Manage. Execut.* **1996b,** *10*, 8–16.

Hall, D. T. *Protean Careers In and Out of Organizations.* Sage: Thousand Oaks, CA, 2002.

Hall, D. T. The Protean Career: A Quarter-century Journey. *J. Vocat. Behav.* **2004,** *65*, 1–13.

Hall, D. T.; Mirvis, P. H. The New Career Contract: Developing the Whole Person at Mid-life and Beyond. *J. Vocat. Behav.* **1995,** *47*, 269–289.

Hall, D. T.; Chandler, D. E. Psychological Success: When the Career is a Calling. *J. Organ. Behav.* **2005,** *26*(2), 155–176.

Hansen, L. S. *Integrative Life Planning: Critical Tasks for Career Development and Changing Life Patterns.* Jossey-Bass: San Francisco, CA, 1997.

Hess, N.; Jepsen, D. M. Career Stage and Generational Differences in Psychological Contracts. *Car. Dev. Int.* **2009,** *14*(3), 261–283. DOI:10.1108/13620430910966433.

Inkson, K.; Thorn, K. Mobility and Careers. In: *The Psychology of Global Mobility*; Springer: New York, 2010; pp 259–278.

Kim, E.; Jyung, C. The Hierarchical Linear Relationship among Protean Career Attitudes, Individual Characteristics, and Organizational Characteristics of Office Workers in Large Corporations. *Agric. Educ. HRD* **2011,** *43*(2), 171–189.

King, Z. New or Traditional Careers? A Study of UK Graduates' Preferences. *Hum. Resour. Manage. J.* **2003,** *13*(1), 5–27.

Kniveton, B. H. The Influences and Motivations on Which Students Base their Choice of Career. *Res. Educ.* **2004,** *72*, 47–57.

Kuchinke, P.; Park, U. The Self-directed Career in an Era of Economic Instability: Opportunities and Limitations in Cross-cultural Comparison. *J. Knowl. Econ. Knowl. Manage.* **2012,** *7*, 1–8.

Levinson, D. J. *The Seasons of a Man's Life.* Knopf: New York, 1978.

Littler, C. R.; Wiesner, R.; Dunford, R. The Dynamics of Delayering: Changing Management Structures in Three Countries. *J. Manage. Stud.* **2003,** *40*(2), 225–256.

Lloyd, J. The Truth about Gen Y. *Mark. Mag.* **2007,** *112*(19), 12–22.

Markus, H. R.; Kitayama, S. Culture and the Self: Implications for Cognition, Emotion, and Motivation. *Psychol. Rev.* **1991,** *98*(2), 224–253.

Meier, J.; Austin, S. F.; Crocker, M. Generation Y in the Workforce: Managerial Challenges. *J. Hum. Resour. Adult Learn.* **2010,** *6*(1), 68–78.

McDonald, R. P. Semiconfirmatory Factory Analysis: The Example of Anxiety and Depression. *Struct. Equat. Model.* **2005,** *12*, 163–172.

Ng, E. S. W.; Burke, R. J.; Fiksenbaum, L. Career Choice in Management : Findings from US MBA Students. *Car. Dev. Int.* **2008,** *13*(4), 346–361. DOI:10.1108/13620430810880835.

Peiperl, M.; Jonsen, K. Global Careers. *Handbook of Career Studies*; 2007; pp 350–372.

Reitman, F.; Schneer, J. A. The Promised Path: A Longitudinal Study of Managerial Careers. *J. Manage. Psychol.* **2003,** *18*(1), 60–75. DOI:10.1108/02683940310459592.

Rousseau, D. M. Psychological and Implied Contracts in Organisation. *Employ. Responsib. Rights J.* **1989,** *2*, 121–139.

Sargent, L. D.; Domberger, S. R. Exploring the Development of a Protean Career Orientation: Values and Image Violations. *Car. Dev. Int.* **2007,** *12*, 545–564.

Smith, F. Workspace: Companies Will Keep Top Staff. *Austr. Financ. Rev.* **2008,** *50*, 33–45.

Sturges, J.; Conway, N.; Guest, D.; Liefooghe, A. Managing the Career Deal: The Psychological Contract as a Framework for Understanding Career Management, Organizational Commitment and Work Behavior. *J. Organ. Behav.* **2005,** *26*(7), 821–838.

Sullivan, S. E.; Arthur, M. B. The Evolution of the Boundaryless Career Concept: Examining Physical and Psychological Mobility. *J. Vocat. Behav.* **2006,** *69*, 19–29.

Sullivan, S. E.; Baruch, Y. Advances in Career Theory and Research: A Critical Review and Agenda for Future Exploration. *J. Manage.* **2009,** *35*(6), 1542–1571.

Terjesen, S.; Frey, R. Attracting and Retaining Generation Y Knowledge Worker Talent. In: *Smart Talent Management: Building Knowledge Assets for Competitive Advantage;* Vaiman, V., Vance, C. M., Eds.; Edward Elgar Publishing, Inc.: Cheltenham, UK, 2008.

Valcour, M.; Tolbert, P. Gender, Family and Career in the Era of Boundarylessness: Determinants and Effects of Intra- and Inter-organizational Mobility. *Int. J. Hum. Resour. Manage.* **2003,** *14*(5), 768–787. DOI:10.1080/0958519032000080794.

Vigoda-gadot, E.; Grimland, S. Values and Career Choice at the Beginning of the MBA Educational Process. *Car. Dev. Int.* **2008,** *13*(4), 333–345. DOI:10.1108/13620430810880826.

Volmer, J.; Spurk, D. Protean and Boundaryless Career Attitudes: Relationships with Subjective and Objective Career Success. *J. Labour Mark. Res.* **2010,** *43*, 207–218. DOI:10.1007/s12651-010-0037-3.

Yahoo! HotJobs/Robert Half International. *What Millennial Workers Want: How to Attract and Retain Gen Y Employees,* 2008. Retrieved from http://www.hotjobsresources.com/pdfs/MillennialWorkers.pdf.

CHAPTER 2

WHAT MOTIVATES THE MILLENNIALS: A PERSPECTIVE

PARUL WASAN*

Manav Rachna International University, Faridabad, Haryana, India

*E-mail: parul.wasan@yahoo.co.in

CONTENTS

ABSTRACT

Generation Y or The Millennials (born between 1980 and 2000) account for over half of the population in India. They have been defined variously as "echo boomers" because they are the children of parents born during the baby boom (the baby boomers) of the United States of America, Millennials, echo boomers, internet generation, iGen, net generation, etc. They will also form 50% of the global workforce by 2020. The purpose of this survey was to (a) understand their career aspirations and attitudes about work, which will eventually define the culture of the 21st century workplace, to answer (b) what motivates of the Gen Y in India, and (c) to suggest policy modifications HR managers. A survey close to 150 subjects from both the factory workers and the corporate communities in Gurgaon and Faridabad, Haryana was conducted. *Overviews of Gen Y motivators in Canada and USA, Sweden, and China have then been briefly discussed, before discussing the country context of India.*

The survey results suggest that the Millennials in India value two aspects, namely, "the pay and respect for me as a person" the most within their departments and companies. This indicates that there is a high sense of self-worth among them. They are hence impatient and critical of system that asks them to put in a requisite amount of time to the job at hand. Gen Y considers it appropriate to be ambitious and therefore asks for immediate results, in other words "if something can be changed for the better, then we better do it now and not wait for things to happen at some later date." The survey thus concludes that Gen Y is all about self-actualization an aspect that is demonstrated in their "inside-out" attitude toward how they perceive their career, Gen X and the world in general. They want to solve problems that plague them and they do not hesitate to be creative about the solutions. They are often critical of Gen X which they think is holding them back. They are adaptive and are not adverse to change.

2.1 INTRODUCTION

Gen Y or The Millennials (born between 1980 and 2000) account for over half of the population in India. They will also form 50% of the global workforce by 2020. According to a 2011 report of Pricewaterhouse Coopers International Limited, "millennials matter because they are not only different from those that have gone before, they are also more numerous than any since the soon-to-retire Baby Boomer generation." The report then goes on to assert that their knowledge of new technologies, career aspirations, and attitudes about work will define the culture of the 21st century workplace.

To understand, a survey close to 150 subjects from both the factory workers and the corporate communities in Gurgaon and Faridabad, Haryana was employed. The tool used for the study is "Work Motivation Checklist" of This and Lippit (1999).

2.2 THEORETICAL BACKGROUND AND LITERATURE REVIEW

Defining Gen Y: Generation Y has been defined variously as "echo boomers" because they are the children of parents born during the baby boom (the baby boomers) of United States of America, Millennials, echo boomers, internet generation, iGen, net generation, etc. Children born during these years have had unprecedented access to technology, namely, computers and cell phones in their youth. They are according to Kotler and Keller (2005) edgy, urban focused, and idealistic. This generation has observed the war on terror, rise of the information age, global recession, stock market crashes, political turmoil, increasing rates of suicides across all strata of the society, and uncertain job markets, as a result they are more apt to "leave their jobs sooner as compared to three to five years of Generation X" (The Canberra Times, March 29, 2006). According to Sai and Swati (2012), parents of Gen Y provided them all those material aspects that they (parents) did not have while they were growing up. Parents thus gave Gen Y an environment in which Gen Y could focus on the higher order of needs of Maslow's Hierarchy such as Self Actualization instead of stability and security, which was the prime need of their parents. Ilene Siscovick (2015) a partner in the New York-based global consulting firm Mercer, is of the opinion that in the absence of the need of stability and security, "Gen Y doesn't have the same allegiances or affinities as other generations," and also that "while they're devoted to their career, they want to have meaningful work and a sense that their company's products or services are somehow improving conditions in the broader society."

Shinjini Das (2015) has characterized Gen Y as having—(a) lots of energy and a fresh perspective, (b) thirst for challenges and a hunger to grow, (c) desire to be a hero and create a profound impact, (d) keen sense of adaptability, and (e) quest for mentors and brutally honest feedback. Not only are they Tech savvy, they are also family oriented, ambitious, team players, good communicators, and they like to be loved. Main (2013) characterized Gen Y or the Millennials as having both negative and positive definitions. On the negative side, he has highlighted as to how Gen Y has been various described "as lazy, narcissistic and prone to jump from job to job. He further to quotes the 2008 book 'Trophy Kids'" by Ron Alsop, where "Alsop discusses how many young people have been rewarded for minimal accomplishments (such as mere participation) in competitive sports, and have unrealistic expectations of working life." To reiterate this school of thought, he then goes on to summarize the more positive views regarding Gen Y as "being generally as more open-minded, and more supportive of gay rights and equal rights for minorities." He also states that Gen Y has been given positives adjectives that describe them as being receptive to new ideas and ways of living, upbeat, confident, self-expressive, and liberal.

Asghar (2014) quotes Jamie Gutfreund of the Intelligence Group in defining Gen Y as having a different concept of authority, different set of motivations, desire for a different set of work environment, and a different concept of the "progress on the project" status.

Dorsey (2010), an authority on Millennials and Gen Y, has identified the following Gen Y characteristics in his book "Y-Size your Business"—

A.   The feeling of entitlement is not true for everyone in Gen Y because there are both focused and strugglers among the Millennials.

B.   Contrary to popular perception, Gen Y is "not tech savvy," instead they are "tech dependent" and they cannot live without it. Often they do not know how the technology works in actual terms.

C.   Gen Y does not wish to wait in line for its turn to come. They always "never have time and patience."

D.   Although Gen Y always knows what it wants, it seldom knows and value steps that are needed to be taken to reach its destination. They do not understand linearity of a process. They are outcome driven, therefore they often understand the process backward from the expected result to the beginning.

2.2.1   CONCEPT OF SELF ACTUALIZATION

Goldstein (1939) in Stahnisch and Hoffmann (2010) explained self-actualization as "the tendency by which an individual, as much as possible, optimizes his individual capacities" in the society. He further said that this tendency determines the life of an organism. Additionally, Goldstein stated that self-actualization cannot be understood as a kind of goal to be reached sometime in the future, because at any given moment, an organism can actualize all its capacities and its whole potential, in any situation in the world under the given circumstances. He further stated that a "normal society is a type of organization through which the fullest possible actualization on the part of all individuals is assured...." He then goes on to say "If we acknowledge and utilize social organization as an instrument by means of which all individuals may actualize themselves to an optimal degree, then a genuine social life becomes possible. Only under these conditions is a social organization capable of doing justice to every individual; only this makes it a real organization and secures its duration."

Under the influence of Goldstein, Abraham Maslow developed a hierarchical theory of human motivation in Motivation and Personality (1954).

Maslow (1943) wrote a paper called "A Theory of Human Motivation," wherein Maslow first proposed "Human Hierarchy of Needs." This theory says that an individual can achieve greater needs only after his/her basic human needs are met.

Maslow's Hierarchy of Needs

Maslow suggested that basic needs such as breathing, food, water, rest are the foundations of an individual's development. After meeting these core requirements, a human being seeks out the attainment of other, higher needs.

According to Maslow, the next step on the pyramid relates to the safety and the security of one's resources, family, health, and property. Progressing upward, an individual would then seek love and friendship. Maslow says that for some individuals, it is very difficult to progress further than these first three levels of need (physiological, safety, and love) and therefore they remain forever trapped in these three states. However, an individual could desire creativity, spontaneity, morality, problem solving, and acceptance, and thus move toward seeking self-actualization. This is the pinnacle of Maslow's pyramid and according to him, it is the ultimate human goal.

This concept of self-actualization, put forth by Maslow and Goldstein, was extended in 1959 by Carl Rogers and Herzberg.

Carl Rogers (1959)—People will flourish and reach their potential if their environment is good enough, that is, they get an environment that has genuineness (openness and self-disclosure), acceptance (being seen with unconditional positive regard), and empathy (being listened to and understood). Through such an environment, people tend to fulfill one's potential and achieve the highest level of "human-beingness." Rogers believed that when people achieve their goals, wishes, and desires in life, they undergo self-actualization. Roger (1961) called such people fully functioning persons. Such persons satisfy five factors. Rogers identified five characteristics of the fully functioning person. Such people are

A. **Open to experience:** Such persons accept both positive and negative emotions. They do not deny negative feelings and work through them without resorting to "ego."
B. **Existential living:** These people are open to all the experiences as they occur. They avoid prejudging and preconceptions about various phenomena of life. They live and fully appreciate the present. Although they may, at times, look back to the past or speculate on the future.
C. **Trust feelings:** They trust their feelings, instincts, and gut reactions. They trust themselves to take the right decisions and make the right choices.

D. **Creativity:** They take risks and think out of the box. They do not always play safe. They can therefore adapt to any situation and are open to new experiences.

E. **Fulfilled life**: Although such people are happy and satisfied with their life, they are always on the lookout for new challenges and experiences. Such people are constantly growing and changing. Roger mentioned that we behave as we do because of the way we perceive our situation. "As no one else can know how we perceive, we are the best experts on ourselves."

Herzberg (1959) in Gawel (1997) during the same period as Rogers and Maslow, put forth a two-dimensional paradigm of factors or motivators which affect people's attitude toward work. These factors are (a) achievement, (b) recognition, (c) the work itself, (d) responsibility, and (e) advancement. Herzberg further stated that these factors/motivators are needed for an employee to remain satisfied in the organization for a long time. Except "salary," which he concluded is a short-term Hygiene factor and which may lead to job dissatisfaction. Additionally, the salary factor could also mar his/her relationship to the context or environment in which she or he performs the job, although he/she may be satisfied with the work that they do. Additionally, it is also possible that Millennials may not relate their work to the situation in which this work is being done. Table 2.1 describes the summary of the concepts.

TABLE 2.1 Summary of the Concepts.

S. No.	Name of the author	Concepts
1	**Goldstein (1939)**	The tendency by which an individual, as much as possible, optimizes his individual capacities
		…A normal society is a type of organization through which the fullest possible actualization on the part of all individuals is assured
2	**Maslow (1943)**	…When basic needs such as breathing, food, water, and rest are the foundations of an individual's development. After meeting these core requirements, a human being seeks out the attainment of other, higher needs like safety and the security of one's resources, family, health, and property

TABLE 2.1 *(Continued)*

S. No.	Name of the author	Concepts
3	**Rogers (1959)**	People will flourish and reach their potential, if their environment is good enough, that is, they get an environment that has genuineness (openness and self-disclosure), acceptance (being seen with unconditional positive regard), and empathy (being listened to and understood). Through such an environment, people tend to fulfill one's potential and achieve the highest level of "human-beingness"
		…When people achieve their goals, wishes, and desires in life, they undergo self-actualization. Such people are called fully functioning persons
4	**Herzberg (1959)**	…factors or motivators which affect people's attitude toward work—achievement, recognition, the work itself, responsibility, and advancement … these factors/motivators are needed for an employee to remain satisfied in the organization for a long time
		…Except "salary," which was concluded as a short-term Hygiene factor and which may lead to job dissatisfaction

2.3   EARLIER RESEARCH ON MILLENNIAL MOTIVATION

The year 2010 saw the publication of a seminal work of Ng, Schweitzer et al. where they studied 23,413 Millennial graduate students across Canada in order to explore their motivational factors. They concluded that "Gen Y rated opportunities for career-growth as the most important after work related attribute." Millennials also wish to work with good people and to report to, they want good training opportunities for skills development and always want a good work life.

In 1999, Montana's and Lenaghan's based their research on Leslie This and Gordon Lippitt (Montana & Lenaghan, 1999). They identified that both Generation X and Generation Y values the same six motivational factors, namely, (a) steady employment, (b) respect for me as a person, (c) good pay, (d) chance for promotion, (e) opportunity for self-development and improvement, and (f) large amount of freedom on the job. They concluded that Generation X and Generation Y value similar motivational factors and that both these generations have very little difference between them. However, later in 2008, Montana with Petit expanded his research to highlight the fact that Generation Y considers "respect for me as a person," "good pay," "getting along well with others on the job," "chance for promotion," "opportunity to do interesting work," and "opportunity for self-development and improvement" as prime motivators of the job (Montana & Petit, 2008).

In another report, Barford and Hester (2011) indicated that motivational differences exist between Generation X and Generation Y, they opined that they are not motivated by the same factors.

Researchers Leschinsky and Michael (2004) modified the "This and Lipit" survey methodology. Instead of using students and young worker across industry, they used case blue-collar production employees in the wood products industry. Their results of the mean-score ratings show that good pay and having steady employment are most sought after among the workers. They also point out that, "respect for me as a person" and "opportunity to do interesting work" were considered as the most important factors by the blue collared workers.

In 2006, a study conducted by Steven Rumpel and John W. Medcof cited Tower Perrin, while discussing the importance of "Pay" for nontechnical job categories. They also demarcated the motivational aspect of the technical work force from nontechnical workers by citing other studies which highlighted work environment, learning, and development as the number one motivating factor. Towers Perrin study used O'Neal's reward frame work to highlight the strategic importance of pay (direct financial items, namely, base pay, variable pay, incentives, stock and equity sharing programs, etc.) in "motivating and retaining" an nontechnical employee. On the other hand, technical employees consider how work content and affiliation fit under work environment, and learning and development. They consider pay and financial benefits secondary to work environment.

2.4 GLOBAL CONTEXT

2.4.1 CANADA AND UNITED STATES

A study carried out by the Dale Carnegie Training in 2014 among 300 Dale Carnegie graduates in Canada and the United States tried to answer how Millennials compared to the older members of corporates (Dale Carnegie, 2014). Of these 300, 50% were older and non-Millennials. This study specifically tried to understand whether the drivers of employee engagement are different across the Gen Y and the Gen X and whether management and communication styles should therefore be adapted accordingly. This study raised some interesting insights that point toward essential factors for engagement and what it takes to engage the Gen Y. It pointed toward the fact that (a) Millennials differ from Gen X in that they are more likely to be engaged in their jobs than Gen X, (b) although they feel less valued they are confident and are connected, (c) they believe in charting their own path and multiple levels, and (d) like Gen X, they would also recommend their workplace (if they like it) to their friends for the purpose of doing business. This is indicative of their desire for a balance between individualism and loyalty to their friends. This engagement is additionally linked to (a) a sense of fulfillment (which comes from helping others find a good business partner) and b) loyalty and personal gain (derived from recognition for bringing business to their employer). The study also states that Gen Y proactively seeks training opportunities in their work place in order to overcome the feeling of inadequacy and low value. This helps them to feel more confident, valued, and connected. They look forward to doing courses in leadership, self-confidence, public speaking, and team management. Owing to their work schedule, they prefer flexible timings for online training programs with simple and engaging content. Additionally, the study pointed toward the importance of emotional and functional qualities that were very important to the Millennial like a positive workplace environment, that is, limited oversight, trust, flexible work hours, etc., and a supervisor with whom they can relate by the virtue of trust, respect, support, leading by example, etc. They look for two specific attributes in their leader in the work space (a) interpersonal connectedness, and (b) professional competence as it is about. The study therefore goes on to recommend the following requirements for Millennial engagement—(1) effective employer–employee communication; (2) need for a system of incentives, employee perks, awards, and recognition; (3) value addition in compensation and benefits packages of the employee; (4) develop challenging tasks to keep for team members engaged; (5) build supportive environment; and (6) provide regular training and development opportunities.

2.4.2 SWEDEN

Reputation Institute (2013) ranked Sweden second (behind Canada) in global rankings. This ranking measures the international perceptions of the country in terms of environment, economy, and government. Despite the worldwide recession of past years, many internationally successful start-ups have emerged from Sweden, especially in the technology industry which is being spearheaded by Gen Y. The Millennials in Sweden have responded by optimistically to the employment market, by exhibiting keen interest in personal development opportunities and by taking on work for the betterment of the society. Not only do they want to experience various facets of the work that they are involved in, they are also opting out of their current jobs with speeding regularity to work for new employers. Gen Y in Sweden don't like to use the term "leader" when it comes to describing their boss at the work place. However, like their counter parts in the United States, Canada, and Asia, they wish their boss to be their coach and mentor, whom they consider to be their peer and not a boss in the traditional sense of the term. The quality that they most look for him is the ability to motivate them to achieve newer heights in their career. Gen Y in Sweden has given a miss to the culture of collectivism and team ethos, which is the standard workplace culture in Sweden, and is now looking toward entrepreneurship. Experts call this phenomenon a reaction to the high youth unemployment rate in the country. Although in Sweden, one of the few countries in the world where entrepreneurial spirit is extremely low, popularity of the start-ups like *Sotify* and *Sound Cloud* in the domain of technology have become extremely popular with the Gen Y. The arche-typal vision of leadership of Ingvar Kamprad, the founder of the country's most successful company, IKEA, resonates across various start-ups across the country. The business culture of Gen Y companies is in consonance with the national business culture, which is all about ease of communica-tion, low internal competition, and absence of hierarchy within the organi-zation structures. Gen Y maintains the traditional expectations of leaders who are mentors who display humility. Such leaders are valued over direc-tive leaders. A strong commitment to education coupled with free educa-tion inculcates in them the passion for training and development. Gen Y in Sweden wishes to learn something new in order to progress rather than to earn more money, an aspect that makes them standout from their contempo-raries across the world. According to Hays' 2014 report, Gen Y in Sweden remains in the educational institutions far longer than the global average. They are usually very adept at learning new languages and hence they are forever in demand for international projects. As mentioned earlier, money is not a motivation for Gen Y (average wages are very high in Sweden). They seek interesting work more than personal wealth, this gives them satisfaction unlike their peers in the other parts of the world where the primary motivation is money. Gen Y like their predecessors define career success as achieving work–life balance. However, very much like the Gen X in Sweden, they (Gen Y) like to complete their working hours between 3 and 5 pm. They therefore prefer to work for an employer that can provide them fun and social interaction woven around the office work. They thus do they seek long hours and overtime. As a country, Swedes are highly computer literate and have extensive and very fast broadband connection throughout the country. This also makes them a country extremely that is connected. In tune with this national obsession, the Gen Y also keeps itself connected with the social media at all times. It can thus be said that unlike Gen Y, elsewhere in the world, the Swedes have a different take on few Millennial obsessions—(a) Swedes do not consider money as important as Gen Y round the world; (b) they are not a great fan of working overtime, and (c) learning for them is a lifelong process and is not connected to career advancement.

2.4.3 CHINA

Hays Report (2013) into the workplace needs, attitudes, and aspirations of Gen Y in China surveyed 1000 respondents and concluded that China's Gen Y has come out from years of collectivist political and economic isolation, to see their country transform into the emerging global market super power. They therefore tend to demonstrate a desire to exercise initiative and independence now that restrictions on private enterprises have eased. However, those who chose to work in offices and corporations are more likely to stick to their jobs, this is in sharp contrast to Gen Y in the Western or other Asian countries. Gen Y in China prefer job security. They are just like their peers around the world, better educated, tech-savvy, individualistic, multitasking, and open-minded generations. They are connected like their western counterparts to the social networking websites, mobile, and the internet. Additionally, they also want that their superiors and their organizations should have similar communications channels. There has been a rapid rise on internet use in China, and Gen Y regards social media, *eSinaWeibo*, the Chinese equivalent of Twitter, and *RenRen*, the equivalent of Facebook, as a central part of their lives. According to Hays report, and China Internet Network Information Centre, there are over 591 million internet users, and a 10% rise in the numbers since last year. This means that over 44% of the country's population now uses the web and other net services. The center reports that over 464 million citizens now have access to the net through their smartphones or other wireless devices. The hold of the social media is such that Gen Y workers do not wish to work with companies that block the access to the internet because they feel that social media is a helpful tool in their work. Additionally, Gen Y considers social media important in finding a good job despite the need of a two-page CV, which is considered a priority in the Chinese industry. Gen Y considers online tools essential to find jobs. They usually have a LinkedIn profile or a Sino Weibo profile and consider it important to have recommendations on these websites. Unlike their counterparts across the world, this generation is "one-child" policy generation. But like the rest of Gen Y they tend to be similarly inclined toward a life style devoted to freedom and personal satisfaction. They are aware of the western culture more than the westerners know about their culture. They are characterized in their work place as adventurous, impressionable, and achievers. They are also highly sought after by multinational companies as many of them have degrees in engineering and science. They value work–life balance and value quality work. Like all single children, they are pampered and have a sense of entitlement to the families resources. Hays Report (2013) says that major drivers of Gen Y in the work place are "Bonus" and a boss that can "motivate and inspire them," who has integrity and deals fairly with them. Gen Y wants opportunities to learn, acquire knowledge, and expertise at the work place. They therefore prefer organizations that offer such opportunities along with clear personal development strategies. Gen Y in China craves for recognition of their work. When asked what would be their definition of a successful career they gave equal weightage to job titles, gaining public, and professional recognition of their achievement, along with creation of personal wealth. This aspect is also culturally important to them because this reflects the Chinese tradition of maintaining "face." Gen Y in China like their generation aspires to work in clean, pleasant, and modern conditions. While in the office they wish to work in teams and prefer flexible benefits in order to enjoy time socially and with their families. They want a boss with whom they can talk, and discuss both personal as well as professional issues. An ideal Chinese office today has a sociable environment. Leadership is as important and there is a close relationship between boss and the employees. The situation as off now is more toward building a network of a strong interpersonal relationship with your boss who is expected to be both a coach and friend. This relationship, a few steps ahead of the traditional concept of "Guanxi," where the emphasis is only to develop a network of interpersonal relationships with powerful individuals.

2.4.4   INDIAN CONTEXT

According to the 2011 Census of India data, 65% of the population in India is between 15 and 64 years, and it is estimated that by 2020 half of the Indian population will be below 25 years of age. The work force will have grown by 5.6%. Studies conducted in Indian context by Sinha et al. (1994), among 750 students in seven cities of India observed the extent to which the Millennials gave importance to other's (over their own) opinions, desires, and interests. They identified principal motivators that guide Gen Y in India in their jobs, namely, clear job description, no stringent rules and regulations, unclear communication, and gender biases. They also look for tolerant organizations that can help them (a) learn, (b) have flexible HR policies that can help them to work from home and which gives them access social media, (c) they wish to have an informal work culture that can be fun to work in, (d) where the leaders are supportable and approachable, and (e) they desire quick promotions and recognition of their work.

According to Alley and Shah, in an IIM Bangalore, roundtable on multigenerations in workforce in Srinivasan (2012), on the other hand, among the main motivators of Gen Y are equitable pay, responsibility, and independence. They also have a contradiction or a diametric motivation which guides them through their career paths. This contradiction makes them want both responsibility and freedom, risk and stability, and guidance and supervision as well as to be their own boss. They also need a work–life balance as well as instant gratification. They look for well-defined organizational roles and job enrichment. However, Udupa in Srinivasan (2012) is of the opinion that movement of an individual on the Maslow's continuum is dependent upon environment but is independent of time. Therefore, an individual could be on any part of Maslow's continuum at any given point of time depending upon the environment, namely, individual situation, family community, socioeconomic situation, education, exposure, health and technology, and access to both global and local opportunities. Udupa further states that depending upon one's need an individual may exhibit behavior that could be contrary to one's age. He further points out that an individual's movement on the Maslow's continuum could be both forward or reverse depending upon his/her needs, which in turn would be dependent upon the environment which could be either an enabler or act as a constraint. However, Rajesh (2015) in Srinivasan (2012) in the same roundtable emphasized flexibility as a way of life. He then identified four major trends of Millennials from Handy's "Gods of Management." These trends are (a) *My time is my own (meaning they will switch off their official phones after the work day is over and also on the weekends)*, (b) *My work is integral part of my life (this means that weekends could be spent in the office if need be)*, (c) *I have other things to do and work is one part of the "other things,"* (d) *I have an identity that goes beyond work or my work does not define me,* and (e) *I am divinely touched and therefore have no need to throw my weight around.* While speaking on the IIM roundtable, Sripada Chandrashekhar, of IBM, in Srinivasan (2012), stated the need of reverse mentoring wherein a much younger colleague mentors a senior colleague on variety of issues such as use and handling of the social media like and Twitter handles and FB accounts. She also discussed issues pertaining to knowledge management and shadow teams and collaboration between generations. She further emphasized though companies take a programmatic approach to compensation and benefit. They are in at best a temporary solution to the needs of the Millennials in India. The primary requirement of Gen Y for all time is Compensation, Career Advancement, Competence (learning) and Care.

The above observations from Canada, USA, Sweden, China, and India therefore directed my study toward realizing the following:

2.5   RATIONALE AND OBJECTIVES

Panwar (Hindustan Times, September, 2015), observed that "today's youth is all about living in the moment." This aspect was reinforced when one comes across quotes like—"I like to break away from the pack. My goal in life is simple: I want to do things that I want to do, and not things that are expected of me. Life is abort. I want my life to be happy and healthy, not long and boring." And "Worrying is worrysome; it's stressful, and stress will kill you. Instead of fretting about getting everything done, why not simply accept that being alive means having things to do?"

We also came across a survey conducted by a leading newspaper of India, which stated that around 41.2% of the Millennials said that they were (a) willing to take lots of risks to succeed in life and also (b) they will most likely to use multiple sources of information before applying for a job. This survey further said that Millennials in India have an ability to stay positive and to take tough situations head-on. As a researcher, I found this surprising that this challenge that makes them extremely satisfied with their work life.

It is against this back drop that the researcher decided to implement the "Work Motivation Checklist" of This and Lippit (1960–1970)[1], in order to gain an insight into to Gen Y motivators.

This study was carried out the industrial city of Faridabad and cyber city of Gurgaon in the state of Haryana, India. It is an industrial town that employs a very large Millennial population in engineering and fabrication in factories that build tractors, automobile parts, etc. Gurgaon in the state of Haryana on the other hand is a knowledge city populated with Fortune 500 companies. It employs Millennials in IT professions and as knowledge workers. Both the cities employ two different profiles of Millennial workers (a) Faridabad has more mechanical factory workers, (b) Gurgaon has more workers in financial, IT, consultancy, and call centers areas.

It was hoped that the sample would throw interesting insights into the Millennial mindset.

---

[1] Leslie This and Gordon Lippitt studied 6000 managers and 500 representatives of different companies and government agencies. They asked them to rank six of 25 motivational factors that allowed them to do their best work.

Accordingly, researcher have tried to understand the Millennials through the following questions:

(1) What motivates the Gen Y in India?
(2) To suggest policy modifications HR managers.

In order to find answers to the about questions, the sample has been divided into two groups:

(1) 18–25 years,
(2) 26–33 years.

My rationale of this division was twofold. First, 18–25-year olds have just started to work on their first jobs, while 26–33-year olds have by now changed their jobs and they are now in their second or third job.

Second, their motivations are different at these stages. Earning money and exhibiting independence is of prime importance to 18–25-year olds, while 26–33-year olds have by now a very clear vision of what exactly are they looking for in their jobs besides pay.

2.6 METHODOLOGY

The "This and Lippit" checklist was discussed with 150 Gen Y members. Some of them have been working on regular basis for at least 2–5 years or have 1-year internship experiences at various organizations.

Extensive one-on-one interviews were also carried out during the interactions in order to understand the logic, their thought process and the background of the interviewees. Care was taken to establish a rapport with the subjects so that they are at ease with the checklist and the interviewer. They were given time to assimilate and discuss the checklist with the interviewer for clarification.

2.7 OBSERVATIONS

Principal motivators as were identified have been given in the table format:

Principal Motivators of Gen Y	
(In the order of Preference)	
1.	Good pay
2.	Respect for me as a person
3.	Chance of promotion
4.	Opportunity of self-Development
5.	Feel my Job is important
6.	Steady employment

"Pay" and "Respect for me" as a person/professional matters to these young men and women not only for economic reasons but also as a symbol of their worth and status. "A chance for promotion and self-development" is also important to them so that they can move ahead of their peers. They are hence impatient and critical of system that asks them to put in a requisite amount of time to the job at hand. Hence, they often hop jobs. However, only a half of those who answered the questionnaire said that "their jobs are important to their organizations because of which their organization gives them additional benefits" ("We are a bunch of youngsters in the office who take care of the social media activities of the company."). Then they were who said that it doesn't matter what job they are in so long as they get paid for it and therefore the "steadiness of the employment" is all that is there on their minds which is why it is ranked a close sixth with on their list of motivators.

When queried further regarding their need for steady employment, we got the following answers:

18–25 years

I need this job, so that I can pay for my MBA.

Nobody tells me how am I doing at my job ... all they tell me how bad am I...

My job gives me money to spend on myself.

I felt good ... when I bought a gift for my sister of my own money .... Every year my father gave me money to buy gifts....

I saw people in office work late hours ... so that they get incentives ... I have also started doing that.

I am learning every day at office ... yesterday my supervisor praised me.

Yesterday I handled an extremely angry customer ... no senior was there ... I was scared but I tried to do my best ... later the customer called my supervisor and praised me ... it was great feeling.

My parents no longer feel that I am irresponsible ... now that I have started working.

In other words, it could be concluded that money (salary and incentives) along with a sense of entitlement for this age group are prime motivators for this age group.

26–33 years

I have just married; I have to take care of my wife.

My father has just retired; I am therefore the sole earning member of the house.

I have to contribute to my sister's wedding this summer.

My family has just bought a house and I want to pay for my portion, which is one story of the of the house.

I need this job so that I can develop my skills and experience.

I can make a difference in my workplace ... I have started to improve my process.

I get weekends off ... it is good to finish the work before the start of next week.

Work is work.... It has to be done ... its ok if I have to sit late in the office a few times a week.

I have got a promotion at 27 years I will celebrate it by buying a car and give it to my family.

It can thus be concluded that self-worth, money, and achievement are the main motivators of this age group.

2.8   CONCLUSIONS AND INFERENCES

As can be deduced from the above observations, the Millennials in India value two aspects, namely, "the pay and respect for me as a person" the most within their departments and companies. This indicates that there is a high sense of self-worth among them. As people Millennials exhibit a sense of entitlement, which makes them assume that people will come to them instead of they going to people for anything. This aspect is seen within the workplace where most often do not go out of their way to create workplace networks or make efforts to get along well with others on the job (they actually start networking a little late then Gen X). When asked about this reason they often say that "we are here to work and *not suck up* to the boss." There is an attitudinal aspect to their work-place behavior. Additionally, they often assert themselves in family and work-related decisions; in other words, they often declare their independence from the systems both at home and outside and thus pride themselves of their independent thinking. This aspect is often criticized by Gen X which perceives them as thankless, ungrateful, and unwilling to give credit to any suggestion or advice. At work place, supervisors are often at the receiving end of their confusing behavior wherein they exhibit extreme docility and aggression in quick succession. Supervisors at the shop floor have to be constantly on the lookout for any harm that may befall them as they are always known to be preoccupied on their phones in some manner or another and therefore they have a tendency to exhibit a careless behavior.

The second observation that was made regarding the Millennials between the age group of 26–33 years was that self-actualization in its myriad forms is sought by this group through (a) search for jobs, (b) definition of the responsibilities and duties, and (c) impact of their job on the company in order to achieve image enhancement. The Millennials prefer to scour various company websites for a better opening; additionally, they carry out thorough search on a company using multiple sources like LinkedIn, Google jobs, and other social media sites before applying. They are also not averse to utilizing their extensive personal network and talent communities before they consider a company for their next job change even if it involves relocating in any part of the world. By this stage in their life, they are aware their skill sets and capabilities to a larger extent, they thus look for job postings that clearly define their roles, duties, and responsibilities within the company. They do not like confused and amorphous job postings and usually they give those a wide berth. As they are well aware of their worth in a job, they often look for a job that could help them shine in front of their peers and family. They therefore value aspects like reward systems, perks, and facilities like laundry, cheap home loans, club/gym memberships, etc. which are often considered important by them. They therefore not only want a very clear understanding of the job, company, and the profile, but they also want to feel the power and importance that the job profile gives them. There inherent sense of entitlement, however, clearly demarcates the personal and professional life and they clearly like to keep it that way. However, the power and importance of their jobs keeps them and their team at their desks late in in the evening and often over the weekends. In all the above aspects, importance of money and various perks is never diminished, and clearly they feel money is important for (a) increased social standing, (b) buying property, (c) increased respect within the family, (d) better marriage prospects, and (e) enhanced self-worth that gives them a sense of "arrival" in the society.

2.9 DISCUSSION

In 2013 and 2015 institutions like Price Waterhouse Cooper (PwC), University of Southern California, and London Business School, and Futurestep, a Korn Ferry (NYSE:KFY) company carried out studies into the motivations of Gen Y and revealed aspects like—balance and workload, work/life imbalance, impact of the workload, etc. This report went on to state that "perceived pay equity" is an important factor for retaining Millennials. While the Futurestep of (2015) revealed that 13–18% of the Millennials consider pay to be an important factor besides clear path of advancement.

Our understanding of the "pay phenomenon" is different from these reports, and it is our perception that all aspects being similar, Indian Millennials prefer good-paying jobs over other aspects that have been highlighted by both the above-mentioned reports. The expressions of this phenomenon have been mentioned in the earlier part of our write-up.

Millennials have been brought up in secure, loving, and comfortable environment and therefore do not know deprivation and danger. Families are small today, Millennials therefore have never stayed, for most part of their lives away from their families. They therefore have a constant need for gratification and expect the outside world to cater to their whims as their families would.

This sense of idealism carried forward from their secure homes leads them to try and change the world, they are actually willing undergo all sorts of hardships, danger, and loneliness to create their desired world. According to Maslow "it's easier for them to make sacrifices especially when they have never faced chronic deprivation." Financial safety and security is not a priority, since they have never known a life without it. Self-development and job hopping are indicative of their hierarchy needs for gaining self-actualization (self-worth) quicker. Gen X often perceives this characteristic as impatience and non-willingness to "dig-in for a long haul." Gen Y on the other hand considers it appropriate to be ambitious and therefore asks for immediate results, in other words "if something can be changed for the better, then we better do it now and not wait for things to happen at some later date." Gen Y is all about self-actualization an aspect that is demonstrated in their "inside-out" attitude toward how they perceive their career, Gen X and the world in general. They want to solve problems that plague them and they do not hesitate to be creative about the solutions. They are often critical of Gen X which they think is holding them back. They are adaptive and are not adverse to change.

2.10 POLICY IMPLICATIONS FOR THE HR MANAGERS OF TODAY

2.10.1 TECHNOLOGY

Today technology influences the rapidly evolving HR environment and therefore there is a gap that requires an in-depth integration of technology in the areas like recruitment, retention, and nurturing the Millennials through training as future company leaders. There is a growing consensus among the researchers, human resource executives, and consultants on some of the strategies for engaging the Millennials. These strategies tend to rely on the use of open social-media-style platforms that stress on transparency along with the state of the art technology. Additionally, they also purposefully highlight an organization's core values, and various approaches taken toward cross-generational mentoring. They also point to the fact that the workplace has fun to be fun and where the training and collaborative efforts need to be reinvented as games or have inbuilt tangible rewards. HR executives around the globe are of the view that companies do not have to make massive changes in their business plans. Millennials are happy as long as their company uses technological tools with which they are comfortable in the company processes.

2.10.2 *SOCIAL IMMEDIACY*

In the past decade, a lot of companies have reworked their HR processes of worker feedback and recognition. They have made efforts to make worker commendations more open and accessible by sharing it on public sites like Facebook. The optimum way to go about this process is to engage departments like marketing and the IT to create websites of their own wherein encouragement from the peers and feedback from the management can be instantaneous. According to Kathy Weaver, Senior vice president of organizational excellence at the Dallas based global tax-services firm Ryan, "We wanted something that would engage, and bring feedback out for everyone to access," … Therefore through Facebook and other social-media sites, colleagues across the company can see each posting and add their own comments." Ryan thus created a website called Ryan-PRIDE which has, over the last three years, been upgraded to promote more feedback and encouragement from peers. The site's motto: "If you see it, say it."

Companies like Accenture, mindful of the Gen Y need to advance rapidly in their career, have created a career marketplace on the lines of LinkedIn as a way to encourage of Gen Y to shop of jobs in-house. Senior HR managers at Accenture are aware that "The biggest thing is career to Generation Y." Since they spend a lot of time on other social networks, including those that focus on various job opportunities, they are not necessarily loyal to the company.

Another company, Futurestep has formulated an online recruitment initiative called the "talent community" that focuses on the Millennials.

Neil Griffiths, a London-based global practice leader, believes that the talent communities often connect on a regular basis. He also says that companies like Kraft Foods have created talent community for recruiting military veterans so that a prospective worker and a company could be matched to an opening. This initiative engages young employees through websites through which prospective employees can collect and collate information about the company and develop a connection with the company brand.

2.10.3 MENTORING EACH OTHER: IT WORKS BOTH WAYS

Traditional view of mentoring is that it is a more knowledgeable person guides a less experienced person toward developing a certain area of expertise. This view holds that the more knowledgeable person is usually an older person, both in terms of age and experience.

This concept has modified itself especially when it comes to the use of technology, social media, and current trends. It is now called reverse mentoring, wherein Gen Y mentor's Gen X; in this way, they upgrade older employees in areas of technology and its consumption. These areas are often second nature to Gen Y employees, whose lives are intensively integrated with computers and the Web. According to Gina D'Ambra, human resource manager of Thermo Fisher Scientific, a Waltham, Massachusetts-based life sciences firm, "Connecting millennial employees to more older tenured employee, makes them feel more connected, and helps them map their future career path." This connection, more than anything else, is often the reason of Millennial retention in the company. Additionally, it also provides feedback on where they stand "good, bad, indifferent," within the department. This aspect is therefore an important factor in their stay within the company. Companies like Futurestep and PwC conclude that, Gen Y finds "development and ongoing feedback" among the issues that matter most to young workers, behind "ability to make an impact on the business" and having "a clear path for career advancement." Indian Millennials show a very similar need for career advancement.

2.10.4   MAKE IT PAY: PLAY GAMES

Reward is a powerful incentive in every area, and reward pays for an employee is an additional incentive to connect to the company. Simply put, the elements of a total rewards program include aspects that an organization uses to attract and retain employees; this includes salary, bonuses, incentive pay, benefits and employee growth opportunities such as professional development and additional training. Most of the incentives featured in the above list are provided to the employees as a matter of routine. However, Millennials need something extra—HR managers have come up with the concept of *gamification*. This provides a number of advantages to company and encourages Millennials to develop positive relationships with employers. The concept of gamification encourages Gen Y to pitch for rewards. These rewards could be in the form of stand-alone incentives, additional leaves, cash, facilities like company subsidized holiday very early in career, a film on company time, an additional day off while on an out of town assignment, certifications that have one time encashment build into it, and also a Knowledge pay, which encourages the Millennials to acquire additional qualifications. HR managers also try to build in an element of playfulness to assessments, evaluations, or business task by bringing a sense of competition. The participants are then given points with which they can reward their colleagues, which can later be redeemed by the awardee from an from an online catalogue with more than 2500 items that has accessories, trips, and electronics and appliances.

Mercer, warehousing and shipping company, uses gamification as a useful tool to engage this generation that grew up playing video games like "Call of Duty." HR managers at Mercer have been proactive in developing games that are aimed at profiling and recruiting candidates for its vast sales force. They also ask the game developers to build sophisticated psychometric tests. HR managers ask a potential candidate to take part in these games for the purpose of hiring.

2.10.5   LEARN AND CONNECT: WHAT ARE OTHER PEOPLE DOING IN MY COMPANY?

"Will this company help me grow?" is the most common refrain that can be heard from the Millennials. HR all over the world has to keep this aspect in mind, and at the same time, they have to keep in mind the restless and constantly shifting focus that defines Millennials. Large national and multinationals have therefore launched cloud learning centers that cater 24×7 to the need of Gen Y. Accenture in 2014 launched a program called "Connected Learning." This program in unlike any traditional learning program because it provides real-time information though the company's global network and is also accessible on mobile phones. Through this facility, the learning is round the clock as the employees can access write-ups, blogs, opinions, e-book, videos, and also interact with their colleagues and peers around the vast global network of Accenture.

KEYWORDS

- **Gen Y**
- **India**
- **millennials**
- **motivators**
- **managerial implications**

REFERENCES

Asghar, R. *Gen X is from Mars, Gen Y is from Venus: A Primer on How to Motivate a Millennial*, 2014 (online). Available from <http://www.forbes.com/sites/robasghar/2014/01/14/gen-x-is-from-mars-gen-y-is-from-venus-a-primer-on-how-to-motivate-a-millennial/> (accessed on 12 September 2015).

Barford, I. N.; Hester, P. T. *Analysis of Generation Y Workforce Motivation Using Multi-attribute Utility Theory*, 2011 (online). Available from <http://sigite2012.sigite.org/wp-content/uploads/2012/08/session07-paper01.pdf> (accessed on 8 October 2015).

Bunch, W. *Engaging Gen Y*, 2015 (online). Available from <http://www.hreonline.com/HRE/view/story.jhtml?id=534358521> (accessed on 4 September 1015).

Bunton, T. E.; Brewer, J. L. *Discovering Workplace Motivators for the Millennial Generation of IT Employees*, 2012 (online). Available from <http://sigite2012.sigite.org/wp-content/uploads/2012/08/session07-paper01.pdf> (accessed on 7 October 2015).

Das, S. 5 *Valuable Gifts of the Millennial Workforce*, 2015 (online). Available from <http://www.huffingtonpost.com/shinjini-das/5-valuable-gifts-of-the-millennial_b_7996110.html?ir=India&adsSiteOverride=in> (accessed on 12 September 2015).

Dorsey, J. *Y-size Your Business*, 2010 (online). Available from <http://jasondorsey.com/millennials/the-top-gen-y-questions-answered/> (accessed on 12 September 2015).

Futurestep. *Futurestep Survey Finds Compensation One of the Least Important Factors for Recruiting Millennial Talent*, 2015 (online). Available from <http://www.futurestep.com/news/futurestep-survey-finds-compensation-one-of-the-least-important-factors-for-recruiting-millennial-talent/> (accessed on 10 October 2015).

Gawel, J. E. Herzberg's Theory of Motivation and Maslow's Hierarchy of Needs. *Pract. Assess., Res. Eval.* **1997,** *5*(11) (online). Available from <http://PAREonline.net/getvn.asp?v=5&n=11> (accessed on 18 February 2016).

Gen Y and the World of Work. A Report into the Workplace Needs, Attitudes and Aspirations of Gen Y China, 2013 (online). Available from <http://social.hays.com/wp-content/uploads/2013/11/Read-the-full-report-on-Gen-Y-China-and-The-World-of-Work1.pdf> (accessed on 10 October 2015).

Gen Y and the World of Work. A Report into the Workplace Needs, Attitudes and Aspirations of Gen Y Sweden, 2014 (online). Available from <http://www.hays.se/cs/groups/hays_common/documents/digitalasset/hays_1253311.pdf> (accessed on 10 October 2015).

The Canberra Times. Generation Y takes over the Workplace, March 29, 2006, *The Canberra Times*.

Griffin, E.; Ledbetter, A.; Sparks, G. *A First Look at Communication Theory*, McGraw Hill Publication, USA, 2015.

Hindustan Times, New Delhi, Wednesday, September 16, 2015, *Hindustan Times*, p 15.

Dale Carnegie. *Igniting Millennial Engagement*, 2014 (online). Available from <www.dalecarnegie.com/assets/1/7/Engage_millennials_wp_031815.pdf> (cited 8 October 2015).

Kotler, P.; Keller, K. L. *Marketing Management* (12th ed.); Prentice Hall: Upper Saddle River, NJ, 2005; pp 81, 252–254.

Leschinsky, R. M.; Michael, J. H. Motivators and Desired Company Values of Wood Products Industry Employees: Investigating Generational Differences. *For. Prod. J.* **2004,** *54*(1), 34–39.

Main, D. *Who are the Millennials?* 2013 (online). Available from <http://www.livescience.com/38061-millennials-generation-y.html> (accessed on 12 September 2015).

McLeod, S. *Updated 2014 Carl Rogers*, 2007 (online). Available from <http://www.simplypsychology.org/carl-rogers.html> (cited 18 September 2015).

Siscovick, I. *Inside Employees' Minds (Survey)*, 2015. <http://www.mercer.com/newsroom/mercer-survey-finds-us-workers-satisfied-with-retirement-and-health-benefits-but-fears-of-future-affordability-rise-dramatically.html> (cited 13 October 2016).

Millennials at Work Reshaping the Workplace, 2011 (online). Available from <http://www.pwc.com/gx/en/managing-tomorrows-people/future-of-work/assets/reshaping-the-workplace.pdf> (accessed on 12 September 2015).

Montana, P. J.; Petit, F. Motivating and Managing Generation X and Y on the Job While Preparing for Z: A Market Oriented Approach. *J. Bus. Econ. Res.* **2008,** *6*(8), 35–39.

Montana, P. J.; Lenaghan, J. A. What Motivates and Matters Most to Generations X and Y. *NACE J.* **1999,** *59*(4), 27–30.

PwC. *PwC's NextGen: A Global Generational Study: Evolving Talent Strategy to Match the New Workforce Reality Summary and Compendium of Findings*, 2013 (online). Available from <http://www.pwc.com/gx/en/hr-management-services/pdf/pwc-nextgen-study-2013.pdf> (accessed on 10 October 2015).

Reputation Institute. *2013 Country RepTrak Topline Report in GEN Y AND THE WORLD OF WORK (2014): A Report into the Workplace Needs, Attitudes and Aspirations of Gen Y Sweden*, 2013 (online). Available from <http://www.hays.se/cs/groups/hays_common/ documents/digitalasset/hays_1253311.pdf> (accessed on 10 October 2015).

Reverse Mentoring. *Definition*, n.d. https://www.techopedia.com/definition/28107/ reverse-mentoring.

Ryan. *MyRyan—Careers*, 1999 (online). Available from <http://www.ryan.com/Careers/ OurCulture/myRyan> (cited 10 October, 2015).

Sai, P. V. S.; Swati, P. S. Managing Gen Y—Some Issues, Challenges and HR Initiatives. In: *Today's HR for a Sustainable Tomorrow*; Mishra, R. K, Sarkar, S.; Singh P., Eds.; Allied Publishers: New Delhi, 2012.

Sinha, J. B.; Daftuar, C. N.; Gupta, R. K.; Mishra, R. C.; Jayseetha, R.; Jha, S. S; Vijayakumar, V. S. R. Regional Similarities and Differences in People's Beliefs, Practices and Preferences. *Psychol. Dev. Soc.* **1994,** *6*(2), 131–149.

Srinivasan, V. Multigenerations in the Workforce: Building Collaboration. *IIMB Manage. Rev.* **2012,** *24*(1), 48–66.

Stahnisch, F. W.; Hoffmann, T. Kurt Goldstein and the Neurology of Movement During the Interwar Years Physiological Experimentation. *Clin. Psychol. Early Rehab.* **2010** (online) Available from <www.th-hoffmann.eu/texte/stahnisch.hoffmann.2010-goldstein_movement.pdf> (cited 18 September 2015).

The Economic Times, New Delhi, Tuesday, October 6, 2015, *The Economic Times*; p 10.

This, L. E.; Lippit, G. L. *Work Motivations Checklist 1A*. Development Publications: Washington, DC, 1999.

Determine What Motivates You:

Please indicate the six factors below which you believe are the most important in motivating you to do your best work.

1. _____	Steady employment
2. _____	Respect for me as a person
3. _____	Adequate rest periods or coffee breaks
4. _____	Good pay
5. _____	Good physical working conditions
6. _____	Chance to turn out quality work
7. _____	Getting along well with others on the job
8. _____	Having a local employee paper
9. _____	Chance for promotion
10. _____	Opportunity to do interesting work

11. _____   Pensions and other security benefits

12. _____   Not having to work too hard

13. _____   Knowing what is going on in the organization

14. _____   Feeling my job is important

15. _____   Having an employee council

16. _____   Having a written job description

17. _____   Being complimented by my boss when I do a good job

18. _____   Getting a performance rating

19. _____   Attending staff meetings

20. _____   Agreement with the organization's objectives

21. _____   Opportunity for self-development and improvement

22. _____   Fair vacation arrangements

23. _____   Knowing I will be disciplined if I do a bad job

24. _____   Working under close supervision

25. _____   Large amount of freedom on the job (chance to work without direct or close supervision).

Work Motivation Checklist, a copyrighted instrument, by Leslie E. This and Gordon L. Lippitt, *Work Motivations Checklist 1A*, Development Publications: Washington, DC.

CHAPTER 3

EMPLOYEE RETENTION STRATEGIES IN SERVICE INDUSTRIES: OPPORTUNITIES AND CHALLENGES

RABINDRA KUMAR PRADHAN*, LALATENDU KESARI JENA, and RANJAN PATTNAIK

Department of Humanities and Social Sciences, IIT Kharagpur, Kharagpur, West Bengal, India

Corresponding author. E-mail: rkpradhan@hss.iitkgp.ac.in

CONTENTS

ABSTRACT

Present day organizations are found to be taking all preventive measures to arrest attrition of their employee resources, but sadly they are unable to focus on retaining them. When we have introspected to explore the reason behind it, we found that action and attitude of bosses, managers, and the organization influences more than the salary, incentive, perks, etc. On the other hand, research findings have confirmed that today's knowledge workers employed in booming service sectors have a deep desire to feel that they are succeeding and their talents and capabilities are used in the best possible way—creating a difference to their career and organization. Therefore, the onus lies on HR to identify what resonates with employees and accordingly devise policies and strategies to meet the preferences and interests of them. This kind of proactive approach is believed to augment retention. Progressive countries of contemporary times are lined up for robust hiring plans to cater the young and skilled workforce in service industries. The biggest challenge is to retain these young talents as losing them will have a huge impact on organization. This study is an attempt to visit the factors that affect employee retention in one hand and document some of the best practices followed by selected service organizations. The implications for practitioners' are also discussed in light of the propositions.

3.1 INTRODUCTION

Hiring employees is just a start toward creating a strong workforce. Retaining and keeping them motivated and interested for optimal productivity, and success is most important. Research findings have brought out the fact that, the cost of replacing old employees with new ones is anticipated around twice the employee's annual salary. Precisely, an estimation of the total cost of losing a single position ranges from 30% of annual salary of the position for junior management to 70%, and the cost can be as high as 150% for senior leadership positions across industries (Hewitt Associates, 2004). Replacing one's high potential and high performing employees places a big dent in companies' budget. It is very much frustrating for an organization to go through the entire process of hiring and training, only to find employees leaving after 2 months or sometimes even quitting just after the training period is over (Dibble, 1999). In Indian context, given the high percentage of employees who plan to seek new employment opportunities as the job markets are rebounding; warrants HR professionals and business leaders to understand turnover's costly impact and focus on ways to keep their best employees on board (Sanghavi, 2015).

One of the foremost challenges for HR professional is to identify the interests and preferences of employees as the labor force of today's market is much different. The age old practices like "incentivizing financially" though required, but it is no longer a key attribute of retention management. Venkatanarayanan, President-HR, Rane Group, Chennai, India says that in this technology age by the time the organization thinks of giving a raise, employees have been already accepting better offers elsewhere (Venkatanarayanan, 2015). This warrants HR professionals to think beyond incentivizing to focus for creating a meaningful employee experience that can provide challenging assignments, while encouraging learning and career growth. Therefore, this study is attempting to visit the factors that affect employee retention in one hand and document some of the best practices followed by selected service organizations.

3.2 CHALLENGES FROM EMPLOYER'S PERSPECTIVE

Today's employers are most likely managing three generation of employees which creates various value sets within an organization. The first generations born during 1950–1970s are very much loyal to their employers and are prepared to work their way to the top. The second generations born during 1971–1980s are probably looking for a shortcut to the top. This generation entered the workplace when terms like downsizing, negative growth, and jobless recovery were catch phrases. The third generation, born during 1980s onward, understands that knowledge is power; their learning style is extremely adaptable, and they demand flexibility, have confidence in their ability, and expect quick promotions (Arnold, 2005). Employers can perceive these generations as fickle, selfish, and transient.

As the first generation employees retire and the young third generation enter the workforce, this massive demographic shift is causing big problems for even the most successful companies. Especially, these young third generations are highly sought after for their technological savvy, energetic work ethos, and young hip attitudes that can help companies to connect with young customers (Cotton & Tuttle, 1986). Many employers are able to recruit the young third generation employees effectively, but end up alienating and losing them shortly thereafter. It is observed by many employers that despite their good qualities, they do not always share the traditional values of boomers with whom they often come into conflict. Disenchanted by this attitude of young generations, the first generations are found giving up their parent employers and look for opportunities outside the corporate world (Hytter, 2007). This high turnover rate among all these generations who are to be replaced is costing companies billions of dollars every year (Whitt, 2006). However, it is imperative that one cannot afford to let generational differences stand in the way of getting things done. Nor can one afford to alienate one generation by favoring another.

Retaining talented employees is one of the topmost priorities of employers today. Each and every good employee is found to be quitting for better jobs, and it stands as an important concern amongst Indian PSUs and Private MNCs. Therefore, for HR professional, the challenge is not only to attract the best talent but also to retain them throughout their professional career.

Predominantly, rising opportunities for career development, life-style decisions, frequent job changing, work–life imbalance, poor mentoring, and stress are some of the factors which influence an individual's decision to continue or to quit (Krishnan & Singh, 2010; Rosse, 1988). For employers, retaining talent takes efforts to work out with aforesaid factors diligently. Wermeling (2009) in his study has stated that, companies that utilize effective employee retention strategies know that wages and benefits are not the most crucial factors in determining whether an employee will stay or leave. Instead employee wants to feel that they are valued contributors. Contemporary organizations have realized the fact that loosing talented employees are not only costly, but it also impacts morale and productivity and can have a negative effect on valued customers.

The employees stay or leave their jobs and organizations for various different reasons. The terminated employees leave the organizations because management wants them to do so, but those who leave organizations voluntarily are a matter of great concern. A survey conducted by McKinsey and Company (2012) found that the most critical factors that affect the attraction and retention of managers and executives are value and culture of the organization, the management of the organization, the freedom and autonomy in the job, career advancement and growth opportunities, the compensation and incentives that they receive, the geographic location they work in as well as respect they earn from their peers and colleagues. Lagunas (2015) in his study on retention factors across organizations in international scenario has opined that, "Employees don't quit jobs;" instead "They quit managers." His findings have estimated that 80% of turnover is driven by the environment a manager creates for an employee (compared to 20% resulting from issues with company culture). On the other hand, there are also umpteen personal issues like health problems, mortgage, divorce, death, etc. causing distraction from work and affecting the abilities of employees to continue working with an organization. Therefore, these kind of sensitive issues stand as a challenge for an employer to address (Boswell et al., 2012).

For our research, we have found more than 1500 keywords from scholarly works on employee retention from Google database agreeing that turnover is a crisis and searching the right talent is a challenge. Every kind of literature in this domain have impressed on the fact that organizations must have the best talent in order to succeed in the hypercompetitive and increasingly complex global economy. Along with the understanding of the need to hire, develop, and manage, organizations need to take necessary steps to retain talented people. However, a handful of researchers' have stated that that there is lack of enthusiasm for investing resources to maintain top talent. The question that arises at this point of time is, if talent is hard to find, then why not employers do the whole lot of thing for retaining them?

3.3   FACTORS AFFECTING RETENTION IN SERVICE-BASED INDUSTRIES

Retention is ideally an intentional move by an organization to construct an environment which engages employees in the long run (Chaminade, 2007). The main reason for retention is to avoid the loss of experienced employees from the organization as this could have adverse effect on organizational productivity. Most of the time employees migrate to competing organization with the acquired knowledge and trade secrets that they have learned from their former employers especially in service industries (Abassi & Hollman, 2000). Empirical studies in this connection have brought the fact that in this era of globalization employees usually go for a change in every 6 years (Pandit, 2007). This stands as a challenge for management to identify the reasons for such a frequent change of employment. Chary (2002) in his book on "Business Guru Speaks" have cited a quote of N. R. Narayana Murthy, Chief Mentor of Infosys, India in the context of employee retention as, "when our key assets, that is, employees walk out every evening, our net worth is reduced to zero. Our challenge is to ensure that they are coming back next day rejuvenated, refreshed and energized."

Keeping in view of the large cost associated with employee turnover, characterized by downsizing and layoffs, the onus is on HR managers to work out with sound policies that will enable them to retain their talented employees (Steel et al., 2002). These practices are reflected under the term "retention management." Johnson (2000) has defined retention management as, "the ability to hold onto those employees one want to keep for longer than its competitors." In the literature, numerous factors are put forward as important points to check in while devising plans for retention management. These factors vary from purely financial inducement to so-called new age benefits. Basically, these inducements are grouped into four major categories: (1) financial incentives, (2) career development opportunities, (3) job content, (4) social atmosphere, and (5) work–life balance (e.g., Horwitz et al., 2003; Roehling et al., 2000; Ulrich, 1998).

First, *financial rewards* or the provision of an attractive remuneration package are one of the most widely discussed retention factors in service sectors and especially in IT industries across the globe. They not only fulfill financial and material needs, but also have a social meaning with the salary level providing an indication of the employee's relative position of power and status within the organization. However, research shows that there is much interindividual variability in the importance of financial rewards for employee retention (Pfeffer, 1998; Woodruffe, 1999). For instance, a study conducted by the "Institute for Employment Studies" reveals that only 10% of people who had left their employer gave dissatisfaction with pay as the main reason for leaving (Bevan, 1997). However, despite the fact that many studies show financial rewards to be poor motivating factors, it remains a tactic used by many organizations to commit their employees to the organization by means of remuneration packages (Mitchell et al., 2001; Woodruffe, 1999). For instance, in a recent study made by Horwitz et al. (2003) states that the most popular retention strategies reported by HR managers of knowledge firms engaged in service sectors are related to compensation.

Second, *opportunities for career development* are considered as one of the most important factors affecting employee retention. It is suggested that an organization that wants to strengthen its bond with its employees must invest in the development of its human resource (Hall & Moss, 1998; Hsu et al., 2003; Steel et al., 2002). Other factors relating to career development are the provisions of mentoring or coaching to employees, the organization of career management workshops and the setup of competency management programs (Roehling et al., 2000). For instance, in a recent study conducted by Allen et al. (2003) found that employees perceptions of growth opportunities offered by their employer reduced turnover intentions. Steel et al. (2002) have also reported through an empirical data showing that lack of training and promotional opportunities were the most frequently cited reasons for high performers to leave the company.

The third category of retention factors relates to *employees job content*, more specifically the provision of challenging and meaningful work. It builds in the assumption that people do not just work for the money but to create a purpose and satisfaction in their life through their job (Mitchell et al., 2001; Pfeffer, 1998). According to Woodruffe (1999), employees in addition to a strong need to deliver excellent results also wants to take on difficult that are relevant for the organization if they find the job meaningful. However, if their work consists of the routine based performance of tasks, the likelihood of de-motivation and turnover gets to be relatively high. By thinking carefully about which task to include in which jobs, companies can affect their retention rates (Steel et al., 2002). Though, employers use to focus on rewards as it plays a critical role in employee attraction however, recognition has a huge role in aspects like employee engagement and retention.

The *social atmosphere*, that is, the work environment and the social ties with in this environment, is the fourth retention factor considered by many researchers. Cappelli (2001) states that loyalty to the organization is a thing of the past, but loyalty to one's colleagues may act as an effective means of retention. When an employee decides to leave the organization, this also means the loss of a social network. Some research suggests that social contacts between colleagues and department are an important factor for retaining talent. Organizations can contribute to the creation of a positive social atmosphere by stimulating interaction and mutual cooperation among colleagues and through open and honest communication between management and employees (Roehling et al., 2000).

3.4   ADDRESSING THE RETENTION AGENDA

According to the study report carried out by Bernthal and Wellins (2001) in Development Dimensions International stating that the turnover rate for nonmanagement employees were almost twice as high as for management employees (19.3% vs. 10.3%). This has proved that it is more difficult to retain frontline and professional staff in service industries, while there is an increase in turnover rates. The report has stated about top five reasons on employees leaving their most recent positions. They were quality of relationships with immediate supervisors, ability to balance work and home life, the feeling of making a difference, level of teamwork and cooperation with coworkers and level of trust in the workplace.

Since HR functionaries play a vital role in employee retention, a well-coordinated and market-driven strategy is warranted to be followed by them. This requires a fundamental change in understanding the nuances of employee retention. It is proposed that organizations can arrest the employee attrition through selective hiring, ensuring work–life balance and creating a healthy organizational climate. Some of the Indian and International best practices have been cited to enrich the discussion on these agendas.

3.4.1   SELECTIVE HIRING

An effective retention strategy begins at the early stages of the recruitment and selection process. This is true because most of the employees' turnover happens due to chemistry or bad fit. Research indicates that most of the people leave organizations due to mistakes made during the hiring phase (Trevor et al., 1997). For this reason, some smart companies are adopting the strategy of "hire for attitudes and train for skill." They have realized that it is easier to develop the skills and capabilities that an employer needs than to attempt to change the employee's personality or mindset.

For instance, Infosys, India receives about twice the number of applications from its competitors. However, it is very much selective in recruiting employees in its board. Infosys basically looks for candidates with good analytical ability, team work and leadership potential, besides innovation skills and structured approach for problem solving. Some companies go to even extraordinary lengths before hiring some body.

According to Hewitt, a leading Indian company was selecting a senior marketing manager after the candidate had gone through several rounds of interviews. The company had identified the person and thought he was the right one for the job. Hewitt associates were on the verge of making an offer. The HR head took him out for dinner and during the meal the prospective marketing manager was particularly rude to a waiter in front of his potential employer and peer. The company reversed its decision to recruit him because if he could be rude to the waiter in such a setting, how could this behavior reflect the company image and culture if he were in charge of marketing?

Besides getting the right people in the door, recruiting has an important symbolical aspect. When a prospective employee undergoes rigorous selection process, the person feels that he or she is joining an elite organization. This creates high expectations and gives an impression that people matter most to the organization (Ramlall, 2003; Wells & Thelen, 2002). When companies recruit people, they often focus on attracting precisely those people who will be the most difficult to retain. Peter Capelli, Professor of Wharton School of Business, advocates that companies must shift their sights to workers who can do the job, but are not in great demands and hence organizations may be able to shelter themselves from market forces. In India, few companies are found to be not going to premier business schools for recruitment, rather they visit second-tier technical and management institutions for campus recruitment in order to improve their selection process.

Since compatibility is the key to long-term relationships, cultural issues need to be addressed throughout the selection process, which begins with recruitment. To determine cultural fitness, companies must look beyond the information in job descriptions, while assessing potential candidates and making hiring decisions (Tett & Meyer, 1993). For instance, Southwest Airlines looks for those who are fit especially with company culture. During the interview of a pilot, interviewers probe for attitudes, beliefs, and behaviors that would give clues about how the potential employee might treat flight attendants or peers, how he might deal with conflicts at work and what mattered most to him. While exploring the reasons behind it, we have found an interesting piece of a paper on "why does Southwest care about an employee's sense of humor, especially the Pilots?" Southwest managers basically test the sense of humor of prospective employees in many ways during series of interviews because it indicates that the company is truly looking for a fit between the way works get done at Southwest Airlines and the personality of the prospective employees.

3.4.2   WORK–LIFE BALANCE

Work–life balance has an important consequence for employee attitudes toward their organizations as well as for the lives of employees. The work–life boundary may be especially significant in managing highly skilled knowledge workers, such as round the clock service professionals whose commitment and loyalty to adhere to different time zones presents a challenge to employees. Some of the Indian and international surveys and studies have shown that work–family initiatives have a very encouraging impact in areas such as employee motivation, satisfaction, commitment, and stress levels. We have tried to bring some of the study findings that buttress the return on investment on work–family initiatives:

(a) Mercer Inc. (2003) in its survey with more than 800 organizations has found from 86% of its surveyed employees stating that, "the companies cannot remain competitive without addressing work–life and diversity issues." In addition to it, more than 50% of the respondents said that work–life programs had a positive impact on employee productivity, morale, and attendance of employees.

(b) In a survey of employees and customers conducted by First Tennessee Bank (2015) revealed that business units run by managers, who were rated highest by employees in supporting work–family initiatives, had a 7% higher customer retention rates than those run by other managers.

(c) Nearly 60% of 3000 employees surveyed at Hoechst Celanese (1997) reported that the ability to balance work–family responsibilities was of great importance in their decisions to remain with the company.

(d) A recent survey carried out by Ceridian employer services confirmed what savvy managers have learned by experience: Flexible work arrangements are "highly successful" in retaining employees. Nearly two-third of Ceridian's respondents felt that virtual teams, flexible work plans, and telecommuting were effective in boosting retention.

In order to attract and retain talent, smart organizations are addressing the issues of work–life balance. For instance, the work–life at Infosys, India is tailored around the personal lives of its employees. The company positions itself as an extension of family and takes care of every individual need, from seeing an employee through a close relative's illness to celebrating special occasions together. A number of special occasions are organized regularly where both the employees and their family participate. Besides other programs, annual special day like "Pretit Infocian" is held where children of employees participate and have an opportunity to explore their parents' workplace.

The work–life of Hewlett-Packard, India focuses around helping employees in meeting the company's objectives as well. HP appreciates personal life needs of an employee; beyond providing statutory leaves, an additional leave often days are provided for employees who are getting married. Another 10 days, if the employee is preparing for a competitive exam, 8 weeks for legally adopting a child, 5 days for employees who suffer bereavement in the family; 5 days for new fathers. Further depending upon the need, the employees can take leave beyond stipulated time. All special events from employee's birthday to business achievements are celebrated with sounds of success.

Retaining employees in the hospitality industry was vital as the cost of recruiting and training new employees was very high. Marriott operated in an industry where every day counted and weekends and holidays generated more business than weekdays. Customer service had to be provided on a 24/7/365 basis. The implications was that employees had to go through a hectic work schedule, with an average work week lasted for more than 50 h. With the increasing workload due to rising customers in late 1990s, several key managers at Marriott left. They wanted to devote more time in their personal lives and their jobs at Marriott were not helping the cause. Facing this challenge, Marriott launched a new program called "Management Flexibility" on a pilot basis at three of its hotel chain. The aim was to assist Marriott's managers in balancing their professional and personal lives, without negatively affecting customer service or the company financials. As a result of this the company was able to avoid unnecessary meetings and formal procedures. The working hours of managers at these hotels were reduced to five days per week. The emphasis was given more to work being done properly rather than the number of hours worked. The results were pretty encouraging. Feedback revealed that after implementing the program, managers felt their jobs are less strenuous and that burnout was significantly lower.

3.4.3 BUILDING AN ADAPTIVE ORGANIZATIONAL CULTURE

Building an adaptive and strong organizational culture is a prerequisite for employee engagement and satisfaction, which in turn improves retention. Companies with strong and adaptive cultures earn significant premium over competitors, making the nurturing of such cultures exceedingly desirable. A study of 207 companies in 22 countries found that adaptive companies get revenue four times greater than that of nonadaptive companies achieving 12 times the stock price performance (Kotter & James, 1992). In a research conducted by Collins and Porras (1994), organizations that are built to last characterize having a strong organizational culture. They have revisited the dimensions of culture and proposed an adaptive culture on the ground of employee retention. They have stated culture as

- A deep seated powerful ideology or set of core beliefs in which employees are thoroughly indoctrinated with an insistence on "walking the talk."
- It acts like glue that holds the members together.

Whereas adaptive culture should embrace:

- The ability to embrace change—through gradual evolution or placing big strategic bets.
- Encouraging trials and error, opportunism of "purposeful accidents."
- A penchant for self-improvement, even if they are already best in class.

Retention is basically aligned with a high-performance organization culture and is built around elements such as innovative, purpose, leadership, and trust. However, the core element which acts as a dynamo is "people" in service industries. This is because in service-oriented industries, people brings performance, process do not. Therefore, people management practices need to be revamped and synthesized to form high-performance work systems. High-performance work systems are a kind of management approach in achieving optimal fit between people, technology, work, and information (Vandenberg et al., 1999) in service-based industries. The characteristics of high-performing work systems that augment employee retention are as follows:

- Flexible organizational settings where work rules and job descriptions are fluid and employee friendly.
- Minimal formal or hierarchical structure.
- High level of employee involvement and participation in the operation of organization policies and decision making.
- Highest level of continuous learning and development.
- High level of transparency and shared communication.

National Thermal Power Corporation of India, India practices on recruitment, training and development, mentoring, comprehensive benefits, and employee welfare establishes a culture of high performance making the company the second most efficient power generator in the world. A survey commended by Mercer HR consulting in 2014 revealed that the company's leadership commitment to employees, the alignment of its HR policies with corporate culture and its ability to create work–life balance has achieved a low attrition rate of 1.7% (for 2014).

Nicholas Piramal India Ltd., a pharmaceutical company with a long history of merger and acquisitions, is an amalgamation of many cultures, most of being multinational in nature. As a result, the company imbibed the best of all cultures and created its own which reflects unity in its diversity. The company's work culture is focused on high performance, entrepreneurship, and empowerment and is based on mutually beneficial personal development that understands and helps employees to manage their dreams and goals. The HR system is designed with transparency and feedback as the primary pivots of employee evaluation and growth. The company through dedicated and regular training programs runs throughout the year, providing organizational learning and development opportunities to all its employees. Communication is a priority for top management. It is evident that the company's chairman regularly shares success and triumphs with the company as a whole through personalized meeting and digital in-house journals which reach majority of the employees. Nicholas Piramal has set up a culture where in talent at all levels with leadership potential can be spotted quickly and potential leaders were presented with the opportunity to grow. This has been seen from the fact that the attrition levels are very much lower than the industry norms.

Tata Steel is a company which has a high-value system and a philosophy of "total return to the community." Tata Steel has been a pioneer in introducing many HR initiatives like 8-h day, free medical aid, establishment of welfare departments at plant levels, paid leave, provident fund, work-men's accident compensation, technical training, maternity benefits, gratuity, social audit, pension schemes, etc. which later became the statutory provisions of Govt. of India. Tata Steel has taken several initiatives on retention management through improving the motivational level of employees, ensuring their participation, enhancing decision making abilities at all levels. A formalized system of personal development laying emphasis on training of the employees, talent review, job rotation systems, performance linked compensations, formal rewards and recognition systems, linking of knowledge management systems with appraisal, leadership opportunities in quality circles, continuous improvement programs, formal and documented moral and social conduct code and ethical norms, etc. are some of the feature of such initiatives. The company also has a provision of "an extremely transparent and credible multi-path communication system" using both formal and online tools.

3.5   IMPLICATIONS FOR HR PRACTITIONERS

Service-oriented organizations that are serious about managing and retaining talents need to develop a formalized strategy. The key to having an effective strategy is to fully understand the real reasons that people choose and leave an organization. Turnover can have a significant impact on a company's bottom line (Mitchell et al., 2001). While the cost of turnover varies from company to company, experts estimate that the loss of an employee can range from 30% to up to 150% of that person's salary. Unfortunately, most organizations do not have a formal retention strategy to address this key business issue (Mohamed et al., 2006). Instead, they approach retention on as need basis rather than developing a comprehensive, coordinated, and preventive strategy.

Before the problem of retention can be addressed, one needs to know where the problem exists. The first step toward developing a retention strategy is to quantify and measure turnover rates for the entire organization, including specifics for each department and supervisor (Rosse & Hulin, 1985; Winterton, 2004). The second step is to analyze the competition. What companies are competing with for employees? Do the organization's compensation package, working conditions, and reputations match or exceed those of the competitors?

Since the relationships between supervisor and employee are a vital factor in retention, the strategy should include a plan for selecting and developing leadership talent (Kaira, 1997). Most people who become supervisors never receive training on managing employees until they are in the job for a significant amount of time or in some cases, not at all. One of the key measures of supervisor performance should be employee satisfaction and turnover (Brockbank, 1999; Döckel et al., 2006). It is found that the more positive the work environment, the higher is the retention rate. Most organizations spend more time criticizing and reprimanding than they do praising and recognizing (Kochanski & Ledford, 2001). Workers who are emotionally engaged in their work are less likely to change jobs. A good retention strategy addresses each employees need to understand his or her importance to the company and its mission. Finally, the strategy should establish open communications based on trust and integrity throughout the company.

Why an employee stays with an organization is a strategic issue for HR managers in order to workout retention policies and practices that are effective at individual and organizational levels. Literature survey indicates that career development is the most important retention factors since offering good opportunities for career development not only prevents employees from leaving the company, but also contributes in a positive way to their loyalty to the organization (Maurer & Lippstreu, 2008; Weng et al., 2010). HR managers should better take into account what their employees values and how they evaluate their organizations efforts toward retention management, if they are to contribute in a cost-efficient way to the strategic objectives of the organization. If HR managers are to be effective in their retention management this means that they should take into account this subjectivity instead of departing from generally agreed upon views on what is important to employees in general. This in turn, should contribute to their role in the company as a strategic partner given that the attraction and retention of talented employees will stay an important factor of competitive advantage for service-oriented organizations both in times of economic downturn and upheaval.

KEYWORDS

- selective hiring
- work–life balance
- adaptive culture
- reward
- recognition
- retention
- best practices
- retention strategies

REFERENCES

Abassi, S. M.; Hollman, K. W. Turnover: The Real Bottom Line. *Publ. Pers. Manage.* **2000,** *2*(3), 303–342.

Allen, D. G.; Shore, L. M.; Griffeth, R. W. The Role of Perceived Organizational Support and Supportive Human Resource Practices in the Turnover Process. *J. Manage.* **2003,** *29*(1), 99–118.

Arnold, E. Managing Human Resources to Improve Employee Retention. *Health Care Manager* **2005,** *24*(2), 132–140.

Bernthal, P. R.; Wellins, R. *Retaining Talent: A Benchmarking Study.* Development Dimensions International: Pittsburgh, PA, 2001, https://www.ddiworld.com/DDIWorld/media/client-results/reducingturnover_rr_ddi.pdf?ext=.pdf (accessed on 7 June 2015).

Bevan, S. Quit Stalling, *People Manage.* **1997,** *1*(1), 32–35.

Boswell, W. R.; Zimmerman, R. D.; Swider, B. W. Employee Job Search: Toward an Understanding of Search Context and Search Objectives, *J. Manage.* **2012,** *38*(1), 129–163.

Brockbank, W. If HR Were Really Strategically Proactive: Present and Future Directions in HR's Contribution to Competitive Advantage. *Hum. Resour. Manage.* **1999,** *38*(4), 337–352.

Cappelli, P. A Market-driven Approach to Retaining Talent. In: *Harvard Business Review on Finding and Keeping the Best People*; Harvard Business School Press: Boston, MA, 2001; pp 27–50.

Chaminade, B. *A Retention Checklist: How Do You Rate?* 2007. www.humanresourcesmagazine.co.au (accessed on 9 June 2015).

Chary, S. N. *Business Gurus Speak.* McMillan India Ltd.: New Delhi, 2002.

Collins, J. C.; Porras, J. I. *Built to Last.* Harper Collins Publishers: New York, 1994.

Cotton, J. L.; Tuttle, J. M. Employee Turnover: A Meta-analysis and Review with Implications for Research. *Acad. Manage. Rev.* **1986,** *11*(1), 55–70.

Dibble, S. *Keeping Your Valuable Employees: Retention Strategies for Your Organization's Most Important Resource.* Wiley: New York, 1999.

Döckel, A.; Basson, J. S.; Coetzee, M. The Effect of Retention Factors on Organizational Commitment: An Investigation of High Technology Employees. *SA J. Hum. Resour. Manage./SA Tydskrif vir Menslikehulpbronbestuur.* **2006,** *4*(2), 20–28.

First Tennessee Bank. *Frost & Sullivan Honors First Tennessee Bank with 2015 Best in Class CIO Impact Award,* 2015. Retrieved from http://www.ciceroinc.com/frost-sullivan-honors-first-tennessee-bank-with-2015-best-in-class-cio-impact-award/ (accessed on 9 June 2015).

Hall, D. T.; Moss, J. E. The New Protean Career Contract: Helping Organizations and Employees Adapt. *Organ. Dyn.* **1998,** *26*(3), 22–37.

Hewitt Associates. *Employee Engagement Higher at Double Digit Growth Companies. Research Brief.* Hewitt associates LLC, 2004.

Hoechst Celanese Survey. *Hoechst Celanese Ties Work/Family Programs to Employee Retention,* 2015. Retrieved from http://www.thefreelibrary.com/Hoechst+Celanese+Survey+Ties+Work%2FFamily+Programs+to+Employee...-a019180457 (accessed on 9 June 2015).

Horwitz, F. M.; Heng, C. T.; Quazi, H. A. Finders, Keepers? Attracting, Motivating and Retaining Knowledge Workers. *Hum. Resour. Manage. J.* **2003,** *13*(4), 23–44.

Hsu, M. K.; Jiang, J. J.; Klein, G.; Tang, Z. Perceived Career Incentives and Intent to Leave. *Inf. Manage.* **2003,** *40*(1), 361–369.

Hytter, A. Retention Strategies in France and Sweden. *Irish J. Manage.* **2007,** *28*(1), 59–79.

Impact of Work and Non-work Related Issues. *J. Manag. Issues* **2006,** *18*(4), 512–529.

Johnson, M. W*inning the People War, Talent and the Battle for Human Capital.* Copyright Licensing Agency: London, 2000.

Kaira, S. K. Human Potential Management: Time to Move Beyond the Concept of Human Resource Management? *J. Eur. Ind. Train.* **1997,** *21*(5), 176–180.

Kochanski, J.; Ledford, G. 'How to Keep Me': Retaining Technical Professionals. *Res. Technol. Manage.* **2001,** *44*(3), 31–38.

Kotter, J. P.; James, L. H. *Corporate Culture and Performance.* Free Press: New York, 1992.

Krishnan, S. K.; Singh, M. S. Outcomes of Intention to Quit of Indian IT Professionals, *Hum. Resour. Manage.* **2010,** *49*(3), 421–437.

Lagunas, K. *5 Employee Retention Strategies for a High Performance Environment,* 2012, Retrieved from: http://www.tlnt.com/2012/07/17/5-employee-retention-strategies-for-a-high-performance-environment/ (accessed on 8 June 2015).

Maurer, T. J.; Lippstreu, M. Who Will be Committed to an Organization that Provides Support for Employee Development? *J. Manage. Dev.* **2008,** *27*(2), 328–347.

McKinsey and Company. McKinsey Quarterly: Is There a Payoff from Top Team Diversity? *McKinsey and Company Report,* 2012.

Mercer Inc. Mercer Study Raises Red Flags for Employer Pay and Benefits Plans (Findings of the 2002 People at Work Survey). *Human Resource Department Management Report,* 2003; pp 8–15.

Mitchell, T. R.; Holtom, B. C.; Lee, T. W. How to Keep Your Best Employees: Developing an Effective Retention Policy. *Acad. Manage. Execut.* **2001,** *15*(4), 96–109.

Mohamed, F.; Taylor, G. S.; Hassan, A. Affective Commitment and Intent to Quit: The Pandit, Y. V. L. Talent Retention Strategies in a Competitive Environment. *NHRD J.* **2007,** *Hyderabad,* 27–29.

Pfeffer, J. Six Myths about Pay. *Harv. Bus. Rev.* **1998,** *76,* 38–57.

Ramlall, S. Managing Employee Retention as a Strategy for Increasing Organizational Competitiveness. *Appl. Hum. Resour. Manage. Res.* **2003,** *8*(2), 63–72.

Roehling, M. V.; Cavanaugh, M. A.; Moynihan, L. M.; Boswell, W. The Nature of the New Employment Relationship: A Content Analysis of the Practitioner and Academic Literatures. *Hum. Resour. Manage.* **2000,** *39*(4), 305–320.

Rosse, J. G. Relations Among Lateness, Absence, and Turnover: Is There a Progression of Withdrawal? *Hum. Relat.* **1988,** *41*(1), 517–531.

Rosse, J. G.; Hulin, C. L. Adaptation to Work: An Analysis of Employee Health, Withdrawal and Change. *Organ. Behav. Hum. Dec. Process.* **1985,** *36*(2), 24–47.

Sanghavi, N. Steps to Employee Retention. *Bus. Manager* **2015,** *17*(11), 29–30.

Steel, R. P.; Griffeth, R. W.; Hom, P. W. Practical Retention Policy for the Practical Manager. *Acad. Manage. Execut.* **2002,** *16,* 149–162.

Tett, R. P.; Meyer, J. P. Job Satisfaction, Organizational Commitment, Turnover Intention and Turnover: Path Analysis based on Meta-analytic Findings, *Pers. Psychol.* **1993,** *45*(1), 259–293.

Trevor, C. O.; Gerhart, B.; Boudreau, J. W. Voluntary Turnover and Job Performance: Curvilinear and the Moderating Influences of Salary Growth and Promotions. *J. Appl. Psychol.* **1997,** *82*(1), 44–61.

Ulrich, D. Intellectual Capital = Competence × Commitment. *Sloan Manage. Rev.* **1998,** *39*(2), 15–26.

Vandenberg, R. J.; Richardson, H. A.; Eastman, L. J. The Impact of High Involvement Work Processes on Organizational Effectiveness: A Second-order Latent Variable Approach. *Group Organ. Manage.* **1999,** *24*(3), 300–339.

Venkatanarayanan, R. People Retention. *Bus. Manager* **2015,** *17*(11), 36–37.

Wells, M.; Thelen, L. What Does Your Workspace Say About You? The Influence of Personality, Status and Workspace on Personalization. *Environ. Behav.* **2002,** *3*(1), 300–321.

Weng, Q.; McElroy, J. C.; Morrow, P. C.; Liu, R. The Relationship Between Career Growth and Organizational Commitment. *J. Vocation. Behav.* **2010,** *77*(2), 391–400.

Wermeling, L. Social Work Retention Research: Three Major Concerns. *J. Sociol., Soc. Work Soc. Welfare* **2009,** *3*(1), 1–21.

Whitt, W. The Impact of Increased Employee Retention on Performance in a Customer Contact Center. *Manuf. Serv. Oper. Manage.* **2006,** *8*(3), 235–252.

Winterton, J. A Conceptual Model of Labour Turnover and Retention. *Hum. Resour. Dev. Int.* **2004,** *7*(3), 371–390.

Woodruffe, C. *Winning the Talent War: A Strategic Approach to Attracting, Developing and Retaining the Best People.* John Wiley & Sons: Chichester, UK, 1999.

CHAPTER 4

GENDER EQUALITY AT DECISION-MAKING POSITIONS IN ORGANIZATIONS

URBAN ŠEBJAN and POLONA TOMINC*

Faculty of Economics and Business, Razlagova 14, University of Maribor, Maribor, Slovenia

Corresponding author. E-mail: polona.tominc@um.si

CONTENTS

ABSTRACT

The gender equality at all areas of life and work is the focus of European Union and its fundamental value. Gender balance is important from both, macroeconomic and microeconomic perspective, where, among other aspects, the positive correlation between the share of women at the companies' decision-making positions and the companies' performance exists. But causes for the underrepresentation of women in decision-making processes and positions within organizations are multidimensional and complex. Purpose of the present research is to analyze the gender equality at decision-making positions within organizations in Slovenia in more depth, with three major objectives: (i) to analyze the gender difference regarding the perceived equality of genders in the organizations at the decision-making positions; (ii) to analyze the gender differences regarding reconciliation of private and professional life of female and male managers; and (iii) to investigate the opinion of female and male managers regarding possible/feasible measures to achieve gender equality in decision-making positions in the economy. A survey with random sample of female and male managers of medium-sized and large companies in Slovenia was conducted. Research results suggest that the sociocultural attitudes toward women, that put women in the center of the family with all responsibilities in relation to that, seem not to have the negative effect on gender equality at decision-making positions in organizations in Slovenia. Research results suggest that system-imposed barriers are more influential and refer especially to those at the company level, such as unclear or vague career advancement criteria leading to biased decision regarding career promotions and opinions regarding women's career potential. Research brings several policy implications for policy makers as well as for business practice.

4.1   INTRODUCTION

The gender equality at all areas of life and work has been the focus of European Union (EU) and its fundamental value since the Treaty on EU in 1957 (consolidated version of the Treaty on EU, 2012). Despite the sustained emphasis and activities in this field, the success is only partial and limited. As gender equality index shows, there have been visible, albeit marginal, improvements in the last observed time period in EU (EIGE, 2015). Since gender equality is a multidimensional and complex concept, the gender equality index covers eight domains, with six of them (work, money, knowledge, time, power, and health) being combined in a core index, and two additional domains (violence and intersecting inequalities). From the view point of the present research, it is important to note that "most pronounced, although marginal improvements are evident in the domain of work and money, reflecting the EU's focus on economic and labor market policy" (EIGE, 2015, p. 3). Although the domain of work measures the extent to which women and men can benefit from equal access to employment and appropriate working conditions (participation), the extent of segregation, and quality of work, it can be viewed as a rough estimate of gender equality at decision-making positions in the economy as well. Results show that not only are women less likely to participate in the labor force in all EU member states, they also tend to work fewer hours and are likely to spend fewer years in work than men overall. Women also represent the strong majority of employees in predominantly feminized sectors (education, health services, and social work).

On top of that, research results show that inequality is also present in the field of payments (European Commission, 2014). Women's average earnings in male-dominated occupations are approximately as high as men's average earning, while the situation is quite different in integrated occupations with the gap being even deeper in female-dominated occupations (Tominc, 2002). Statistical data show (European Commission, 2015) that for the economy as a whole (http://ec.europa.eu/eurostat/statistics-explained/index.php/Gender_pay_gap_statistics) in 2013, women's gross hourly earnings were on average 16.4% below those of men in the EU and 16.6% in the euro area. Across Member States, the gender pay gap varied by 26.7 percentage points, ranging from 3.2% in Slovenia to 29.9% in Estonia (Eurostat, 2015). As expected, economic sectors with even negative gender gaps were in 2013 male-dominated sectors. Thirteen Member States registered negative gender pay gaps in the water supply, sewerage, waste management, and remediation activities and 12 Member States in the construction industry.

Evidence from the literature suggests that gender equality is important for economic and social development, with a bidirectional relationship between economic development and improving the ability of women to access the constituents of development, especially in low-income countries (Duflo, 2012), but in the EU Member States as well: the strong positive relationship between the Gender Equality Index and the gross domestic income across EU Member States is noticeable (EIGE, 2013).

Statistical data show that women are underrepresented in decision-making positions, particularly in politics and business across the EU Member States. Although the situation differs between EU countries, on average women represented only 28% of members of the single or lower houses of parliaments in the EU countries in November 2014. In business organizations the gap regarding leadership positions is even deeper. At the end of 2014, women accounted for only one-fifth of board members of the largest publicly listed companies registered in the EU countries and less than one-tenth of president positions (European Commission, 2015).[1]

Purpose of the present research is to analyze the gender equality at decision-making positions within organizations in Slovenia in more depth. This research investigates whether the differentiation between females and males at decision-making positions in organizations exists—this gender gap can be reflected in the perceived attitudes and support of family regarding the coordination of family responsibilities, in satisfaction with employment, current job position, and career, in the perceived gender equality in the organization and in economy in general, in satisfaction with the quality of work and life—these are domains analyzed in the present research, as well as in other aspects. The objective of this research is also to test and investigate the opinion of female and male managers regarding different aspects of measures and actions aimed at achieving the equality of genders at decision-making positions in organizations.

Objectives of the present research are

(i)   to analyze the gender difference regarding the perceived equality of genders in the organizations at the decision-making positions;
(ii) to analyze the gender differences regarding reconciliation of private and professional life of female and male managers; and
(iii) to investigate the opinion of female and male managers regarding possible/feasible measures to achieve gender equality in decision-making positions in the economy.

---

[1]The companies covered are the largest publicly listed companies in each country. Publicly listed means that the shares of the company are traded on the stock exchange. The "largest" companies are taken to be the members (max. 50) of the primary blue-chip index, which is an index maintained by the stock exchange covering the largest companies by market capitalization and/or market trades. Only companies which are registered in the country concerned (according to the ISIN code) are counted (European Commission, 2015).

4.1.1 THEORETICAL BACKGROUND AND HYPOTHESES FORMED

Equality in decision-making positions is one of the five priority areas (others being equal economic independence, equal pay for equal work and work of equal value, dignity, integrity, and an end to gender-based violence, gender equality in external action) in Women's Charter (European Commission, 2010a) and European Commission's Strategy for Equality between Women and Men 2010–2015 (European Commission, 2011c). The strategy highlights the contribution of gender equality to economic growth and sustainable development, and support the implementation of the gender equality dimension in the Europe 2020 Strategy (European Commission, 2010b).

The importance of the gender balance in EU strategic documents is not surprising since it has effect at both, macro and micro levels. From the macroeconomic view-point, research results report that increase of women's empowerment in the labor force is correlated with economic growth particularly in low-income countries (Duflo, 2012), but in the EU as well (Lofstrom, 2009), although the causality between the two is unknown. In developed economies, the gender balance in the economy may play an important role in addressing the problem caused by aging of population and the burden of pensions (European Commission, 2011a).

At the microlevel perspective, an increasing number of reports indicate that a positive correlation between the share of women at the companies' decision-making positions and the companies' performance exists, although, again, results do not prove any causality (European Commission, 2011a). Also, research results of Ittonen et al. (2013) suggest that firms with female audit engagement partners are associated with smaller abnormal accruals, thereby implying that female auditors may have a constraining effect on earnings management. Their research results support the view that the behavioral differences between women and men may have important implications for the quality of auditing and financial reporting. Moreover, research results suggest (Palvia et al., 2014) that banks with female CEOs hold more conservative levels of capital after controlling for the bank's asset risk and other attributes. Furthermore, while neither CEO nor Chair gender is related to bank failure in general, the strong evidence that smaller banks with female CEOs and board Chairs were less likely to fail during the financial crisis was found.

Again, these results support the view that gender-based behavioral differences may affect corporate decisions.

Also gender diversity among members of management boards contributes to board effectiveness. Perrault (2015, p. 149) discusses "that through real and symbolic representations, women enhance perceptions of the board's instrumental, relational, and moral legitimacy, leading to increased perceptions of the board's trustworthiness which in turn fosters shareholders' trust in the firm."

On the other hand, causes for the underrepresentation of women in decision-making processes and positions are multidimensional and complex (European Commission, 2015). Traditional gender roles and stereotypes are still present in societies, suggesting that household activities and child care continued to be seen as women's tasks. With the purpose to analyze the gender difference regarding the perceived equality of genders in the organizations at the decision-making positions (objective (i) of the present research), hypotheses H1–H3 were formed:

H1—Regarding the perceived gender equality at decision-making positions within organizations significant differences between female and male managers exist.

H2—Regarding the perceived satisfaction with the work and life significant differences between female and male managers exist.

H3—Regarding the perceived satisfaction with the employment position and career significant differences between female and male managers exist.

The lack of support for women and men to balance care responsibilities are also often present. Work–life balance is affected by lack of access to care services for children, elderly, and disabled persons, inadequate maternity-leave schemes and not flexible working arrangements. The work–life balance and gender differences regarding reconciliation of private and professional life of female and male managers were analyzed within the objective (ii) and hypotheses H4 and H5 were formed:

H4—Regarding the perceived family attitudes and support in the balance of work and life, significant differences between female and male managers exist.

H5—Regarding the carried out share of domestic and family obligations, the significant differences between women and men managers exist.

Barriers are found also in many political and corporate cultures where traditional gender roles prevail and lead to the vertical segregation of women, who are underrepresented in management positions that lead to the top management positions. Despite the fact that many international corporations have adopted and implemented diversity policies, women are still seriously underrepresented in decision-making roles, which is reflected in statistical data of share of female presidents or members of the Management Boards. The objective (iii) of the present research is therefore to investigate the opinion of female and male managers regarding possible/feasible measures to achieve gender equality in decision-making positions at the corporate level, with the purpose to identify the most suitable ones.

H6—It is possible to identify possible/feasible measures to achieve gender equality in decision-making positions at the corporate level, based on the opinions of mangers.

4.1.2 SLOVENIA—COUNTRY CHARACTERISTICS

Slovenia is a small open economy and an EU country that has been quite significantly affected by the global economic crisis. Historically, Slovenia was a part of the Austrian–Hungarian Monarchy. It also has the experience of socialism and of a communist history, since it spent also almost half of the century as part of the former Yugoslavia. Slovenia has been a member of the EU since 2004.

The transition process from socialism into the market economy has brought profound changes to the labor market, also regarding the female participation in the labor force, the average level of education, the gender wage gap, etc. The transition process has affected both men and women with a loss in job security and employment costs, but it seems women took over a large share of the adjustment costs (Ruminska-Zimny, 2003). Moreover, it seems that transition changes have also had important and often negative effects on women's position in society where men and women under the communist regime were supposedly equal in all aspects of society (Stoyanovska, 2001). However, with the fall of the communist regime, structural inequalities between men and women became evident (Tominc, 2002). A common characteristic of labor market developments during the transition process, which is also strongly present in Slovenian economy, is gender asymmetry, seen in employment, sectorial changes of employment, income and wages, access to jobs in the private sector, etc. The not completely finished transition process is lately combined with the effects of global economic crisis.

As already mentioned, the statistical data about the extent of the gender imbalance at decision-making positions in companies show that women are still underrepresented (European Commission, 2011a–c). In EU, in general, the participation of women in labor market has grown to nearly 50% (to 44.9% in Slovenia in May 2015) as well has grown the share of women among university graduates to 60% (to 61.1% in Slovenia in May 2015). But unfortunately, the corresponding increase of use of women talents at the decision-making positions in companies is not observed: In business organizations, the gap regarding leadership positions is deep. In October 2014, women accounted for just 20.2% of board members of the largest publicly listed companies registered in the EU countries and only 7% of president positions (European Commission, 2015). There are substantial differences across countries: in France, Latvia, Sweden, and Norway more than 25% of board members are female, while in the Czech Republic and Malta this share is less than 5%. Slovenia is in this respect the average EU country with 20% of females as board members. In NACE sectors from A to F (agriculture, mining, quarrying, and manufacturing), the share was considerably higher—28%, while in NACE sectors from G to K (wholesale, retail, and services) the reported share is lower, only 16%. Regarding the share of female chairpersons of the highest decision-making body in the company, Slovenia is below EU average, with 5%, while 12 out of 28 Member States reported less than 5%.

4.2　DATA AND METHODOLOGY

With the purpose to achieve the above-described objectives of the present study, the random sample of female and male managers in Slovenia was obtained, using the questionnaire.

Design of the questionnaire was conducted in three steps. In the first step, we reviewed the literature and concluded that established and standardized way of measuring these issues in the academic literature does not yet exist as a whole, only partial solutions until now can be used (European Commission, 2011b; Judge et al., 1994; Karatepe & Baddar, 2006; Lyness & Brumit Kropf, 2005; Orser & Leck, 2010; Shapiro & Olgiati, 2002; Singhapakdi et al., 2013). Therefore, we performed also the in-depth individual interviews. The purpose of in-depth interviews was to gather relevant facts, which reflect the personal opinion of the persons who are directly integrated into the decision-making process in companies. Their views represent an additional illumination of the studied areas, helping in the process of designing a questionnaire that will provide the reliability and validity of measurement in the case of random sample used. With this purpose in-depth individual interviews with three female and three male managers in five companies in Slovenia were conducted in February 2014. Two companies were large and three were medium sized according to the number of employees. The aim of this part of qualitative research was therefore to obtain information that helped using the design of specific domains that constitute the multidimensional variables used to analyze these complex phenomena.

In the second step, the on-line questionnaire was prepared and pretested with three male and three female managers. This step resulted in the measurement scales of multidimensional variables, as described in Table 4.1.

With the purpose to test the hypotheses H1–H4, the multidimensional variables are represented in the questionnaire by several statements. We used the following scale: 1 = strongly disagree, 2 = disagree, 3 = mildly disagree, 4 = neither agree nor disagree, 5 = mildly agree, 6 = agree, and 7 = strongly agree. Respondents marked the level of agreement with the statements.

TABLE 4.1 The Measurement Scales for Multidimensional Variables.

Perceived gender equality at decision-making positions within organizations
[GE1]: I feel the same willingness to bid for the top managerial positions in comparison with the opposite gender.
[GE2]: I believe I have the necessary abilities and skills to take over the leading position in comparison with the opposite gender.
[GE3]: I think I have less freedom because of the family obligations, compared with the opposite gender.
[GE4]: I believe that people trust me (as a manager) equally in comparison with the opposite gender.

TABLE 4.1   *(Continued)*

Perceived gender equality at decision-making positions within organizations

[GE5]: I believe that I have the same opportunities to be at the managerial position as the opposite gender.

[GE6]: I believe that I have the same responsibilities for the management of organizations like the opposite sex.

[GE7]: I believe that my rights to take over a managerial position in the business society are treated equally in comparison with opposite sex.

[GE8]: Opinions are divided regarding the gender balance on the boards of organizations.

[GE9]: The law enables equal gender representation on boards of organizations.

[GE10]: In the organization, I feel equal in the decision-making process compared with the opposite gender.

[GE11]: In general, I perceive equal influence in the organization compared with the opposite gender.

[GE12]: I believe that I am equally paid in comparison with the opposite gender.

Perceived satisfaction with the work and life

[SWL1]: In general I feel physically safe at the workplace.

[SWL2]: My job provides positive health benefits.

[SWL3]: I do enough for myself to staying good health.

[SWL4]: I'm satisfied with the payment I receive for my work.

[SWL5]: My job has a positive impact on my family.

[SWL6]: At work, I have good friends.

[SWL7]: I have enough free time to be able to enjoy other things in life.

[SWL8]: I feel valued in the business environment.

[SWL9]: People in the organization respect my expertise in my field of work.

[SWL10]: I feel that working on my position allows me to realize my potential overall.

[SWL11]: I feel that I learned new things that can help me to better perform work activities.

[SWL12]: My working position allows me to strengthen my professional skills.

[SWL13]: My work requires a lot of creativity.

[SWL14]: My working position helps me to develop creativity outside of the workplace.

Perceived satisfaction with the employment position and career

[SEC1]: Overall, I am satisfied with my working position.

[SEC2]: I'm often thinking to stop engaging at a managerial position.

TABLE 4.1 *(Continued)*

Perceived satisfaction with the employment position and career

[SEC3]: In general, I am satisfied with the type of work I'm doing at the job.

[SEC4]: Most people in similar workplace are very satisfied with the work situation.

[SEC5]: People working in a senior managerial position often think to quit to pursue a leading position.

[SEC6]: In general, I am satisfied with the success that I have achieved in my career.

[SEC7]: I'm satisfied with the progress in meeting the goals of my entire career.

[SEC8]: I'm satisfied with the progress in meeting the goals regarding my personal income.

[SEC9]: I'm satisfied with the progress in meeting goals for my career promotion.

[SEC10]: In general, I am satisfied with my career.

Perceived family attitudes and support in the balance of work and life

[FA1]: Difficulties of my working situation interfere with my family life.

[FA2]: Due to the time devoted to my work it is hard to meet my family obligations.

[FA3]: Because of the duties associated with the work, I have to often change plans for family activities.

[FA4]: Requirements of the family or spouse (partner) are connected with my work situation.

[FA5]: At work place, I have to often postpone my work obligations because of the time I need for my family.

[FA6]: Due to demands of my family, I often cannot do at home the work related to my managerial positions that I should do.

[FA7]: My family life hinders my responsibility in the workplace, for example, timely arrival to work, completing everyday tasks, overtime work, etc.

For testing the hypothesis H5, the question about the share of domestic and family obligations that are the responsibility of the respondent was included into the questionnaire and the values were indicated by respondents as numerical values (in %).

For testing H6, several proposed measures to achieve gender equality in decision-making positions at the corporate level, were included into the questionnaire, as presented by Table 4.2. Again, the 1–7 Likert scale of agreement with the statements was used.

The questionnaire included also three questions gathering demographics, that is, gender, level of education, and managerial position of the respondent in the organization.

In the third, final step, the questionnaire was used in the survey. The sampling frame consisted of medium-sized and large companies (more than 50 employees) in Slovenia. From this frame, 150 companies were randomly selected. The questionnaire was sent to the e-mail addresses of male and female managers and executive managers (president, president of the Management Board, members/member of the board, supervisors, managers, partners, directors, owner, etc.), representing top management, who manage an organization. We received $n = 32$ completely filled in questionnaires, representing 21.3% response rate.

TABLE 4.2   Measures to Achieve Gender Equality in Decision-making Positions at the Corporate Level.

Measures
[M1]: To achieve gender equality in decision-making positions, professional and public debates should be implemented.
[M2]: Slovenia should implement an annual conference on gender equality in decision-making positions.
[M3]: Organizations should provide appropriate staff workshops to raise awareness on gender equality, for men and for women.
[M4]: Measures to achieve gender equality in decision-making positions should be created at EU level.
[M5]: The state should raise awareness on gender equality in the workplace with the help of the national media.
[M6]: Country should draw up a national program to achieve the gender equality in decision-making positions.
[M7]: Any organization that have developed an internal good practice on gender equality in decision-making positions, should receive the certificate/award.
[M8]: Each organization must develop the internal legislation/act regarding the gender equality in decision-making positions.
[M9]: It would be necessary to establish and continuously monitor in-depth statistics on gender equality in decision-making positions at the level of organization, at national level, by sector, etc.)
[M10]: Organizations should continuously carry out surveys on gender equality in decision-making positions.
[M11]: The principles of gender equality in decision-making positions should be included in the general guidelines and strategic objectives of the organization.

TABLE 4.2 *(Continued)*

Measures
[M12]: Organizations should use the intranet as a means of providing greater transparency on key issues and practices in the field of gender equality.
[M13]: Organizations should assess executive employees based on the achievement of the goals of gender equality.
[M14]: To achieve gender equality in decision-making is necessary to improve the knowledge and surveillance in connection with the relevant social legislation.
[M15]: Organizations should provide employees family services (maternal and paternal services).
[M16]: Organizations should provide services to employees and allowances for child care.
[M17]: Organizations should provide staff training to facilitate entry into the workplace after parental leave.
[M18]: Problems and awareness on gender equality in the society should be included into all levels of the school system.

The basic characteristics of the sample structure are described in Table 4.3.

TABLE 4.3 Sample Characteristics.

Characteristics	Relative frequency (%)
Gender	
Male	56.25
Female	43.75
Education	
Secondary school	6.25
High school	6.25
University graduates	59.37
Master of science/doctorate	28.13
Position in the company	
President of the Management Board	3.20
Member of the management Board	9.36
Partner, manager of the company, regional manager	40.64
Other managerial positions	
	46.80

Data analysis was performed using univariate and multivariate statistical methods. For constructing the multidimensional variables, the factor analysis was used. The analysis of the data set was based on exploratory factor analysis (Hair et al., 2010) in which the principal component analysis and Varimax method were used. The Bartlett's test of sphericity (BTS), the Kaiser–Meyer–Olkin statistics (KMO > 0.5) and the significance level ($p < 0.10$) were calculated. The reliability and validity of the measurement instrument was tested, keeping in mind the Cronbach's alpha ($\alpha > 0.7$). We have examined factor loadings ($\eta \geq 0.5$), communalities of variables ($h > 0.4$) and eigenvalues of factors ($\lambda \geq 1.0$).

For testing gender differences, the t-test was used at 0.10 significance level.

4.3   RESULTS

4.3.1   VALIDITY AND RELIABILITY ANALYSIS OF THE MEASUREMENT SCALES—FACTOR ANALYSIS

The results of the factor analysis indicate that it is meaningful to use it regarding the BTS and KMO statistics, while the results lead to two factors solution for each multidimensional variable, as presented by Table 4.4. With the purpose to obtain as high as possible share of variance explained by factors of meaningful structure, items with lower factor loadings and communalities were excluded. The second and the third iteration of the factor analysis led to the eight items for "Perceived gender equality at decision-making positions within organizations," eight items for "Perceived satisfaction with the work and life," eight items for "Perceived satisfaction with the employment position and career," and five items for "Perceived family attitudes and support in the balance of work and life." All four obtained measurement scales proved high reliability (Cronbach's alpha > 0.7).

All factor weights of the items included are higher than the value of 0.5. A total of 76.58% of total variance is explained by the "Perceived gender equality at decision-making positions within organizations," factors. According to the factor loadings factor 1 describes "perceived current gender equality at decision-making positions in organization," while factor 2 represents "perceived legislative options and personal willingness to participate at the highest decision making positions in the organization."

A total of 72.66% of total variance is explained by "Perceived satisfaction with the work and life" factors. Factor loadings allow us to explain factor 1 as "perceived satisfaction with the attitude of work colleagues and with payments." This factor includes several aspects: satisfaction with payment, appreciation and respect of colleagues, professionalism, and creativity. Factor 2 describes "perceived friendship and physical safety."

TABLE 4.4 Factor Analysis Results—Communalities and Factor Loadings—and Reliability Analysis Results.

Items of "Perceived gender equality at decision-making positions within organizations"	Comm.	Factor 1—"perceived current gender equality"—loadings	Factor 2—"legislative options and personal willingness"—loadings
GE1	0.736		0.810
GE4	0.684	0.825	
GE5	0.782	0.846	
GE6	0.766	0.861	
GE7	0.747	0.831	
GE9	0.832		0.905
GE10	0.818	0.814	
GE12	0.761	0.709	

KMO: 0.819; BTS: Chi-square = 174.236, $p < 0.001$; Cronbach's alpha: 0.911

Items of "Perceived satisfaction with the work and life"	Comm.	Factor 1—"perceived attitude of work colleagues, payments"—loadings	Factor 2—"perceived friendship and physical safety"—loadings
SWL1	0.803		0.896
SWL4	0.515	0.718	
SWL6	0.814		0.874
SWL8	0.836	0.869	
SWL9	0.799	0.799	
SWL10	0.869	0.926	
SWL12	0.606	0.771	
SWL13	0.571	0.665	

KMO: 0.767; BTS: Chi-square = 147.737, $p < 0.001$; Cronbach's alpha: 0.861

TABLE 4.4 *(Continued)*

Items of "Perceived gender equality at decision-making positions within organizations"	Comm.	Factor 1—"perceived current gender equality"—loadings	Factor 2—"legislative options and personal willingness"—loadings
Items of "Perceived satisfaction with the employment position and career"	Comm.	Factor 1—"general perceived satisfaction—employment" —loadings	Factor 2—"perceived aversion to negative attitudes"—loadings
SEC1	0.862	0.846	
SEC2	0.933		−0.962
SEC3	0.714	0.819	
SEC6	0.855	0.893	
SEC7	0.894	0.945	
SEC8	0.690	0.829	
SEC9	0.798	0.893	
SEC10	0.848	0.883	
KMO: 0.773; BTS: Chi-square = 246.635, $p < 0.001$; Cronbach's alpha: 0.875			
Items of "Perceived family attitudes and support in the balance of work and life"	Comm.	Factor 1—"perceived deprivation of family life"—loadings	Factor 2—"perceived obstacles in fulfilling job obligations"—loadings
FA1	0.782	0.876	
FA2	0.825	0.900	
FA3	0.747	0.844	
FA6	0.812		0.887
FA7	0.822		0.795
KMO: 0.548; BTS: Chi-square = 67.896, $p < 0.001$; Cronbach's alpha: 0.739			

Constructs obtained by the factor analysis reveal that using the factor analysis is meaningful, and that the factors—constructs explain sufficiently high proportion of variance of original variables. All constructs obtained with the factor analysis were used for testing hypotheses H1–H4.

For the "Perceived satisfaction with the employment position and career" a total of 82.44% of total variance was explained with the two factors. From the values of factor loading in Table 4.4, it can be concluded that while factor 1 covers several different aspects of this variable and can be described as "general perceived satisfaction with employment position and career," factor 2 covers "perceived aversion to negative attitudes towards engaging at managerial position."

A total of 79.76% of the total variance for "Perceived family attitudes and support in the balance of work and life" is explained by the two factors obtained. Factor loading suggest that factor 1 describes "perceived deprivation of family life due to the complexity of work," while factor 2 deals with "perceived obstacles in fulfilling job obligations because of family obligations."

4.3.2   HYPOTHESES TESTING

Table 4.5 shows the results of testing H1 on the level of individual items within the multidimensional variable "perceived gender equality at decision-making positions within organizations," while Table 4.6 brings the result using the two factors obtained with the factor analysis. Among all items, female managers assessed the highest the perception of having less freedom because of the family obligations, compared with the opposite gender (5.56 ± 0.984); male managers on general do not perceive that this is true (2.57 ± 1.399). Female managers are on general also more self-confident regarding the necessary abilities and skills needed for leading positions as compared to male managers (women: 5.72 ± 0.895; men: 5.57 ± 1.555) but the gender difference is not significant. On the other hand, male managers assessed the highest statement regarding bidding for the top managerial positions in comparison with the opposite gender (6.14 ± 0.949), but female managers assessed the agreement with this statement high as well (5.17 ± 1.425), nevertheless the difference is significant. It is important to notice, that the beliefs to have the same opportunities to be at the managerial position as the opposite gender are significantly lower by female managers as compared to their male working colleagues (women: 4.22 ± 1.555; men: 5.57 ± 1.399). At all other statements, male managers assessed the agreement with them on average higher, with the difference being significant at six out of remaining eight statements. These results show the support for H1, which is confirmed also at the level of both factors, as presented in Table 4.6—Hypothesis H1 that regarding the perceived gender equality at decision-making positions within

organizations significant differences between female and male managers
exist is therefore supported.

TABLE 4.5   Results of Testing H1 (Individual Items).

Items	Gender	Mean	Std. Dev.	Sig.
[GE1]: I feel the same willingness to bid for the top managerial positions in comparison with the opposite gender	F	5.17	1.425	$p < 0.05$
	M	6.14	0.949	
[GE2]: I believe I have the necessary abilities and skills to take over the leading position in comparison with the opposite gender	F	5.72	0.895	$p > 0.10$
	M	5.57	1.555	
[GE3]: I think I have less freedom because of the family obligations, compared with the opposite gender	F	5.56	0.984	$p < 0.001$
	M	2.57	1.399	
[GE4]: I believe that people trust me (as a manager) equally in comparison with the opposite gender.	F	5.11	1.278	$p > 0.10$
	M	5.71	1.069	
[GE5]: I believe that I have the same opportunities to be at the managerial position as the opposite gender	F	4.22	1.555	$p < 0.05$
	M	5.57	1.399	
[GE6]: I believe that I have the same responsibilities for the management of organizations like the opposite gender	F	5.06	1.514	$p > 0.10$
	M	5.93	1.385	
[GE7]: I believe that my rights to take over a managerial position in the business society are treated equally in comparison with opposite gender	F	4.89	1.367	$p > 0.10$
	M	5.64	1.216	
[GE8]: Opinions are divided regarding the gender balance on the boards of organizations	F	3.78	1.263	$p < 0.05$
	M	4.86	1.512	
[GE9]: The law enables equal gender representation on boards of organizations	F	5.11	1.132	$p < 0.05$
	M	5.93	0.997	
[GE10]: In the organization, I feel equal in the decision-making process compared with the opposite gender	F	4.44	1.580	$p < 0.001$
	M	6.14	0.663	
[GE11]: In general, I perceive equal influence in the organization compared with the opposite gender	F	4.50	1.543	$p < 0.001$
	M	5.86	0.770	
[GE12]: I believe that I am equally paid in comparison with the opposite gender	F	4.39	1.819	$p < 0.001$
	M	5.93	1.141	

TABLE 4.6   Results of Testing H1 (Factors).

Factor	Gender	Mean	Std. Dev.	Sig.
1: "Perceived current gender equality at decision-making positions in organization"	F	−0.288	1.053	$p < 0.10$
	M	0.370	0.820	
2: "Perceived legislative options and personal willingness to participate at the highest decision making positions in the organization"	F	−0.334	0.997	$p < 0.05$
	M	0.429	0.855	

Tables 4.7 and 4.8 bring results of testing hypothesis H2, first at the level of individual items and also at the level of both factors for multidimensional variable "Perceived satisfaction with the work and life". The highest assessment by female managers was obtained regarding the feeling of physically safe at the workplace (5.78 ± 1.003), which was assessed even higher by male managers (5.86 ± 1.027), while the difference between genders is not significant. Item that was assessed the highest by male managers is that they perceive that their work requires a lot of creativity (6.21 ± 0.975), where the difference with female managers (5.28 ± 1.274) is statistically significant. Few more significant differences between the two genders are found. Male managers on general feel significantly more highly valued by the business environment as compared to female managers (women: 4.72 ± 1.320; men: 5.64 ± 0.842); they also feel that their expert work is respected in organization to the higher extent as compared by female managers (women: 5.00 ± 1.138; men: 5.86 ± 0.535). Research results suggest that hypothesis H2 is not entirely supported. This conclusion is suggested also based on the results at the level of factors, presented in Table 4.8.

TABLE 4.7   Results of Testing H2 (Individual Items).

Items	Gender	Mean	Std. Dev.	Sig.
[SWL1]: In general I feel physically safe at the workplace	F	5.78	1.003	$p > 0.05$
	M	5.86	1.027	
[SWL2]: My job provides positive health benefits	F	4.11	1.367	$p > 0.05$
	M	3.86	1.406	
[SWL3]: I do enough for myself to stay in good health	F	4.28	1.742	$p > 0.05$
	M	4.36	1.336	

TABLE 4.7   *(Continued)*

Items	Gender	Mean	Std. Dev.	Sig.
[SWL4]: I'm satisfied with the payment I receive for my work	F	4.56	1.464	$p > 0.05$
	M	5.21	0.802	
[SWL5]: My job has a positive impact on my family	F	4.39	1.243	$p > 0.05$
	M	4.71	1.326	
[SWL6]: At work, I have good friends	F	4.94	1.349	$p > 0.05$
	M	5.21	1.051	
[SWL7]: I have enough free time to be able to enjoy other things in life	F	3.56	1.542	$p > 0.05$
	M	4.00	1.301	
[SWL8]: I feel valued in the business environment	F	4.72	1.320	$p < 0.05$
	M	5.64	0.842	
[SWL9]: People in the organization respect my expertise in my field of work	F	5.00	1.138	$p < 0.05$
	M	5.86	0.535	
[SWL10]: I feel that working on my position allows me to realize my potential overall	F	4.56	1.381	$p > 0.05$
	M	5.21	0.893	
[SWL11]: I feel that I learned new things that can help me to better perform work activities	F	5.61	0.916	$p > 0.05$
	M	5.86	0.864	
[SWL12]: My working position allows me to strengthen my professional skills	F	5.22	1.060	$p > 0.05$
	M	5.43	0.852	
[SWL13]: My work requires a lot of creativity	F	5.28	1.274	$p < 0.05$
	M	6.21	0.975	
[SWL14]: My working position helps me to develop creativity outside of the workplace	F	4.44	1.756	$p < 0.05$
	M	5.57	0.756	

TABLE 4.8   Results of Testing H2 (Factors).

Factor	Gender	Mean	Std. Dev.	Sig.
1: "Perceived satisfaction with the attitude of work colleagues and with payments"	F	−0.313	1.125	$p < 0.05$
	M	0.403	0.646	
2: "Perceived friendship and physical safety"	F	−0.082	1.060	$p > 0.10$
	M	0.106	0.945	

While the significant gender differences regarding factor 1 are found, testing differences regarding factor 2 revealed no significant result. Hypothesis H2 that regarding the perceived satisfaction with the work and life significant differences between female and male managers exist is partly supported.

Results of testing H3 are presented in Tables 4.9 and 4.10. Table 4.9 presents results of testing gender differences regarding individual items of variable "perceived satisfaction with the employment position and career."

TABLE 4.9 Results of Testing H3 (Individual Items).

Items	Gender	Mean	Std. Dev.	Sig.
[SEC1]: Overall, I am satisfied with my working position	F	5.22	1.263	$p > 0.10$
	M	5.50	0.855	
[SEC2]: I'm often thinking to stop engaging at a managerial position	F	2.72	1.447	$p > 0.10$
	M	2.64	1.277	
[SEC3]: In general, I am satisfied with the type of work I'm doing at the job	F	5.39	1.037	$p > 0.10$
	M	5.79	0.699	
[SEC4]: Most people in similar workplace are very satisfied with the work situation	F	4.56	0.984	$p < 0.10$
	M	5.14	0.663	
[SEC5]: People working in a senior managerial position often think to quit to pursue a leading position	F	3.11	1.323	$p > 0.10$
	M	3.50	1.454	
[SEC6]: In general, I am satisfied with the success that I have achieved in my career	F	4.94	1.349	$p < 0.05$
	M	5.86	0.663	
[SEC7]: I'm satisfied with the progress in meeting the goals of my entire career	F	4.78	1.396	$p < 0.10$
	M	5.43	0.646	
[SEC8]: I'm satisfied with the progress in meeting the goals regarding my personal income	F	4.33	1.572	$p < 0.05$
	M	5.36	0.745	
[SEC9]: I'm satisfied with the progress in meeting goals for my career promotion	F	4.56	1.381	$p < 0.05$
	M	5.50	1.019	
[SEC10]: In general, I am satisfied with my career	F	5.11	1.183	$p > 0.10$
	M	5.64	0.745	

Significant differences are found especially regarding satisfaction with different aspects of achieved career, although women and men assess these items relatively high. Lower assessments as compared to males were achieved by female managers regarding the satisfaction with success achieved in career (women: 4.94 ± 1.349; men: 5.86 ± 0.663), regarding meeting goals of the entire career (women: 4.78 ± 1.396; men: 5.43 ± 0.646), as well as regarding personal income (women: 4.33 ± 1.572; men: 5.36 ± 0.745) and career promotion (women: 4.56 ± 1.381; men: 5.50 ± 1.019). The overall satisfaction with the career does not differ significantly between genders (women: 5.11 ± 1.183; men: 5.64 ± 0.745). Significant differences regarding items describing employment positions were not found.

TABLE 4.10   Results of Testing H3 (Factors).

Factor	Gender	Mean	Std. Dev.	Sig.
1: "General perceived satisfaction with employment position and career"	F	−0.297	1.178	$p < 0.05$
	M	0.382	0.539	
2: "Perceived aversion to negative attitudes towards engaging at managerial position"	F	0.0495	1.089	$p > 0.10$
	M	−0.064	0.908	

Again, research results suggest that H3 is partly supported. That is confirmed also based on the gender differences at the level of factors that are presented in Table 4.10, where significant difference was found regarding the general perceived satisfaction with employment position and career, while women and men do not differ significantly regarding the perceived aversion to negative attitudes toward engaging at managerial position. Therefore, hypothesis H3 that regarding the perceived satisfaction with the employment position and career significant differences between female and male managers exist is partly supported.

Test results of testing H4 regarding the "Perceived family attitudes and support in the balance of work and life" variable, are presented in Table 4.11 at the level of individual items and in Table 4.12 at the level of both factors.

TABLE 4.11 Results of Testing H4 (Individual Items).

Items	Gender	Mean	Std. Dev.	Sig.
[FA1]: Difficulties of my working situation interfere with my family life	F	4.00	1.782	$p > 0.10$
	M	4.14	1.657	
[FA2]: Due to the time devoted to my work it is hard to meet my family obligations	F	4.11	1.410	$p > 0.10$
	M	4.43	1.651	
[FA3]: Because of the duties associated with the work I have to often change plans for family activities	F	4.50	1.581	$p > 0.10$
	M	3.86	1.956	
[FA4]: Requirements of the family or spouse (partner) are connected with my work situation	F	2.06	1.056	$p > 0.10$
	M	2.50	1.092	
[FA5]: At workplace, I have to often postpone my work obligations because of the time I need for my family	F	2.67	1.572	$p > 0.10$
	M	2.79	1.311	
[FA6]: Due to demands of my family, I often cannot do at home the work related to my managerial positions that I should do	F	1.89	1.183	$p > 0.10$
	M	2.00	1.617	
[FA7]: My family life hinders my responsibility in the workplace, for example. Timely arrival to work, completing everyday tasks, overtime work, etc.	F	3.06	1.514	$p > 0.10$
	M	2.86	1.834	

Mean assessments of individual items are on general relatively low by both genders, also none gender differences were found. There is no clear pattern of differently perceived aspects regarding family attitudes and support in the balance of work and life between men and women. Results are similar when testing H4 at the level of factors, as presented in Table 4.12.

TABLE 4.12 Results of Testing H4 (Factors).

Factor	Gender	Mean	Std. Dev.	Sig.
1: "Perceived deprivation of family life due to the complexity of work"	F	0.017	0.946	$p > 0.10$
	M	−0.022	1.102	
2: "Perceived obstacles in fulfilling job obligations because of family obligations"	F	0.018	0.830	$p > 0.10$
	M	−0.024	1.218	

No significant gender differences are observed, neither regarding the perceived deprivation of family life due to the complexity of work (factor 1), nor regarding perceived obstacles in fulfilling job obligations because of family obligations (factor 2). Therefore, hypothesis H4 that regarding the perceived family attitudes and support in the balance of work and life significant differences between female and male managers exist is not supported.

But significant gender differences were found regarding the carried out share of domestic and family responsibilities, when testing hypothesis H5, as presented in Table 4.13.

TABLE 4.13   Results of Testing H5.

Variable	Gender	Mean	Std. Dev.	Sig.
The carried out share of domestic and family responsibilities	F	65.00	22.622	$p < 0.001$
	M	37.14	11.387	

Female managers reported that they carry out on average 65% of all domestic and family responsibilities, while male managers much less, 37.14%. The gender differences are also statistically significant; therefore, hypothesis H5 regarding gender differences in the carried out share of domestic and family responsibilities is supported.

In Table 4.14, the possibility and feasibility of measures to achieve gender equality in decision-making positions at the corporate level, based on the opinions of mangers are presented. Measures are ranked regarding the mean values of assessment, separately for female and male managers, gender differences were tested as well.

TABLE 4.14   Measures to Achieve Gender Equality in Decision-making Positions within Organizations.

Measure	Gender	Mean	Std. Dev.	Sig.
[M1]: To achieve gender equality in decision-making positions professional and public debates should be implemented.	F	4.56	1.294	$p > 0.10$
	M	4.36	1.499	
[M2]: Slovenia should implement an annual conference on gender equality in decision-making positions.	F	4.56	1.199	$p > 0.10$
	M	4.43	1.604	

TABLE 4.14   *(Continued)*

Measure	Gender	Mean	Std. Dev.	Sig.
[M3]: Organizations should provide appropriate staff workshops to raise awareness on gender equality, for men and for women	F	4.61	1.501	$p > 0.10$
	M	4.14	1.460	
[M4]: Measures to achieve gender equality in decision-making positions should be created at EU level	F	5.00	1.283	$p > 0.10$
	M	4.86	1.351	
[M5]: The state should raise awareness on gender equality in the workplace with the help of the national media	F	5.00	1.237	$p > 0.10$
	M	4.79	1.424	
[M6]: Country should draw up a national program to achieve the gender equality in decision-making positions	F	5.00	1.414	$p > 0.10$
	M	4.50	1.557	
[M7]: Any organization that have developed an internal good practice on gender equality in decision-making positions, should receive the certificate/award	F	5.28	1.447	$p < 0.10$
	M	4.64	1.393	
[M8]: Each organization must develop the internal legislation/act regarding the gender equality in decision-making positions	F	4.72	1.526	$p < 0.05$
	M	3.86	1.512	
[M9]: It would be necessary to establish and continuously monitor in-depth statistics on gender equality in decision-making positions at the level of organization, at national level, by sector, etc.	F	4.89	1.367	$p > 0.10$
	M	4.43	1.505	
[M10]: Organizations should continuously carry out surveys on gender equality in decision-making positions.	F	4.56	1.580	$p > 0.10$
	M	3.71	1.773	
[M11]: The principles of gender equality in decision-making positions should be included in the general guidelines and strategic objectives of the organization	F	5.22	1.517	$p > 0.10$
	M	4.64	1.499	
[M12]: Organizations should use the intranet as a means of providing greater transparency on key issues and practices in the field of gender equality	F	4.72	1.602	$p > 0.10$
	M	4.14	1.351	
[M13]: Organizations should assess executive employees based on the achievement of the goals of gender equality	F	5.17	1.339	$p < 0.10$
	M	4.00	1.922	
[M14]: To achieve gender equality in decision-making is necessary to improve the knowledge and surveillance in connection with the relevant social legislation	F	5.00	1.188	$p > 0.10$
	M	4.43	1.399	

TABLE 4.14 *(Continued)*

Measure	Gender	Mean	Std. Dev.	Sig.
[M15]: Organizations should provide employees family services (maternal and paternal services)	F	5.22	1.478	$p < 0.05$
	M	3.93	1.385	
[M16]: Organizations should provide services to employees and allowances for child care	F	5.17	1.339	$p < 0.05$
	M	3.57	1.399	
[M17]: Organizations should provide staff training to facilitate entry into the workplace after parental leave	F	5.00	1.609	$p > 0.10$
	M	4.29	1.437	
[M18]: Problems and awareness on gender equality in the society should be included into all levels of the school system	F	5.22	1.396	$p > 0.10$
	M	5.21	1.122	

Although the assessments between both genders mainly do not differ significantly, female managers assess all measures higher than male managers. The highest assessment by female managers is reported regarding the suggestion that any organization that have developed an internal good practice on gender equality in decision-making positions, should receive the certificate/award (women: 5.28 ± 1.447; men: 4.64 ± 1.393), followed by the opinion that the principles of gender equality in decision-making positions should be included in the general guidelines and strategic objectives of the organization (women: 5.22 ± 1.517; men: 4.64 ± 1.499), the opinion that organizations should provide employees family services (maternal and paternal services) (women: 5.22 ± 1.478; men: 3.93 ± 1.385) and opinion that problems and awareness on gender equality in the society should be included into all levels of the school system (women: 5.22 ± 1.396; men: 5.21 ± 1.122). It can be concluded that it is possible to identify possible/feasible measures to achieve gender equality in decision-making positions at the corporate level based on the opinions of mangers (H6 is supported).

4.4   CONCLUSIONS AND DISCUSSION

Our research results show that several gender differences and inequalities exist regarding the balance at the decision-making positions within organizations in Slovenia.

Research results suggest that both aspect of perceived gender equality, namely, perceived gender equality at decision-making positions in the respondent's organization (factor 1), as well as perceived gender equality ensured by legislation and perceived personal willingness to participate at the highest decision-making positions (factor 2), are perceived significantly differently by female and male managers. Female managers perceive the gender equality on general significantly lower than males, including the lower assessment of personal willingness to bid for the top managerial positions in comparison with the opposite gender. Since female managers are self-confident into necessary skills and abilities needed for decision-making positions in organization there are obviously other reasons for underrepresentation of women at top managerial positions in organizations. Among the reasons, there is undoubtedly the general organizational culture in organizations, since female managers perceive having significantly lower opportunities to be at the top managerial position as the opposite gender. Hypothesis H1 that regarding the perceived gender equality at decision-making positions within organizations significant differences between female and male managers exist is therefore supported.

While the friendship relationships of managers in organizations as well as physical safety are perceived on general equally by female and male managers (factor 2), the picture is quite different regarding the perceived attitudes of work colleagues (factor 1), which are assessed significantly lower by female managers as compared to males. The business climate in organizations on general does not provide equal respect and equal evaluation of managers, since female managers perceive these attitudes on general significantly lower. If these results are added to previous findings of perceiving lower opportunities to be at the top managerial position as males, it seems that female managers are facing different treatment in comparison with their male colleagues. Hypothesis H2 that regarding the perceived satisfaction with the work and life significant differences between female and male managers exist is partly supported.

Research results suggest that female managers are on average significantly less satisfied with the progress in their careers as compared to males. The significantly lower satisfaction by female managers is found regarding all components of the perceived career goals: women are less satisfied regarding the achieved career success, regarding meeting career goals, goals regarding personal income, as well as regarding career promotion. This result is expected, taking into account the obstacles and characteristics of organizational culture and attitudes of the organizational and business environment. On the other hand, no significant gender differences regarding concerning to stop working at the demanding decision-making positions were found. This implies that women do not differ with regard to perseverance in carrying out work in demanding decision-making positions, despite the, on average, lower perceived support of organizational and the wider business environment. Hypothesis H3 that regarding the perceived satisfaction with the employment position and career significant differences between female and male managers exist is partly supported.

Research results also suggest that perceived attitudes of family and perceived family support in balance of work and family life do not differ between the two genders. While some aspects were assessed higher by women managers (changing plans for family activities due to work, difficulties in everyday responsibilities due to family obligations), others were assessed lower (like difficulties to fulfill family obligations due to work, need to postpone work obligations due to family obligations) and no significant gender differences were found. Therefore, the evidence to support hypothesis H4 that regarding the perceived family attitudes and support in the balance of work and life significant differences between female and male managers exist was not found.

These results imply that families support women and men at decision-making positions on general equally and that both genders perceive generally positive family attitudes and support in the balance of work and life activities, although female and male managers significantly differ regarding the share of carried out household tasks and domestic liabilities, where female managers bear a significantly higher burden. Hypothesis H5 regarding gender differences in the carried out share of domestic and family responsibilities is supported. But obviously, female managers do not perceive the significantly higher share of carried out household and family work as compared to male managers, as the lack of the family support.

These research results confirm that traditional gender roles in family and everyday life are generally not the obstacle to establish the gender balance at the decision-making positions in organizations. The sociocultural attitudes toward women, that put women in the center of the family with all responsibilities in relation to that, seem not to have the negative effect on gender equality at decision-making positions in organizations in

Slovenia. Research results suggest that system-imposed barriers are more influential. These barriers refer especially to those at the company level, such as unclear or vague career advancement criteria leading to biased decision regarding career promotions and opinions regarding women's career potential.

This research brings several policy implications for policy makers as well as for business practice. According to the results of the analysis performed, it is expected that policy measures that should help closing the male–female gap at decision-making positions in organizations, that are highly assessed by female managers, are aimed toward barriers at the corporate level as nonlegislative instruments: especially like prizes and awards for organization that have developed an internal good practice on gender equality in decision-making positions. The organizational culture should contribute to the gender balance also by including principles of gender equality in decision-making positions into the general guidelines and strategic objectives of the organizations. In this respect, the corporate governance code addressing gender equality at top decision-making position and that rely on peer pressure within organization, pressure from stakeholders and media, can play an important role.

As research results show, female managers nevertheless fell that the big share of domestic tasks lies on their shoulders. The measures regarding providing maternal and paternal services as well as child care services by organizations are so highly ranked.

To achieve the objectives of balance equality in decision-making positions in organizations, the educational system at all levels of education is important, according to the opinions of both, female and male managers. Cultural and social norms are very difficult to change, yet certainly not in a short period of time. From this perspective, the integration of the principles of gender equality in all areas of work and life into all stages of the education system is a measure that can contribute to changes in deep-rooted principles of gender inequality in the cultural and social norms.

The high degree of consensus regarding the need to apply virtually all actions and measures recorded implies that gender equality at the decision-making positions within organizations is important topic for discussion, even more—it calls for action. It is important to highlight good practices and it is important to provide reference material in order to contribute to gender balance.

KEYWORDS

- **gender balance**
- **managers**
- **decision-making position**
- **Slovenia**
- **organizations**

REFERENCES

Duflo, E. Women Empowerment and Economic Development. *J. Econ. Lit.* **2012**, *4*(59), 1051–1079.

European Commission. *Gender Balance in Decision-making Positions*, 2015 [online]. Available from: http://ec.europa.eu/justice/gender-equality/gender-decision-making/index_en.htm (last accessed on 8 August 2015).

European Commission. *Tackling the Gender Pay Gap in the European Union*. Publication Office of the European Union: Luxembourg, 2014.

European Commission. *Report on Progress on Equality between Women and Men in 2010: The Gender Balance in business Leadership*. Publication Office of the European Union, Luxembourg, 2011a.

European Commission. *Women in Decision-making Positions*. Eurobarometer, 2011b [online]. Available from: http://ec.europa.eu/public_opinion/archives/ebs/ebs_376_en.pdf (last accessed on 8 August 2015).

European Commission. *Strategy of Equality between Women and Men 2010–2015*. Publication Office of the European Union, Luxembourg, 2011c [online]. Available from: http://ec.europa.eu/justice/gender-equality/files/strategy_equality_women_men_en.pdf (last accessed on 8 August 2015).

European Commission. *A Strengthened Commitment to Equality between Women and Men A Women's Charter*, 2010a [online]. Available from: http://eur-lex.europa.eu/legal-content/EN/TXT/?uri=celex:52010DC0078 (last accessed on 8 August 2015).

European Commission. *Europa 2020. A Strategy for Smart, Sustainable and Inclusive growth*, European Commission: Brussels, 2010b [online]. Available from: http://ec.europa.eu/europe2020/index_en.htm (last accessed on 8 August 2015).

European Institute for Gender Equality (EIGE). *Gender Equality Index 2015—Measuring Gender Equality in the European Union 2005–2012*. Publication Office of the European Union: Luxembourg, 2015.

European Institute for Gender Equality (EIGE). *Gender Equality Index: Report*. Publication Office of the European Union: Luxembourg, 2013.

European Union (EU). Consolidated version of the Treaty on European Union. *Off. J. Eur. Union* **2012**, *C326*(55), 13–46.

Eurostat. *Gender Pay Gap Statistics*, 2015 [online]. Available from: http://ec.europa.eu/eurostat/statistics-explained/index.php/Gender_pay_gap_statistics (last accessed on 8 August 2015).

Hair, J. F.; Black, W. C.; Babin, B. J.; Anderson, R. E. *Multivariate Data Analysis*. Prentice Hall: New Jersey, 2010.

Ittonen, K.; Vähämaa, E.; Vähämaa, S. Female Auditors and Accruals Quality. *Account. Horizons* **2013**, *27*(2), 205–228.

Judge, T. A.; Cable, D. M.; Boudreau, J. W.; Bretz, R. D. *An Empirical Investigation of the Predictors of Executive Career Success*, 1994 [online]. Available from: http://digitalcommons.ilr.cornell.edu/cgi/viewcontent.cgi?article=1232&context=cahrswp (last accessed on 8 August 2015).

Karatepe, O. M.; Baddar, D. An Empirical Study of the Selected Consequences of Frontline Employees' Work–Family Conflict and Family–Work Conflict. *Tour. Manage.* **2006,** *27*, 1017–1028.

Lofstrom, A. *Gender Equality, Economic Growth and Employment*. Umea University, 2009 [online]. Available from: http://ec.europa.eu/social/BlobServlet%3FdocId%3D3988%26langId%3Den (last accessed on 8 August 2015).

Lyness, K. S.; Brumit Kropf, M. The Relationships of National Gender Equality and Organizational Support with Work–Family Balance: A Study of European Managers. *Hum. Relat.* **2005,** *58*(1), 33–60.

Orser, B.; Leck, J. Gender Influences on Career Success Outcomes. *Gender Manage.: Int. J.* **2010,** *25*(5), 386–407.

Palvia, A.; Vähämaa, E.; Vähämaa, S. Are female CEOs and Chairwomen More Conservative and Risk Averse? Evidence from the Banking Industry During the Financial Crisis. *J. Bus. Ethics* **2014,** *6*, 1–18.

Perrault, E. Why Does Board Gender Diversity Matter and How Do We Get There? The Role of Shareholder Activism in Deinstitutionalizing Old Boys' Networks. *J. Bus. Ethics* **2015,** *128*, 149–165.

Ruminska-Zimny, E. Women's Entrepreneurship and Labour Market Trends in Transition Countries, In: *Women's Entrepreneurship in Eastern Europe and CIS Countries*; 2003; pp 1–16.

Shapiro, G.; Olgiati, E. *Promoting Gender Equality in the Workplace*. European Foundation for the Improvement of Living and Working Conditions: Dublin, 2002 [online]. Available from: http://www.eurofound.europa.eu/pubdocs/2001/61/en/1/ef0161en.pdf (last accessed on 8 August 2015).

Singhapakdi, A.; Sirgy, M. J.; Lee, D.-J.; Senasu, K.; Yu, G. B.; Nisius, A. M. Gender Disparity in Job Satisfaction of Western versus Asian Managers. *J. Bus. Res.* **2013,** *April*, 1–10.

Stoyanovska, A. *Jobs, Gender and Small Enterprises in Bulgaria*. ILO: Geneva, 2001.

Tominc, P. Some Aspects of the Gender Wage Gap in Slovenia. *Druš. istraž.* **2002,** *11*(6), 879–896.

PART II
Technology and Knowledge Management

CHAPTER 5

ORGANIZATIONAL ENVIRONMENT FACTORS AFFECTING KNOWLEDGE MANAGEMENT: RESEARCH STUDY IN SOFTWARE COMPANIES

JELENA HORVAT[1] and SAMO BOBEK[2*]

[1]University of Zagreb, Faculty of Organization and Informatics, Varaždin Pavlinska 2, HR 42000 Varaždin, Croatia

[2]Faculty of Economics and Business, University of Maribor, Razlagova, Maribor, Slovenia

*Corresponding author. E-mail: samo.bobek@um.si

CONTENTS

ABSTRACT

Importance of knowledge management (KM) lies in the fact that it could result in empowerment of individuals and organization itself to accomplish activities effectively through organizing of knowledge. Broad scope of KM and its interdisciplinary nature spans traditional functions and professional boundaries. KM is a major issue for human resources management. Organization, culture, and information technology play crucial enabler for various aspects of KM. Aim of this chapter is to present findings regarding influences of organizational factors on KM. In the empirical part, research in SME IT companies is presented. Data were analyzed using partial least square structural equation modeling techniques.

5.1 INTRODUCTION

Success of companies, especially service companies, depends on knowledge in performing activities. The 21st century is declared as a century of knowledge (Milanović, 2010), whereas competitive advantage is perceived as a link to knowledge and therefore interest in knowledge management (KM) grows on. Environment where companies operate has certain properties, according to which companies differ from each other (Belak et al., 2014, p. 29). Companies are becoming the main actors responsible for the destruction and development of environment (Pučko, 2005). Based on extensive review of literature, Boersma and Stegwee (1996) identified 10 different KM approaches: strategic approach, human resource management approach, learning organization approach, intellectual capital approach, knowledge technology approach, ICT approach, organizational approach, innovation approach, network approach, and quality control approach. KM has been defined and explored in various ways, but generally, it relates to unlocking and leveraging knowledge of individuals to gain appropriate knowledge from appropriate individuals in appropriate time (Hutchinson & Huberman, 1994). Therefore, knowledge becomes available as an organizational resource (Anand & Singh, 2011) and helps individuals to share and apply information with regard to organizational performance (Hutchinson & Huberman, 1994). At its core, KM is trying to harvest insights and experience which make the organization function (Sarayreh et al., 2012). Key factors for KM are considered to be (Moffett et al., 2003) macroenvironment (includes economic, technical, and social agents of change), organizational culture (includes organizational structure, strategy, change management), people or personal contributors (include knowledge roles and skills, motivation, empowerment), informational contributors (include information fatigue, information auditing), and technology or technical contributors (include system standardization and technical usability). Every of these factors has influence on KM and can be linked with its performance.

Therefore, this chapter, through an empirical pilot study, investigates KM in small software companies from four different perspectives: technical contributions, informational contributors, people/personal contributors, and culture/organizational climate, in SME IT companies. For data analysis, the partial least square structural equation modeling (PLS-SEM) is used and for results reporting the SmartPLS software.

5.2   PERSPECTIVES ON KNOWLEDGE

Different authors (Davenport & Prusak, 1998; Hicks et al., 2006) address questions in defining KM by distinguishing data, information, and knowledge also known as the knowledge hierarchy (Davenport & Prusak, 1998; Hicks et al., 2006). Although these three terms are sometimes used interchangeably, they are quite distinct one from another. Data comprise facts, observations, or perceptions, whence information is a subset of data including those data that posse's relevance and purpose (Becerra-Fernandez et al., 2004). Unlike the information, which is visible, independent of various actions, decisions, and environment which can easily be conveyed and duplicated, knowledge is invisible, and closely associated with actions and decisions, it identifies itself with existing environment, is transferable through learning, and cannot be duplicated (Kumar, 2010). The knowledge is information that facilitates action and has been recognized as currency for organizational sustenance and competitive advantage (Osterloh et al., 2002; Wiig, 1997). Knowledge embraces main place among traditional factors of production, such as land, labor, and capital. Its importance has been recognized in the 1990s and after that, literature began to fill with research and analysis regarding knowledge (Jengard, 2010). Knowledge is perceived as the most important intellectual property and asset of the company (Chan et al., 2011; Collins, 2010) and also resource that enhances competitive advantages of each organization. Oxford Dictionary and Thesaurus (2007) defines the term knowledge as "awareness or familiarity gained by experience (of a person, fact, or thing)," "person's range of information," "specific information; facts or intelligence about something," or "a theoretical or practical understanding of a subject." Davenport and Prusak (1998) define knowledge as "fluid mix of framed experience, values, contextual information and expert insight that provides a framework for evaluating and incorporating new experiences and information. It originates from and is applied in the minds of knowers." Chase (1997) describes knowledge as an input to production, as valuable as the product itself and competitiveness of companies becomes dependent on the ability to harness and exploit knowledge.

There are different kinds of knowledge classified according to several criteria. One of them was presented by Alavi and Leidner (2001) who have defined various perspectives on knowledge. First, one is viewing knowledge as a state of mind, which focuses on enabling individuals to expand their personal knowledge and apply it according to organizations' needs. Second perspective is an object that posits knowledge that can be viewed as a thing to be stored and manipulated (i.e., an object). In addition, knowledge can be viewed as a process and as a condition of having access to information. Fourth view of knowledge is as a capability. This suggests that KM perspective is centered on building core competencies, creating intellectual capital and understanding the strategic advantage of know-how. According to some authors, knowledge can be divided into nonempirical (occurred through reflection) and empirical (created through experience) (Jelkić, 2011). It becomes meaningful for the company when it is available and its value increases with respect to its availability.

Drawing on work of Nonaka (1991), two dimensions of knowledge emerge in organizations: tacit and explicit. This typology of knowledge had considerable influence on theory of organizational learning and KM in general (Daud & Yusoff, 2011). Explicit knowledge is accurate, it is easy to articulate and therefore easy to code, document in writing or in pictures, easy to carry, share and communicate—it is generally more accessible and systematic than tacit knowledge and relates to the technical knowledge. In contrast to explicit knowledge, tacit experiential knowledge is subconsciously understood, applied, and difficult to articulate. Tacit knowledge includes an understanding, intuition and arises with experience while explicit knowledge can be transferred. It is very personal, includes intangible factors that are related to beliefs, experiences and values of employees, and working knowledge of employees (Pan &

Scarbrough, 1999). Study of tacit knowledge, which was first introduced by Michael Polanyi (Smith, 2003) can be characterized as "invisible" and difficult. In his papers, Polanyi dealt with tacit knowledge, which evoked great interest. This resulted in significant numbers of papers and studies and above all disagreements and contradictions about this issue.

Transfer of knowledge throughout the organization is considered as a critical driver of its effectiveness (Dinur, 2011, p. 246). Chan et al. (2004, p. 14) raised the question of how interaction between individuals and groups should be for a successful knowledge transfer. Discussing the key processes in knowledge transfer, they emphasize that organizational effectiveness is limited if individuals are not willing to share knowledge with other employees. Organizations can develop knowledge in a variety of technical domains. Due to limited resources, organizations are forced to make a choice. They chose to develop knowledge in one technological domain and so reduce their own possibilities to develop expertise in another field. Such choice affects organization's ability to succeed in the long run (Moorthy & Polley, 2010). Therefore, it is important to study the KM and its factors and their interplay in companies.

5.3   KNOWLEDGE MANAGEMENT

KM is explicit and systematic management of vital knowledge and is associated with creating, gathering, organizing, diffusion, use, and exploitation of the knowledge (Anand & Singh, 2011). It is not only a necessity, but also source of competitive advantage and thus an important strategic resource for business organizations. Popularity of KM has been increasing fast, especially in the late 1990s and it has become a central topic of management philosophy and a management tool (Edvardsson, 2003). Rise of KM is attributed to development of intranets that allowed faster and direct communication within organizations (Srikantaiah & Koenig, 2000, p. 25). In 1995, there were 45 articles on KM in the ABI/information database, 158 in 1998, and 835 in 2002 (Edvardsson, 2003). One of the main goals of KM is to take the tacit knowledge, what people carry around with them, what they observe and learn from experience and what is internalized and therefore not readily available for transfer to another, and turn it into explicit knowledge which has been formalized in people's heads, or documented in books and paper (Srikantaiah & Koenig, 2000, p. 223).

KM is defined as creating, acquiring, storing, sharing, transferring, and utilizing both explicit and implicit forms of knowledge at individual, group, organizational, and community level through harnessing of people, process, and technology (Madhoushi et al., 2010). KM, from an operational perspective, is perceived as systematic process by which "organization identifies, creates and acquires, shares and leverages knowledge" (Chivu & Popescu, 2008). Housel and Bell (2001) summarized four main goals of KM: (1) gathering: bringing information and data into the system; (2) organizing: associating items to subjects establishing context, making them easier to find; (3) refining: adding value by discovering relationships, abstracting, synthesizing, and sharing; (4) disseminating: getting knowledge to people who can use it. Term KM does not imply only on a set of technologies or methodologies, but also practice and discipline that involves interaction of people, processes, and technology. KM is not only a necessity but also source of competitive advantage and thus an important strategic resource for business organizations.

Most definitions of KM include a combination of a management philosophy related to organizational knowledge and a technology-based knowledge gathering and sharing systems. Overall accepted definition defines KM as (Thite, 2004): "Creating, acquiring, storing, sharing, transferring and utilizing both explicit and implicit forms of knowledge at individual, group, organizational and community level through harnessing of people, process and technology." Literature discussing applications of IT to organizational KM initiatives reveals three common applications (Alavi & Leidner, 2001): (1) coding and sharing of best practices; (2) creation of corporate knowledge directories; (3) creation of knowledge networks. Housel and Bell (2001) summarized four main goals of KM: (1) gathering: bringing information and data into the system; (2) organizing: associating items to subjects establishing context, making them easier to find; (3) refining: adding value by discovering relationships, abstracting, synthesizing, and sharing; (4) disseminating: getting knowledge to people who can use it. KM, from an operational perspective, is perceived as systematic process by which "organization identifies, creates and acquires, shares and leverages knowledge" (Chivu & Popescu, 2008).

Existing research and conceptual studies in KM filed are identified as dynamic sets of activities (Mehta, 2008). These are called the KM processes. Knowledge discovery and creation process is defined as development of new tacit or explicit knowledge from data and information or from the synthesis of prior knowledge (Becerra-Fernandez et al., 2004, p. 33). Knowledge capturing is process of retrieving explicit or tacit knowledge that resides within people (individuals or groups), artifacts (practices, technologies, or repositories) or organizational entities (units, organizations, or interorganizational networks) (Becerra-Fernandez et al., 2004, p. 33). Capturing is a process by which expert's thoughts and experience are captured (Awad & Ghaziri, 2004, p. 121) in order to elucidate importance or interlinkage (Staab et al., 2000). Once knowledge is created, it needs to be stored in database for subsequent use by employees in different departments (Storey & Kelly, 2002). Knowledge storage is a process of structuring and storing of knowledge (Massa & Testa, 2009) and is considered as one of the essential elements in the KM process as it helps in the prevention of losing important information (Lee et al., 2013). Refined and stored knowledge enables employees to retrieve and disseminate knowledge conveniently and therefore proving it to be a valuable element for company (Gold et al., 2001). In knowledge creation and storage process, a distinction between individual and organizational memory can be made. Individual memory is based on a person's observations, experiences and actions (Alavi & Leidner, 2001). Collective also known as organizational memory are means by which knowledge from past experience and events influence present organizational activities and it extends beyond the individuals memory (Alavi & Leidner, 2001; Stein & Zwass, 1995). Process of knowledge application depends on the available knowledge and on the whole KM process. The better the processes of knowledge discovery, capture, and storage the higher, the chance that the knowledge needed will be available (Becerra-Fernandez et al., 2004, p. 35).

Knowledge sharing process is a social interactive culture, involving the change of employee knowledge skills and experience through every department in organizations. It comprises set of shared understandings related to providing employees access to relevant information (Hoegl et al., 2003). Such process of sharing organizational knowledge facilitates exchange of working experiences, technical know-how and individual insights between and among individuals. Knowledge sharing has become essential in organizations because it enables enhancement of innovation performance and reduces the redundant learning efforts (Lin, 2007). It can also be concaved as involving supply and demand for new knowledge (Ardichvili et al., 2003). Most authors agree that knowledge sharing depends on individual factors like experience, values, motivation, and beliefs (Connelly & Kelloway, 2003; Lee & Choi, 2003; Lin, 2007).

Knowledge sharing has been identified as a major focus and research area of KM. It provides a link between the level of individual knowledge workers, where knowledge resides and level of organization, where knowledge attains its (economic, competitive) value (Hendriks, 1999). Process of sharing starts at individual level and expands to group and organizational level (Lin, 2007). Most authors agree that knowledge sharing depends on individual factors like experience, values, motivation, and beliefs (Connelly & Kelloway, 2003; Lee & Choi, 2003; Lin, 2007). Many researchers have been examining effectiveness of knowledge sharing from different viewpoints focusing on the problem of transferring tacit and complex knowledge across organization parts, on the nature of informal relationships between two parties to transfer knowledge and on the problem of searching for knowledge (Brauner & Becker, 2006; Connelly & Kelloway, 2003; Fong et al., 2011; Goh & Hooper, 2009; Hoegl et al., 2003; Jalote, 2003; Lee et al., 2011; Martins & António, 2010; Moorthy & Polley, 2010). Effectively sharing knowledge increases the accumulation of organizational knowledge and develops the capability of its employees for better performance of their jobs (Jalote, 2003). Combining knowledge of different employees creates new opportunities and responds to challenges in innovative ways (Mathew et al., 2011). In addition, Lin (2007) argues that the survival of company may be substantially undermined if employees are not willing to share knowledge, by which the ethic foundations can seriously be affected.

Effectively sharing knowledge increases the accumulation of organizational knowledge and develops the capability of its employees for better performance (Jalote, 2003). Organizations can develop knowledge in a variety of technical domains. Due to limited resources, organizations are forced to make a choice. They chose to develop knowledge in one technological domain and so reduce their own possibilities to develop expertise in another field. Such choice affects organization's ability to succeed in the long run (Moorthy & Polley, 2010). Therefore, it is important to study the KM and its factors and their interplay in companies.

5.4   INTERNAL FACTORS AND KNOWLEDGE MANAGEMENT

Aspects of KM have been dominated by two main factors: (1) the supporters of information and communication technology and (2) the human resource (HR) views (Jantz, 2001). Some authors believe that IT is the main driver for KM, though others disagree and believe that KM is mostly about people and culture and not technology (Soliman & Spooner, 2000). Various authors (Attafar et al., 2012; Daud & Yusoff, 2011; du Plessis, 2007b; Mathew et al., 2011; Moffett & McAdam, 2009; Moffett et al., 2003; Scarbrough, 2003) have dealt with KM, aspects, and factors. According to research by Mathew et al. (2011), KM initiative is determined by the organizational culture and technology and people; to the same conclusion came Moffett et al. (2003). They concluded that organizations should have effect on all three factors in order to successfully exploit the knowledge in organization and have defined these factors as companies' internal factors. Research conducted by Daud and Yusoff (2011) suggests that combination of process of KM as well as organizational skills and intellectual capital as a strategic organizational asset enables the increases of organizational effectiveness.

5.4.1 PEOPLE AND TECHNOLOGY IN KNOWLEDGE MANAGEMENT

Knowledge-oriented culture challenges people to share knowledge throughout the company (Davenport & Prusak, 1998). People in an organization conform to the culture of the organization. Changes in organizations' culture places demands on employees to change their mindsets. Through education, remuneration, and knowledge sharing, people influence the company in all areas: quality advantage and speed of development. Therefore, to upgrade the status of the company, first thing is to develop people (employees) (Bahtijarević-Šiber, 1999). Thus, within the field of organizational change resulting from KM, human issues must be considered as a key factor. This consideration has given rise to the *knowledge worker*, and key influences on this concept are increased information technology, a shift in markets away from labor intensive manufacturing and an increase in education (Carter & Scarbrough, 2001; Moffett et al., 2003; Pan & Scarbrough, 1999). Technology is viewed as a key contributor and also enabler to the field of KM (Davenport & Prusak, 1998). This perspective is related to technological ability in capturing data, information, and knowledge that surpasses human capacity in absorbing and analyzing these in a focused manner. Literature regarding KM, particularly, reflects a techno-centric focus which, in essence, regards knowledge that can be captured, manipulated, and leveraged through IT (Pastor et al., 2010). This perception is limited and needs to be enhanced with a human-centric focus. Such focus perceives knowledge as a social creation emerging at the interface between people and information and between people and people themselves (Pastor et al., 2010). From this perspective, KM is concerned with the way organizations create, supplement, and organize knowledge around their activities, within their cultures and develop organizational efficiency by improving the use of employees' talent (Pan & Scarbrough, 1999).

Different researchers (Ahmed & Ahmad, 2012; Attafar et al., 2012; Davenport & Völpel, 2001; Hussock, 2009; Ishak et al., 2010; Lin, 2007; Oltra, 2005; Özbebek & Toplu, 2011; Theriou & Chatzoglou, 2008) have been interested in area of KM and employees, combining it with perspectives of strategic, project, and information management. Modern companies and development of management support the premise of the "knowledge workers." Key influences on this concept are increased information technology, a shift in markets away from labor intensive manufacturing and an increase in third-level education opportunities (Scarbrough, 2003).

5.4.2 KNOWLEDGE MANAGEMENT AND CULTURE

Various factors have influence on KM in companies (e.g., employees, organizational values, infrastructure, culture, technology, macroenvironment, and others). Organizational culture with its values and norms is of essential meaning for ensuring the long-term success of companies. Culture is a set of achievements of human society: all creations, both material and non-material (Belak et al., 2014, p. 29). Companies' culture is perceived as one of the key success factors in MER Model of Integral Management developed by the Institute for Management and Development (MER) in Slovenia (Milfelner & Belak, 2012). Considering MER Model of Integral Management (Belak & Duh, 2012) selected companies key success factors are compatibility, competitiveness, efficiency, culture, credibility, ethics, ecology, entrepreneurship, synergy, and philosophy. Culture of a particular company in this is a wide, complex, and multi-faceted phenomenon, which forms the social system and consequently impacts the environments (Belak et al., 2014, p. 29). More recently, corporate culture has been denoted as encompassing the assumptions, beliefs, goals, knowledge, and values that are shared by organizational members (Belak et al., 2014).

Pučko (from Belak et al., 2014, p. 37) considered companies culture as a system of values, beliefs and usage which are typical for people from the company and are permanently present in the context of strategic planning, implementation, and control. The culture deals with a process that stems from the environment and returns to it (Belak et al., 2014, p. 30). Employees (people) invest in the circumstances of the evolving knowledge society in developing some specific knowledge or skills, which demands from them to take over a significant risk that authorizes them to participate in the governing of their enterprises. The existing tendency of the increasing variable component of employees' remuneration, which is apparent, requires risk sharing between owners and employees (Pučko, 2005). In addition to higher profit, the image of companies is improved through technology used for communication with shareholders (information sharing and marketing) (Babu, 2012). Organizational culture is considered a key element in managing organizational change and renewal.

Culture is regarded as an effective factor on KM success in many researches (Carrión et al., 2004; Davenport & Prusak, 1998; Mason & Pauleen, 2003; Nonaka, 1991; Oltra, 2005; Pan & Scarbrough, 1999). Culture of confidence and trust is required to encourage the application and development of knowledge within an organization (Pan & Scarbrough, 1999). It plays an effective role in KM effectiveness as the identity and foundation of organization (Attafar et al., 2012) and can have direct effect on employees' empowerment and knowledge sharing behavior (Özbebek & Toplu, 2011). Judge and Cable (1997) and Kristof (1996) focused on the importance of a fit between new employees and the organization's knowledge culture. Their research is linked to the person-organizational fit literature within HRM, with emphasis on a fit between organizational culture and hiring of suitable personality, as well as the socialization of individuals into the culture of the company. Culture plays an effective role in KM effectiveness as the identity and foundation of the organization (Attafar et al., 2012) and is principal determinant of successful KM initiatives (Mathew et al., 2011). Knowledge-sharing culture must enable an incentive system to support the participation toward knowledge sharing (e.g., "knowledge sharer" of the month or giving out virtual dollars) (Hussock, 2009). Attafar et al. (2012) state that stronger culture influences stronger KM. According to du Plessis (2007a), the cultural realities in companies should be taken into account when implementing KM. In company where success is measured with billable hours, such systems leaves little time for

KM; therefore, the culture is oriented toward financial measurement and KM is not seen as important (du Plessis, 2007a).

In today's knowledge-intensive organizations, primary objective of information and communication technology is to lead users to the information they need. This includes creating, gathering, storing, accessing, and making available the right information that will result in insight for the organizations' users (Davenport & Prusak, 1998). Information technology is essential; it organizes, communicates, and creates the life-blood of a modern organization: business critical data and eliminates barriers and boundaries—enabling innovation and competitive advantage (Standards Australia, 2012). Technology is expected to play a key role in helping organizations achieve their business objectives; it has an important role in corporate governance of the companies (Chandiramani, 2007).

5.5   RESEARCH MODEL AND RESEARCH METHODOLOGY

Aim of this chapter is to analyze the interrelationship between people, technical aspects, and KM. As stated above, the organizational environment (KM) factors include: technical contributions, informational contributors, people/personal contributors, and culture/organizational climate (see Moffett et al., 2003). Moffett et al. proposed a *MeCTIP* model that portrays five factors (Me—macroenvironment; C—culture; T—technology; I—information, and P—people) which have influence on adoption of KM within organizations. Authors have analyzed and tested this MeCTIP model in various sectors. For the purpose of this research, the relationship between four factors and KM is observed in IT Croatian companies. Accordingly, following hypothesis and conceptual model based on literature findings have been researched:

H1: *Organizational environment factors have a significant association with knowledge sharing.*

The above stated hypothesis has been divided into four:

H1a: *Technical contributors have a significant association with knowledge management.*

H1b: *Informational contributors have a significant association with knowledge management.*

H1c: Personal contributors have a significant association with knowledge management.

H1d: Culture has a significant association with knowledge management.

Later methodology of data collection and analysis and results of the pilot study in Croatia IT companies are presented.

For data collection, survey method was used for small IT companies. Accordingly, online questionnaire was formed in GoogleDocs and sent through e-mail with appropriate cover letter, whereas 77 answers were received. Focus was on KM, which was analyzed through four main factors: culture, people, information, and technology. For culture 9 items were used and for people (personal contribution) 10, for informational climate 5, and technical contribution 6. KM in general was measured using five items. Responses were measured on a 5-point Likert scale ranging from 1 = strongly disagree to 5 = strongly agree. Form the mentioned items; several had to be dropped in order to ensure reliability and validity of the model. The partial least square structural equation modeling (PLS-SEM) technique was employed to analyze the research model constructed in Figure 5.1.

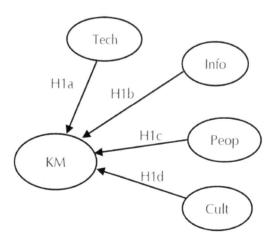

FIGURE 5.1   Conceptual framework. *Note*: KM, knowledge management; Tech, technical contributions; Info, informational contributors; Peop, people/personal contributors; Cult, culture/organizational climate.

Structural equation modeling (SEM) techniques are considered to be a major component of applied multivariate statistical analyses which are used in various sciences (e.g., biology, economy, educational researchers, marketing researchers, medical researchers, and others) (Pugesek et al., 2003). For this purpose, several specialized SEM programs are used in order to simplify the calculations, for example, AMOS, EQS, LISREL, Mplus, Mx, RAMONA, SEPATH, SmartPls (Pugesek et al., 2003). SEM models represent translations of a series of hypothesized cause–effect relationships between variables into a composite hypothesis concerning patterns of statistical dependencies (Pugesek et al., 2003). SEM is a combination of factor analysis and multiple regression (Pugesek et al., 2003). It is used to determine how sets of variables define constructs (i.e., measurement model) and how these constructs are related to each other (i.e., structural model) (Bollen & Long, 1993). With SEM, the relationship between measured variables and the relationship between unmeasured, hypothetical constructs can be modeled. Two different techniques for SEM can be applied (Afthanorhan, 2013): covariance-based technique (CB-SEM) and PLS-SEM. PLS-SEM increases the explained variance of the endogenous latent constructs (dependent variables) and minimizes the unexplained variance (Afthanorhan, 2013). CB-SEM is a covariance-based technique and attempts to minimize the difference between the sample covariance and that predicted by the theoretical model (Pugesek et al., 2003). PLS does not assume the normality of data distribution, and therefore is more suitable for smaller samples. The analysis can also be conducted with several (fewer than three) indicators (items), whereas the CB-SEM assumes that there are more than three indicators (Afthanorhan, 2013).

For purpose of this research, the PLS-SEM is employed and software the SmartPLS 2.0 (Ringle et al., 2005) is used. SmartPLS is one of the main applications for PLS-SEM. This software, developed by Ringle et al. (2005), has a friendly user interface and advanced reporting features and is freely available to academics and researchers.

5.6   RESEARCH RESULTS

The structural model reflecting the research hypothesis H1 (H1a–d) depicted Figure 5.1 was analyzed using the proposed SmartPLS software.

In order to validate the model several indicators regarding factors and KM had to be dropped. So from nine items measuring culture, four had to be dropped. In the final model, 3 indicators explain the information factor, 7 (out of 10) explain the personal contributors, and technology is explained by 4 items, and KM by 3.

The beta values of path coefficient indicate direct influences of predictor upon the predicted latent constructs (see Fig. 5.2). The coefficient of determination is 0.409 for the KM endogenous latent variable. The inner model suggest that personal contributors have the strongest effect on KM (0.514), followed by culture (0.290) and technical contributions (0.135), although the informational contributors have negative effect on KM (−0.275). The hypothesized relationship between KM and personal contributors, culture and technical contributions is statistically significant due to the high-standardized path coefficients which need to be higher than 0.1 (Wong, 2013), though the hypothesized path relationship between informational contributors and KM is not statistically significant. Consequently, the hypothesis (H1) has been partially confirmed showing that technology, people, and culture have a significant association with KM.

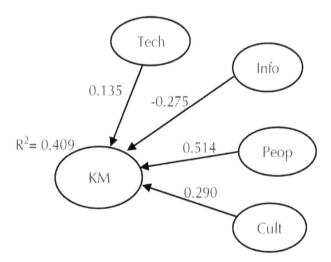

FIGURE 5.2 Structural model with path coefficients. *Note*: KM, knowledge management, Tech, technical contributions; Info, informational contributors; Peop, people/personal contributors; Cult, culture/organizational climate.

The dimensions of KM showed good validity and reliability and thus reflect the overall KM aspects in companies. The four latent variables (personal contributors, culture, technical contributions, and informational contributors) substantially explain 40% of variance of KM. It can be concluded that personal contributors, culture, and technical contributions are substantial predictors of KM, but informational contributors do not predict KM in companies directly.

TABLE 5.1 Reliability and Discriminant Validity Assessment of the Measurement Model.

Items	AVE	Composite reliability	R^2	KM	Cult	Info	Peop	Tech
KM	0.6196	0	0.4092	0.9065				
Cult	0.5216	0.8414	0	0.5143	**0.72221**			
Info	0.6477	0.8461	0	0.347	0.6423	**0.8047**		
Peop	0.501	0.8714	0	0.5868	0.696	0.7412	**0.7078**	
Tech	0.5409	0.8183	0	0.4041	0.3271	0.4134	0.5615	0.7354

Source: Authors' research.

Smart PLS also assesses the psychometric properties of the measurement model and estimates the parameters of the structural model (Yeşil et al., 2013). Results of the reliability and discriminate validity testing of the measurement model are presented in Table 5.1. As shown, the composite reliabilities of measure in the model range from (0.8183) which exceeds the recommended threshold values of 0.70 (Yeşil et al., 2013), though some authors demand a 0.60 minimum level (Wong, 2013). High level of internal consistency reliability is demonstrated among all four reflective variables. To check convergent validity, each latent variable's average variance extracted (AVE) is evaluated. As shown in Table 5.1, it is fund that all of the AVE values are higher than the acceptable threshold of 0.5, and therefore the convergent validity is confirmed. The square root of AVE can be used to establish the discriminate validity of the model (Fornell & Larcker, 1981). These results are presented in Table 5.1 as bolded elements in the matrix diagonal. They are greater, in all cases, than the off-diagonal elements in their corresponding row and column. These results indicate that discriminate validity is well established.

5.7 LINKING KNOWLEDGE MANAGEMENT AND PERFORMANCE APPRAISAL

Human resources management includes decisions that affect the success of business, with aim of achieving long-term company strategy. This specific area of management has been explored throughout last few decades and various authors have defined range of classification and functions of HR management. Human resources practices and functions differ from industry to industry as well from size of the organization. Taking in consideration that there is no universal prescription for HRM policies and practices and everything lays on organization's context, culture, and its business strategy, it is important for every company to find the "best fit." For purpose of this chapter, human resource functions are defined accordingly (Scarbrough, 2003): recruitment and selection, training and development, performance appraisal (management).

Performance management (PM) is continuous activity for evaluating employee work and is engaged for improving employee performance productivity and effectiveness (Chan, 2004). One goal of performance management is gaining information necessary for decision-making related to employee remuneration. Furthermore, tracking performance allows planning of career development which is in the interest of the company and individual employees (Bahtijarević-Šiber, 1999). Performance assessment results with three categories of workers' performance (Tuan, 2011). First, evaluation of employee's abilities and his/her's personal characteristics, then the behavior of employees, and in the end, specific job performance. This is necessary for accurate evaluation in order to develop a remuneration system based on the performance of employees (Tuan, 2011). In assessment it is important to note that all employees do not have same ethical principles, temperament, abilities, skills, and knowledge. Key activities in performance management are performance planning, coaching, and review. Performance planning includes defining job responsibilities and performance expectations, as well as goals and objectives. Second is the performance coaching which occurs during the whole period and it involves monitoring, coaching, and developing the employees. It's mostly based on feedback information. And third key activity is the performance review as the formal stage at the end of the review period. It is based on performance appraisal (Chan, 2004).

Evaluating the performance of work may also contribute to the sharing of knowledge (North, 2008, pp. 149–150). Previous research has shown that motivated employees are willing to share when they think that knowledge sharing will be worth the effort (Lin, 2007). Results of Lam and Lambermont-Ford (2010) research showed that external motivational factors help knowledge sharing in organizations. Some of the previous studies have proven that PM is linked to KM and knowledge sharing (Liu & Liu, 2011). Currie and Kerrin (2003) analyzed performance management systems and stated that the performance management system inhibits knowledge sharing, as much of the conflict between different functions was due to the divergent objectives set out for employees in the performance agreements. Paying attention to needs of employees, giving feedback to them and performance appraisal based on accurate standards is effective on facilitation of KM process (Attafar et al., 2012). Better performance appraisal system can help employees obtain information about the requirements of knowledge sharing (Liu & Liu, 2011). Jaw and Liu (2003) stated that it is important for companies to publish the results of the PM to the employees and through that enforce remedial actions for the underperforming employees. Knowledge sharing can help incentive systems which are measures aimed to increase employee motivation. Some of the incentives can be increase of salary, bonuses, trips, feedback (evaluation) about the performance of work, independence, promotion, etc. Thus, a PM system may serve as a positive pressure in directing employees to develop for better performance, through greater knowledge sharing among other employees (Fong et al., 2011). Therefore, it is important to study effect of performance management and knowledge sharing behavior in companies.

5.8 CONCLUSION

The field of KM continues to grow and expand, bringing new conclusions and discoveries. Individuals, human potentials are in the center of KM, so KM is individuals' management and individuals' management is KM (Davenport & Völpel, 2001). As managerial philosophy, KM is perceivable in practices of different organizations and is not an ultimate tool that solves all information and knowledge problems (Jha, 2011). Broad scope of KM and its interdisciplinary nature spans traditional function and professional boundaries ranging from IT professionals to accountants, marketers, organizational development, and change management professionals (Chivu & Popescu, 2008). Some authors argue that rise and growth of KM is managerial response to various empirical trends associated with globalization and postindustrialism (Scarbrough & Swan, 2001). Value of knowledge tends to perish quickly over time and companies need to speed up innovation and enhance creativity and learning (Kluge et al., 2001). Growing number of organizations adopts team working, organic structures, knowledge-centric cultures (Edvardsson, 2003) through that the importance of knowledge is highlighted in current organizational theory and contemporary organizational trends.

Individuals, human potentials are in the center of KM, so KM is individuals' management and individuals' management is KM (Davenport & Völpel, 2001). If human resources, employees, and their effective managing are essential for company and if people's most valuable resource is knowledge, then HRM and KM are closely interrelated. Importance of KM lies in the fact that it could result in empowerment of individuals and organization itself to accomplish activities effectively through organizing of knowledge (Jantz, 2001).

Presented pilot study was conducted in order to analyze the KM and its factors in software companies. Insight in the factors influencing KM can result with understanding how to improve knowledge collection, encourage knowledge sharing culture, and effectively store and apply the knowledge. Results showed that in software small companies' researched, main effect on KM have people and their characteristics, followed by cultural and technological contributors. Discriminant validity and reliability of the data was confirmed, and SmartPLS was used for data analysis.

Main limitations of this research are the small sample size and the fact that companies are form just one country which limits the generalizability of the findings. Therefore, authors recommend further research in companies from countries with increased sample size. Future studies can also include other variables that may affect performance management and KM in general.

KEYWORDS

- knowledge management
- human resources management
- IT professionals
- IT SME
- organizational environment

REFERENCES

Afthanorhan, W. M. A. B. W. A Comparison of Partial Least Square Structural Equation Modeling (PLS-SEM) and Covariance Based Structural Equation Modeling (CB-SEM) for Confirmatory Factor Analysis. *Int. J. Eng. Sci. Innov. Technol. (IJESIT)* **2013,** *2*(5), 8.

Ahmed, M.; Ahmad, R. R. Human Resource effective Factor for Knowledge Management with IT. In: *International Conference on Technology and Business Management* (vol. 2), 2012. University of Wollongong in Dubai: Dubai. Retrieved from http://www.icmis.net/ictbm/ictbm12/ICTBM12CD/pdf/D2248-done.pdf.

Alavi, M.; Leidner, D. E. Review: Knowledge Management and Knowledge Management Systems: Conceptual Foundations and Research Issues. *MIS Q.* **2001,** *25*, 107–136.

Anand, A.; Singh, M. D. Understanding Knowledge Management. *Int. J. Eng. Sci. Technol.* **2011,** *3*(2), 926–939.

Ardichvili, A.; Page, V.; Wentling, T. Motivation and Barriers to Participation in Virtual Knowledge-sharing Communities of Practice. *J. Knowl. Manage.* **2003,** *7*(1), 64–77. http://doi.org/10.1108/13673270310463626.

Attafar, A.; Soleimani, M.; Shahnazari, A.; Shahin, A. The Relationship between Enhancement of Human Resource Productivity and Knowledge Management in Iranian Libraries: A Correlation Analysis. *Afr. J. Bus. Manage.* **2012,** *6*(8), 3094–3099.

Awad, E. M.; Ghaziri, H. M. *Knowledge Management: International Edition.* Pearson Education International: New York, 2004.

Babu, K. V. S. N. J. *Role of Corporate Governance in Strategic Management* (SSRN Scholarly Paper No. ID 2184235). Social Science Research Network: Rochester, NY, 2012. Retrieved from http://papers.ssrn.com/abstract=2184235.

Bahtijarević-Šiber, F. *Management ljudskih potencijala*; Golden marketing, 1999; pp 507–509.

Becerra-Fernandez, I.; González, A. J.; Sabherwal, R. *Knowledge Management: Challenges, Solutions, and Technologies.* Pearson/Prentice Hall: New York, 2004.

Belak, J.; Belak, J.; Thommen, J.-P. *Integralni management in upravljanje: kultura, etika in verodostojnost podjetja.* MER, 2014.

Belak, J.; Duh, M. Integral Management: Key Success Factors in the MER Model. *Acta Polytechn. Hung.* **2012,** *9*(3), 5–26.

Boersma, J. S. T.; Stegwee, R. A. *Exploring the Issues in Knowledge Management.* University of Groningen: Groningen, 1996. Retrieved from http://som.eldoc.ub.rug.nl/FILES/reports/1995-1999/themeA/1996/96A09/96a09.pdf.

Bollen, K. A.; Long, J. S. *Testing Structural Equation Models.* SAGE, 1993.

Brauner, E.; Becker, A. Beyond Knowledge Sharing: the Management of Transactive Knowledge Systems. *Knowl. Process Manage.* **2006,** *13*(1), 62–71. DOI:10.1002/kpm.240.

Carrión, G. C.; González, J. L. G.; Leal, A. Identifying Key Knowledge Area in the Professional Services Industry: A Case Study. *J. Knowl. Manage.* **2004,** *8*(6), 131–150. http://doi.org/10.1108/13673270410567684.

Carter, C.; Scarbrough, H. Towards a Second Generation of KM? The People Management Challenge. *Educ. + Train.* **2001,** *43*(4/5), 215–224.

Chandiramani, S. *Information Technology and Corporate Governance*, 2007. Retrieved from http://unpan1.un.org/intradoc/groups/public/documents/APCITY/UNPAN026699.pdf.

Chan, P.; Pollard, D.; Puriveth, P. Implementing Knowledge Management. *J. Bus. Econ. Res.* **2004,** *2*(5), 7–17.

Chan, P.; Pollard, D.; Puriveth, P. Implementing Knowledge Management. *J. Bus. Econ. Res. (JBER)* **2011,** *2*(5).

Chase, R. L. The Knowledge-based Organization: An International Survey. *J. Knowl. Manage.* **1997,** *1*(1), 38–49. http://doi.org/10.1108/EUM0000000004578.

Chivu, I.; Popescu, D. Human Resources Management in the Knowledge Management. *Rev. Inform. Econ.* **2008,** *4*(48), 54–60.

Collins, H. *Tacit and Explicit Knowledge.* University of Chicago Press: Chicago, IL, 2010.

Connelly, C. E.; Kelloway, E. K. Predictors of Employees' Perceptions of Knowledge Sharing Cultures. *Leadersh. Organ. Dev. J.* **2003,** *24*(5), 294–301.

Currie, G.; Kerrin, M. Human Resource Management and Knowledge Management: Enhancing Knowledge Sharing in a Pharmaceutical Company. *Int. J. Hum. Resour. Manage.* **2003,** *14*(6), 1027–1045. DOI:10.1080/0958519032000124641.

Daud, S.; Yusoff, W. F. W. How Intellectual Capital Mediates the Relationship between Knowledge Management Processes and Organizational Performance. *Afr. J. Bus. Manage.* **2011,** *5*(7), 2607–2617.

Davenport, T.; Prusak, L. *Working Knowledge.* Harvard Business School Press: Cambridge, 1998.

Davenport, T. H.; Völpel, S. C. The Rise of Knowledge towards Attention Management. *J. Knowl. Manage.* **2001,** *5*(3), 212–222.

Dinur, A. Tacit Knowledge Taxonomy and Transfer: Case-Based Research. *J. Behav. Appl. Manage.* **2011.** Retrieved from http://www.highbeam.com/doc/1P3-2373068391.html.

du Plessis, M. Knowledge Management: What Makes Complex Implementations Successful? *J. Knowl. Manage.* **2007a,** *11*(2), 91–101. http://doi.org/10.1108/13673270710738942.

du Plessis, M. The Role of Knowledge Management in Innovation. *J. Knowl. Manage.* **2007b,** *11*(4), 20–29. http://doi.org/10.1108/13673270710762684.

Edvardsson, I. R. Knowledge Management and Creative HRM. *Occasional Pap.* **2003,** *14*. Retrieved from http://www.strath.ac.uk/media/departments/hrm/pdfs/hrm-pdf-occasionalpapers/media_61995_en.pdf.

Fong, C.-Y.; Ooi, K.-B.; Tan, B.-I.; Lee, V.-H.; Chong, A. Y.-L. HRM Practices and Knowledge Sharing: An Empirical Study. *Int. J. Manpow.* **2011,** 32(5/6), 704–723.

Fornell, C.; Larcker, D. F. Evaluating Structural Equation Models with Unobservable Variables and Measurement Error. *J. Mark. Res.* **1981,** *18*(1), 39. http://doi.org/10.2307/3151312.

Goh, C. H. T.; Hooper, V. Knowledge and Information Sharing in a Closed Information Environment. *J. Knowl. Manage.* **2009,** *13*(2), 21–34. DOI:10.1108/13673270910942673.

Gold, A. H.; Malhotra, A.; Segars, A. H. Knowledge Management: An Organizational Capabilities Perspective. *J. Manage. Inf. Syst.* **2001,** *18*(1), 185–214.

Hendriks, P. Why share knowledge? The Influence of ICT on the Motivation for Knowledge Sharing. *Knowl. Process Manage.* **1999,** 6(2), 91–100.

Hicks, R. C.; Dattero, R.; Galup, S. D. The Five-tier Knowledge Management Hierarchy. *J. Knowl. Manage.* **2006,** *10*(1), 19–31. http://doi.org/10.1108/13673270610650076.

Hoegl, M.; Parboteeah, K. P.; Munson, C. L. Team-Level Antecedents of Individuals' Knowledge Networks. *Decis. Sci.* **2003,** *34*(4), 741–770. http://doi.org/10.1111/j.1540-5414.2003.02344.x.

Housel, T.; Bell, A. H. *Measuring and Managing Knowledge,* 2001. Retrieved from https://calhoun.nps.edu/handle/10945/41074.

Hussock, J. *Defining a Framework for Knowledge Sharing in a Dynamic Sales Oriented Organisation,* Master Thesis, Dublin Institute of Technology, Dublin, September 2009.

Hutchinson, J. R.; Huberman, M. Knowledge Dissemination and Use in Science and Mathematics Education: A Literature Review. *J. Sci. Educ. Technol.* **1994,** *3*(1), 27–47. http://doi.org/10.1007/BF01575814.

Ishak, N.; Eze, U.; Ling, L. Integrating Knowledge Management and Human Resource Management for Sustainable Performance. *J. Organ. Knowl. Manage.* **2010,** *2010*(2010), 1–13. http://doi.org/10.5171/2010.322246.

Jalote, P. Knowledge Infrastructure for Project Management. In: *Managing Software Engineering Knowledge;* Springer, 2003; pp 361–375. Retrieved from http://link.springer.com/chapter/10.1007/978-3-662-05129-0_17.

Jantz, R. Knowledge Management in Academic Libraries: Special Tools and Processes to Support Information Professionals. *Ref. Serv. Rev.* **2001,** 29(1), 33–39. http://doi.org/10.1108/00907320110366778.

Jelkić, V. Kakvo znanje trebamo? *Filozofska istraživanja* **2011,** *31*(2), 255–261.

Jengard, L. *Project Knowledge Management: How to Evaluate Project Knowledge, and Project Knowledge Management Performance,* Student Thesis, Linnaeus University, Faculty of Science and Engineering, School of Computer Science, Physics and Mathematics, 2010.

Jha, S. Human Resource Management and Knowledge Management: Revisiting Challenges of Integration. *Int. J. Manage. Bus. Stud.* **2011,** *1*(2).

Judge, T. A.; Cable, D. M. Applicant Personality, Organizational Culture, and Organization Attraction. *Pers. Psychol.* **1997,** *50*(2), 359–394. http://doi.org/10.1111/j.1744-6570.1997.tb00912.x.

Kluge, J.; Stein, W.; Licht, T. *Knowledge Unplugged: The McKinsey & Company Global Survey on Knowledge Management.* Palgrave Macmillan: Basingstoke, 2001.

Kristof, A. L. Person–Organization Fit: An Integrative Review of Its Conceptualizations, Measurement, and Implications. *Pers. Psychol.* **1996,** *49*(1), 1–49. http://doi.org/10.1111/j.1744-6570.1996.tb01790.x.

Kumar, S. A. Knowledge Management and New Generation of Libraries Information Services: A Concepts. *Int. J. Lib. Inf. Sci.* **2010,** *1*(2), 24–30.

Lam, A.; Lambermont-Ford, J.-P. Knowledge Sharing in Organisational Contexts: A Motivation-based Perspective. *J. Knowl. Manage.* **2010,** *14*(1), 51–66.

Lee, H.; Choi, B. Knowledge Management Enablers, Processes, and Organizational Performance: An Integrative View and Empirical Examination. *J. Manage. Inf. Syst.* **2003,** *20*(1), 179–228.

Lee, V.-H.; Leong, L.-Y.; Hew, T.-S.; Ooi, K.-B. Knowledge Management: A Key Determinant in Advancing Technological Innovation? *J. Knowl. Manage.* **2013,** *17*(6), 848–872. http://doi.org/10.1108/JKM-08-2013-0315.

Lee, W.-L.; Liu, C.-H.; Wu, Y.-H. How Knowledge Cooperation Networks Impact Knowledge Creation and Sharing: A Multi-countries Analysis. *Afr. J. Bus. Manage.* **2011,** *5*(31). DOI:10.5897/AJBM11.522.

Lin, H. F. Knowledge Sharing and Firm Innovation Capability: An Empirical Study. *Int. J. Manpow.* **2007,** *28*(3/4), 315–332. http://doi.org/10.1108/01437720710755272.

Liu, N.-C.; Liu, M.-S. Human Resource Practices and Individual Knowledge-sharing Behavior—An Empirical Study for Taiwanese R&D Professionals. *Int. J. Hum. Resour. Manage.* **2011,** *22*(4), 981–997.

Madhoushi, M.; Sadati, A.; Delavari, H.; Mehdivand, M.; Hedayatifard, M. Facilitating Knowledge Management Strategies through IT and HRM. *Chin. Bus. Rev.* **2010,** *9*(10), 57–66.

Mason, D.; Pauleen, D. J. Perceptions of Knowledge Management: A Qualitative Analysis. *J. Knowl. Manage.* **2003,** *7*(4), 38–48. http://doi.org/10.1108/13673270310492930.

Massa, S.; Testa, S. A Knowledge Management Approach to Organizational Competitive Advantage: Evidence from the Food Sector. *Eur. Manage. J.* **2009,** *27*(2), 129–141. http://doi.org/10.1016/j.emj.2008.06.005.

Martins, J. D. M.; António, N. J. S. The Transfer of Knowledge from the MNEs to their Mozambican Subsidiaries: A Process Based on the Relationship between Source and Recipient. *Afr. J. Bus. Manage.* **2010,** *4*(13), 2615–2624.

Mathew, M.; Kumar, D.; Perumal, S. Role of Knowledge Management Initiatives in Organizational Innovativeness: Empirical Findings from the IT Industry. *Vikalpa* **2011,** *36*(2), 31.

Mehta, N. Successful Knowledge Management Implementation in Global Software Companies. *J. Knowl. Manage.* **2008,** *12*(2), 42–56. http://doi.org/10.1108/13673270810859505.

Milanović, L. Korištenje informacijske tehnologije za upravljanje znanjem u hrvatsim poduzećima. *Zbornik Ekonomskog fakulteta u Zagrebu.* **2010,** *8*(2), 195–211.

Milfelner, B.; Belak, J. Integralni Pristup Kulturi Poduzetništva kao Jednom od Ključnih Faktora Uspjeha u Poduzetništvu. *Ekonomska Istraživanja.* **2012,** *25*(3), 620–643.

Moffett, S.; McAdam, R. Knowledge Management: A Factor Analysis of Sector Effects. *J. Knowl. Manage.* **2009,** *13*(3), 44–59. http://doi.org/10.1108/13673270910962860.

Moffett, S.; McAdam, R.; Parkinson, S. An Empirical Analysis of Knowledge Management Applications. *J. Knowl. Manage.* **2003,** *7*(3), 6–26. http://doi.org/10.1108/13673270310485596.

Moorthy,S.;Polley,D.E.TechnologicalKnowledgeBreadthandDepth:PerformanceImpacts. *J. Knowl. Manage.* **2010,** *14*(3), 359–377. http://doi.org/10.1108/13673271011050102.

Nonaka, I. The Knowledge-creating Company. *Harv. Bus. Rev.* **1991,** *85*(7/8), 162.

Oltra, V. Knowledge Management Effectiveness Factors: The Role of HRM. *J. Knowl. Manage.* **2005,** *9*(4), 70–86. http://doi.org/10.1108/13673270510610341.

Osterloh, M.; Frost, J.; Frey, B. S. The Dynamics of Motivation in New Organizational Forms. *Int. J. Econ. Bus.* **2002,** *9*(1), 61–77. http://doi.org/10.1080/13571510110102976.

Oxford Dictionaries. *Oxford Dictionary and Thesaurus* (2nd ed.). OUP Oxford: New York, 2007.

Özbebek, A.; Toplu, E. K. Empowered Employees' Knowledge Sharing Behavior. *Int. J. Bus. Manage.* **2011,** *3*(2), 69–76.

Pan, S. L.; Scarbrough, H. Knowledge Management in Practice: An Exploratory Case Study. *Technol. Anal. Strat. Manage.* **1999,** *11*(3), 359–374. http://doi.org/10.1080/095373299107401.

Pastor, I. M. P.; Santana, M. P. P.; Sierra, C. M. Managing Knowledge through Human Resource Practices: Empirical Examination on the Spanish Automotive Industry. *Int. J. Hum. Resour. Manage.* **2010,** *21*(13), 2452–2467. http://doi.org/10.1080/09585192.2010.516596.

Pučko, D. Corporate Governance in European Transition Economies: Emerging Models. *Manage.: J. Contemp. Manage. Issues* **2005,** *10*(1), 1–21.

Pugesek, B. H.; Tomer, A.; von Eye, A.. *Structural Equation Modeling: Applications in Ecological and Evolutionary Biology.* Cambridge University Press: Cambridge, 2003.

Ringle, C.; Wende, S.; Will, A. *SmartPLS 2.0 (Beta).* University of Hamburg: Hamburg, Germany, 2005.

Sarayreh, B.; Mardawi, A.; Dmour, R. Comparative Study: The Nonaka Model of Knowledge Management. *Int. J. Eng. Adv. Technol. (IJEAT)* **2012,** *1*(6). Retrieved from http://www.oalib.com/paper/2087615#.VAHbj_l_sgS.

Scarbrough, H. Knowledge Management, HRM and the Innovation Process. *Int. J. Manpow.* **2003,** *24*(5), 501–516. http://doi.org/10.1108/01437720310491053.

Scarbrough, H.; Swan, J. Explaining the Diffusion of Knowledge Management: The Role of Fashion. *Brit. J. Manage.* **2001,** *12*(1), 3–12. http://doi.org/10.1111/1467-8551.00182.

Smith, M. K. Michael Polanyi and Tacit Knowledge, 2003. Retrieved from http://infed.org/mobi/michael-polanyi-and-tacit-knowledge/ (25 May 2014).

Soliman, F.; Spooner, K. Strategies for Implementing Knowledge Management: Role of Human Resources Management. *J. Knowl. Manage.* **2000,** *4*(4), 337–345.

Srikantaiah, T. K.; Koenig, M. E. D. *Knowledge Management for the Information Professional.* Information Today, Inc.: Medford Township, NJ, 2000.

Staab, S.; Schnurr, H.-P.; Studer, R.; Sure, Y. *Knowledge Process and Ontologies,* 2000. Retrieved from http://userpages.uni-koblenz.de/~staab/Research/Publications/isystems-knowledgeprocess.pdf.

Standards Australia. *The Value in Governance of Information Technology,* 2012. Retrieved from http://www.standards.org.au/OurOrganisation/News/Documents/SA-Value-in-Governance-in-IT.pdf.

Stein, E. W.; Zwass, V. Actualizing Organizational Memory with Information Systems. *Inform. Syst. Res.* **1995,** *6*(2), 85–117. http://doi.org/10.1287/isre.6.2.85.

Storey, C.; Kelly, D. Innovation in Services: The Need for Knowledge Management. *Australas. Mark. J. (AMJ)* **2002,** *10*(1), 59–70. http://doi.org/10.1016/S1441-3582(02) 70144-4.

Theriou, G. N.; Chatzoglou, P. D. Enhancing Performance through Best HRM Practices, Organizational Learning and Knowledge Management: A Conceptual Framework. *Eur. Bus. Rev.* **2008,** *20*(3), 185–207. http://doi.org/10.1108/09555340810871400.

Thite, M. Strategic Positioning of HRM in Knowledge-based Organizations. *Learn. Organ.* **2004,** *11*(1), 28–44. http://doi.org/10.1108/09696470410515715.

Tuan, L. T. Human Resource Management in Knowledge Transfer. *Int. Bus. Manage.* **2011,** *2*(2), 128–138.

Vidović, M. Razvijenost prakse upravljanja znanjem u Hrvatskoj. *Zbornik Ekonomskog fakulteta u Zagrebu* **2008,** *6*(1), 275–288.

Wiig, K. M. Knowledge Management: Where Did It Come From and Where Will It Go? *Expert Syst. Appl.* **1997,** *13*(1), 1–14.

Wong, K. K.-K. Partial Least Squares Structural Equation Modeling (PLS-SEM) Techniques Using SmartPLS. *Mark. Bull.* **2013,** *24*, 1–32.

Yeşil, S.; Koska, A.; Büyükbeşe, T. Knowledge Sharing Process, Innovation Capability and Innovation Performance: An Empirical Study. *Proc.—Soc. Behav. Sci.* **2013,** *75*, 217–225. http://doi.org/10.1016/j.sbspro.2013.04.025.

CHAPTER 6

EMPLOYEE ACCEPTANCE OF ERP INFORMATION SOLUTIONS IN SERVICE ORGANIZATIONS: A TAM-BASED RESEARCH

SIMONA STERNAD-ZABUKOVŠEK* and SAMO BOBEK

Faculty of Economics and Business, University of Maribor, Razlagova, Maribor, Slovenia

Corresponding author. E-mail: simona.sternad@um.si

CONTENTS

ABSTRACT

Enterprise resource planning (ERP) solutions have been recently implemented in service organizations. But most of organizations cannot confirm promised benefits of ERP solutions have been achieved. One of the reasons is ERP users who do not accept and use ERP solutions in their whole functionality. Research has showed that users impacts ERP solutions use and acceptance. Technology acceptance model (TAM) developed by Davis (1989) has been most widely used for researching user acceptance and usage of information technologies and solutions. In our study which is based on TAM we have researched ERP solutions use after 1 year of operation in organizations. Because of this we modified TAM with construct extended use instead of actual use. We also present relationship between work compatibility and usefulness. We added new relationship between work compatibility and attitude toward using ERP solution. In majority TAM research papers only limited number of external factors has been researched therefore we added 13 external factors which have influence on ERP extended use. The hypothesized model has been empirically tested using data collected from a survey. Model has been analyzed using PLS.

6.1 INTRODUCTION

Majority of organizations, especially in service sector, have implemented enterprise resource planning (ERP) solutions until today, but it seems that their users don't use ERP solutions in extended way. Because of ERP solutions characteristics, they have huge impact on employees and especially on those who are ERP users. Impact of ERP solutions on their users and their acceptance and degree of ERP usage has been recognized as one of key factors of ERP implementation success.

ERP solutions are transforming business processes and how organizations operate. Although that the most important contributions of business information systems (BIS) are that they significantly reduce the time to complete business processes and help organizations to share information (Lee et al., 2010). They also usually offer a better work environment for their organizational workers as they are given more efficient system to work with. On the other hand, they have been also plagued with high failure rates and inability to realize promised benefits (Kwahk & Lee, 2008). As with any information systems (IS), user perceptions about the IS play a critical role in its usage and eventual success (Saeed et al., 2010). They added that recent studies indicate that evaluating the user's acceptance in the nature of the IS, moderate usage should be an important consideration in evaluating IS usage. If IS is entail with major reorganization of work practices and it is prescribed as mandatory, that it is likely to be trigger with strong user feelings (Saeed et al., 2010). We can say that BIS only give organizations real benefits if users accept and use it extensively during daily tasks. Even more, organizational workers' perceptions play a critical role in determining whether a business solution is implemented effectively and used to its maximum capacity, even that those business solutions are mandate to use. Users can choose to resist using them, use them only minimally or use them in extended way. The most wildly implemented BIS at organizations in last few years are ERP solutions, the chapter will be focused on them.

A primary outcome expectation from the ERP solution is improvement in efficiency. So, much of the success of ERP solutions implementation lies in operational phase of ERP life cycle, which consists of stabilization stage and routine stage (Bradford, 2008; Motiwalla & Thompson, 2009). Stabilization stage is the time from Go-live to about 30–90 days after, or until the number of issues and problems has been reduced to a small, manageable number. In the majority of cases in that stage operational efficiency drops in productivity. Eventually the users' turn-on their ERP systems and they begin to work using ERP systems, instead of having to navigate many different systems. However, at some point in the ERP system's life users begin to see the advantages of the ERP system and then they carefully begin to explore its functions, gradually reaching success. Benefits start to occur as an organization gains more experience in using the ERP system (Gattiker & Goodhue, 2005; Hong & Kim, 2002; Markus & Tanis, 2000; Ross & Vitale, 2000; Saeed et al., 2010). Finally, having mastered the system, the users begin to get creative. This shows that the ERP users have accepted the system and are putting it to extended use.

Moreover, numerous factors influence IS success, especially individual acceptance or resistance (Amoako-Gyampah & Salam, 2004). To improve the efficiency and effectiveness of ERP system use, organizations need to research factors that impact user satisfaction. In this area, technological acceptance model (TAM) is one of the most widely used models for explaining the behavioral intention (BI) and actual usage and can improve our understanding of how influence on actual usage could help increase efficiency and effectiveness of ERP system use (Shih & Huang, 2009). Review of the literature shows that in past few years, a few studies that have examined the users' adoption of ERP systems through TAM have been published (for latest researches see Calisir et al., 2009; Kwak et al., 2012; Lee et al., 2010; Shih & Huang, 2009; Sun et al., 2009; Yang Chan et al., 2013; Youngberg et al., 2009). But all of them examine few contextual factors that have influence on intention to use ERP system or ERP use in stabilization stage or earlier stages of ERP lifecycle. In addition, very few studies have been conducted regarding technology acceptance of ERP systems, especially those dealing with autonomous ERP users (i.e., Sun et al., 2009). Through their scientific work, researchers have recognized that the generality of TAM and researching of small numbers additional factors that have impact on TAM fails to supply more meaningful information on users' opinions about a specific system especially of ERP system that is considered as a strategic IS in organizations. In the ERP context, organizations have to adopt business processes of an implemented ERP system. Inherent business rules behind business processes gives organizational users little choice but to follow strict business process of ERP system. Therefore, there is the need to incorporate additional factors or integrate it with other IT acceptance models for improvement of its specificity and explanatorily utility (i.e., Agarwal & Prasad, 1999; Lu et al., 2003).

6.1.1　TECHNOLOGY ACCEPTANCE FRAMEWORKS

Several theoretical models have been used to investigate the determinants of acceptance and use of new information technology (IT) (Venkatesh et al., 2003). The theory of reasoned action (TRA) (Fishbein & Ajzen, 1975) is a fundamental theory used to explain human behavior (Venkatesh et al., 2003). According to TRA, BI can be explained by the attitude toward behavior and subjective norms (Fishbein & Ajzen, 1975). TRA has been applied into theory of planned behavior (TPB) (Ajzen, 1991) and into the theory of TAM (Davis et al., 1989). Numerous empirical studies have found that TAM consistently explains a substantial proportion of the variance (typically about 40%) in usage intentions and behavior, and the TAM compares favorably with alterative models such as the TRA and TPB (Davis et al., 1989; Legris et al., 2003; Venkatesh & Davis, 2000). TAM has become well established as a robust, powerful, and parsimonious model for predicting user acceptance (Lu et al., 2003; Liu & Ma, 2006; Venkatesh & Davis, 2000) and so among the theoretical models is most widely used by IS/IT researchers (Amoako-Gyampah & Salam, 2004; Davis, 1989; Davis et al., 1989; Lee et al., 2010). The key purpose of TAM is to provide a basis for tracing impact of external factors on internal beliefs, attitudes, and intentions (Davis et al., 1989). TAM posits that two beliefs, perceived usefulness (PU) and perceived ease of use (PEOU) are of primary relevance for computer acceptance behavior (Davis et al., 1989). PU is defined as "degree to which a person believes that using a particular system would enhance his or her job performance" (Davis, 1989, p. 320). PEOU in contrast, refers to "the degree to which a person believes that using a particular system would be free of effort" (Davis, 1989, p. 320). The two central hypotheses in TAM state that PU and PEOU positively influence an individual's attitude toward using a new technology (AT), which in turn influences his or her BI to use it. Finally, intention is positively related to actual use. TAM also predicts that PEOU influences on PU, as Davis et al. (1989, p. 987) put it, "effort saved due to improved perceived ease of use may be redeployed, enabling a person to accomplish more work for the same effort."

6.1.2 THEORETICAL BACKGROUND OF BUSINESS INFORMATION SOLUTIONS ACCEPTANCE BY USERS IN ORGANIZATIONS

One of the main reasons for lack of sophisticated use of ERP systems is, according to many studies, the lack of user acceptance. As several studies (i.e., Umble et al., 2002; Nah et al., 2004) have revealed, a common reason for ERP failures can be attributed to users' reluctance and unwillingness to adopt and to use the implemented ERP system. More researches on TAM stress out that if we want to explain higher proportion of the variance (typically about 40%), we have to include external factors which have impact on cognitive factors (Davis et al., 1989; Legris et al., 2003; Venkatesh & Davis, 2000) or integrate other cognitive factors (i.e., Agarwal & Prasad, 1999; Lu et al., 2003). A literature review of past ERP studies regarding TAM indicates that few studies have investigated ERP user acceptance and usage. Research shows that a small number of researches have been published, and all of them expose small numbers of determinants (external factors) or cognitive factors which could have influence on ERP acceptance and usage in different phases of an ERP system life cycle (see Table 6.1).

TABLE 6.1   ERP Literature Review Regarding TAM.

Reference	Focus	Phase—ERP system life cycle
Nah et al. (2004)	They tested impact of four cognitive constructors (perceived usefulness, perceived ease of use, perceived compatibility, and perceived fit) on attitude toward using ERP system and symbolic adoption	Post-implementation (stabilization phase)
Amoako-Gyampah and Salam (2004)	Their study evaluated the impact of one belief construct (shared beliefs in the benefits of a technology) and two technology success factors (training and communications) on perceived usefulness and perceived ease of use in one global organization	Implementation
Shivers-Blackwell and Charles (2006)	They research student's readiness for change (through gender, computer self-efficacy and perceived benefits of ERP) on behavioral intention regarding the ERP implementation	Implementation
Bradley and Lee (2007)	They investigate through case study the relationship between training satisfaction and the perceptions of ease of use, the perceptions of usefulness, effectiveness and efficiency in implementing an ERP system at a mid-sized university	Implementation
Hsieh and Wang (2007)	They research impact of perceived usefulness and perceived ease of use on extended use	Post-implementation (routine stage)
Kwahk and Lee (2008)	They examined the formation of readiness for change (enhanced by two factors: organizational commitment and perceived personal competence) and its effect on the perceived technological value of an ERP system leading to its use	Post-implementation (stabilization stage)
Bueno and Salmeron (2008)	They develop a research model based on TAM for testing the influence of the critical success factors (top management support, communication, cooperation, training, and technological complexity) on ERP implementation	Implementation
Uzoka et al. (2008)	They extended TAM to research the selection of ERP by organizations using factors: impact of system quality, information quality, service quality, and support quality as key determinants of cognitive response, which ERP system to purchase/use	Selection

TABLE 6.1 *(Continued)*

Reference	Focus	Phase—ERP system life cycle
Sun et al. (2009)	They extended IT usage models to include role of ERP's perceived work compatibility in users' ERP usage intention, usage, and performance in work settings	Post-implementation (routine stage)
Shih and Huang (2009)	Their study attempts to explain behavioral intention and actual use through incorporated additional behavioral constructs: top management support, computer self-efficacy, and computer anxiety	Post-implementation (routine stage)
Calisir et al. (2009)	They examine factors (subjective norms, compatibility, gender, experience, and education level) that affect users' behavioral intention to use ERP system based on potential ERP users at one manufacturing organization	Implementation
Youngberg et al. (2009)	They researched impact of perceived ease of use, result demonstrability, and subjective norm on perceived usefulness and impact of it on usage behavior	Post-implementation (stabilization stage)
Lee et al. (2010)	They examined factor organizational support (formal and informal) on original TAM factors	Post-implementation
Garača (2011)	Study investigates the correlation of the perceived usefulness of the ERP system, perceived ease of using it, and computer anxiety in users to their satisfaction with the used ERP system	Post-implementation
Yang Chan et al. (2013)	They researched three mediating variables, namely perceived ease of use, perceived usefulness, and self-efficacy, and four external variables, namely subjective norms, training support, fairness of the organization's business evaluation system, and belief in the benefit of the IT on factor intention to use in a mandatory environment	Post-implementation

But to assess if TAM is suitable for explaining end-users' acceptance in the ERP context, we need to expose relationships defined by TAM in the ERP context. TAM has been tested primarily on technologies that are relatively simple to use (e.g., e-mail, word processors) and is not mandatory to use by users. Several researchers (Adamson & Shine, 2003; Brown et al., 2002; Nah et al., 2004) have pointed out that TAM needs to be extended or revised in order to explain end-users' acceptance of complex IT in

organizational settings. ERP usage is characterized as mandatory (Nah et
al., 2004; Pozzebon, 2000; Yang Chan et al., 2013) and that one users' tasks
have strong impact on other users' tasks. That's mean that users do not have
the choice not to use the system, regardless of their attitude. All researchers
who examine ERP acceptance and usage through TAM, except Nah et al.
(2004), had focus on external factors which have influence on cognitive
constructs (see Table 6.1). Because of that they exclude relations between
basic constructs of TAM in context of ERP which can be seen in Table 6.2.

At the end of this head in Figure 6.1, the research model with relationships which different researchers had exposed between factors is
presented.

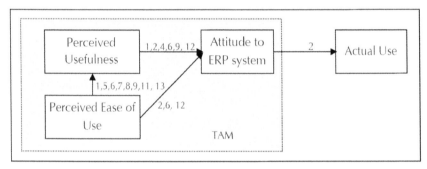

FIGURE 6.1   Relationships between factors in extended TAM for ERP regarding to ERP
researches.

In summary, because of high rate of ERP implementation failure, more
research in the area of technological acceptance is needed (Youngberg et
al., 2009). The original TAM is well established and tested and, furthermore, a variety of extensions has been developed in different IT environments. Regardless of ERP complexity and ERP implementation failure,
very few studies have been conducted regarding technology acceptance,
especially those dealing with autonomous ERP users and including more
cognitive constructs.

6.2   ERP SOLUTIONS ACCEPTANCE MODEL

To examine ERP users' use of ERP systems, we extended the TAM model.
Synthesizing prior researches on TAM and researches on ERP systems, a

TABLE 6.2 Relationship Significance through Cognitive Constructs Regarding Researches.

	Authors	Survey	PEOU → PU	PU → A	PEOU → A	PU → BI	PEOU → BI	AT → BI	BI → Use	AT → Use	PU → Use	EOU → Use
1	Amoako-Gyampah and Salam (2004)	Large global organization that was implemented, ERP system (SAP, 571 respondents)	***	**	n.s.	n.s.		*				
2	Nah et al. (2004)	SAP users of one public institution where ERP was implemented 1999 (229 respondents)		*	**					*** (1)		*** (1)
3	Hwang (2005)	Online survey where 70% users use ERP solutions more than 1 year	n.s.			***	***					
4	Shivers-Blackwell and Charles (2006)	238 students' intention to use ERP system on bases on online newsletter of ERP systems		*	n.s.			*				
5	Hsieh and Wang (2007)	ERP users in large manufacturing company. They use ERP system more than 2 years (79 respondents)	**								* (2)	** (2)
6	Bueno and Salmeron (2008)	Potential users during implementation process (91 responses)	**	***	*	***		**				
	Kwahk and Lee (2008)	Recently finished ERP implementation and had implemented at least two ERP modules (283 usable responses)				* (3)	* (3)					
7	Sun et al. (2009)	Field survey where more than 3 years is implemented ERP system	**	**		**	*		**			
8	Youngberg et al. (2009)	Potential participants from 4 Utah 2-year college	***	***		***			n.s.			

TABLE 6.2 *(Continued)*

Authors	Survey	PEOU → PU	PU → A	PEOU → A	PU → BI	PEOU → BI	AT → BI	BI → Use	AT → Use	PU → Use	EOU → Use
9 Calisir et al. (2009)	Group of potential users at a manufacturing organization (75 respondents)	***	***	n.s.			n.s.				
10 Shih and Huang (2009)	Implement or use ERP system. 99% users have experience of using ERP system at least 12 months (165 respondents)	n.s.			**	**		**		n.s.	
11 Lee et al. (2010)	63.2% users had experience using ERP system in the past	***									
12 Garača (2011)	After winter semester, 180 third year students of the vocational program at the faculty, who may be potential users of the ERP system		**	**			**				
13 Yang Chan et al. (2013)	Defense enterprises, research centers, and program administrators, as mandatory usage environments (342 respondents)	*			*	*					

n.s., Not supported; *$p < 0.05$; **$p < 0.01$; ***$p < 0.001$.

(1) symbolic adoption; (2) extended use; (3) usage intention.

conceptual model that represents the cumulative body of knowledge accumulated over the years from TAM and ERP research has been developed
(see Fig. 6.2). The gray area within the dotted line denotes the original
TAM. Because our research was focused on current usage of ERP system
in routine stage, there was no need to examine the BI on-use; BI was
dropped from the research model.

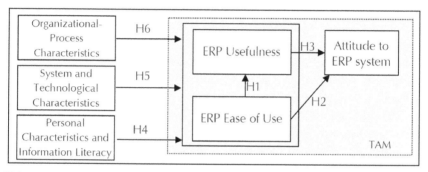

FIGURE 6.2   Conceptual model.

According to Davis (1989) and Davis et al. (1989), PEOU influences
PU, and PU and PEOU influence attitude toward using the system (AT).
Because we examined extended use after 1 year of operation, we can't talk
about PU and PEOU. In our study, we hypothesized:

H1: ERP ease of use has positive and direct effect on ERP usefulness.

H2: ERP ease of use has positive and direct effect on attitude toward
ERP system.

H3: ERP usefulness has positive and direct effect on attitude toward
ERP system.

Research efforts have been devoted to extend to the theory by examining the antecedents of PU and PEOU. As noted by Venkatesh and Davis
(2000), a better understanding of these factors would enable us to design
effective organizational interventions that might lead to increased user
acceptance and use of new IT systems. Over the last two decades, there
has been substantial empirical support in favor of TAM (see Venkatesh
& Bala, 2008). Even though TAM is a model applicable to a variety of
technologies, the constructs of TAM need to be extended by incorporating

additional factors (Calisir et al., 2009). Schwarz (2003) review of identified antecedents to cognitive factors (PEOU and PU) puts factors into
three groups, which are individual variable (such as computer experience,
self-efficacy, and prior experiences), organizational influences (such as
management and external support and perceived resources), and technology characteristics (such as accessibility of the medium and interface
type). On the other hand, Venkatesh and Bala (2008) expose four different
types of determinants of PU and PEOU: individual differences, system
characteristics, social influence, and facilitating conditions. Whereas in the
context of ERP systems, we expose on prior research of external factors
three groups of factors (see Table 6.3): personal characteristics and information literacy (PCIL), system and technological characteristics (STC),
and organizational-process characteristics (OPC).

The problem of TAM researches are that most researchers investigate
small numbers of external factors that have influence on user acceptance
and usage. In context of ERP systems, there are more external factors that
can have influence on users' acceptance and extended usage. Because of
that, conceptualization of higher order factors (in our case second-order
factors), in which more external factors jointly have to be included, have
been researched because we wanted to extent the understanding of user
behavior in ERP settings. On that presumption, we also hypothesized:

H4: There is group of external factors which have influence through
conceptual factor personal characteristics and information literacy on
use of ERP system.

H5: There is group of external factors which have influence through
conceptual factor system and technological characteristics on use of
ERP system.

H6: There is group of external factors which have influence through
conceptual factor organizational-process characteristics on use of ERP
system.

6.3   RESEARCH STUDY

Our hypotheses have been tested empirically using a filed survey of ERP
users using ERP systems in routine stage. Organizations have been selected
using two criteria: (1) the organizations must have implemented one of the

TABLE 6.3 External Factors Mentioned by Authors.

Personal characteristics and information literacy (PCIL)		
External factors	**Authors**	**Description**
Experience with computer	Davis et al. (1989), Venkatesh et al. (2003), Thompson et al. (2006), Venkatesh and Bala (2008), Calisir et al. (2009)	Experience with computer has been found to be important factor for the acceptance of a technology (Calisir et al., 2009)
Computer self-efficiency	Venkatesh and Davis (2000), Venkatesh et al. (2003), Thompson et al. (2006), Shivers-Blackwell and Charles (2006), Venkatesh and Bala (2008), Shih and Huang (2009), Yang Chan et al. (2013)	Computer self-efficiency is defined as the degree to which an individual believes that he or she has the ability to perform a specific task/job using the computer (Shih & Huang, 2009; Venkatesh & Bala, 2008)
Personal innovativeness toward IT	Agarwal and Prasad (1999), Rogers (2003), Yi et al. (2006), Thompson et al. (2006)	Personal innovativeness toward IT represents the degree to which an individual is willing to try out a new IT (Agarwal & Prasad, 1999)
Computer anxiety	Venkatesh et al. (2003), Liu and Ma (2006), Venkatesh and Bala (2008), Shih and Huang (2009)	Computer anxiety represents degree of an individual's apprehension, or even fear, when she/he is faced with the possibility of using computers (Venkatesh et al., 2003)
System and technological characteristics (STC)		
ERP data quality	Venkatesh (1998), Venkatesh and Davis (2000), Gattiker and Goodhue (2005), Kositanuri et al. (2006), Iansiti (2007)	Without accurate and relevant data, an organization is severely constrained in the coordination and task efficiency benefits it can achieve from its ERP system (Gattiker & Goodhue, 2005)
ERP system functionality	Musaji (2002), Somers et al. (2003), Lu et al. (2003), Kositanuri et al. (2006), Iansiti (2007)	System functions are used to measure the rapid response, stability, easy usage, and flexibility of the system

TABLE 6.3   *(Continued)*

Personal characteristics and information literacy (PCIL)

External factors	Authors	Description
ERP system performance	Boudreau (2002), Musaji (2002), Venkatesh et al. (2003), Somers et al. (2003); Kositanuri et al. (2006), Liu and Ma (2006), Iansiti (2007)	ERP system performance refers to the degree to which person believes that a system is reliable and responsive during a normal course of operations (Liu & Ma, 2006)
User manuals (help)	Kelley (2001), Boudreau (2002), Musaji (2002); Kositanuri et al. (2006), Bradford (2008)	The degree to which an individual views inadequate user manuals as the reason for one's unsuccessful ERP performance (Kelley, 2001)

Organizational-process characteristics (OPC)

Social influence	Venkatesh (1998), Venkatesh et al. (2003), Bradford (2008), Calisir et al. (2009), Yang Chan et al. (2013)	*Social influence* joins two factors: subjective norm and social factors. Subjective norm is defined "as a person's perception that most people who are important to him/her think that he/she should or should not perform the behaviour in question" (Venkatesh, 1998). Social factors are "individual's internalization of the reference group's subjective culture, and specific interpersonal agreements that the individual has made with others in specific social situations" (Venkatesh et al., 2003)
Fit with business processes	Amoako-Gyampah and Salam (2004), Nah et al. (2004), Bradley and Lee (2007), Bradford (2008)	Fit with business processes from an end-user's perspective is the degree to which the ERP system perceived by a user to meet his/her organization's needs (Nah et al., 2004)
ERP training and education	Amoako-Gyampah and Salam (2004), Bradley and Lee (2007), Bueno and Salmeron (2008), Yang Chan et al. (2013)	ERP training and education is defined as degree to which user thinks that he/she had enough formal and informal training after ERP implementation

TABLE 6.3 *(Continued)*

Personal characteristics and information literacy (PCIL)

External factors	Authors	Description
ERP support	Boudreau (2002), Lee et al. (2010)	We define ERP support as the degree to which an individual view adequate ERP support as the reason for one's successful ERP usage
ERP communication	Kelley (2001), Musaji (2002), Boudreau (2002), Amoako-Gyampah and Salam (2004), Bueno and Salmeron (2008)	ERP communication problems refer to the lack of communication regarding the ERP applications and their modifications (Kelley, 2001)

two most popular global ERP solutions in Slovenia: SAP or Microsoft Dynamics and (2) the organizations must have used the ERP system for more the 1 year at the time of the study. A total of 122 companies the initial e-mail has been send to each organization to verify if they matched our selection criteria and to explain the purpose of the study. Forty-four organizations agreed to participate in the survey, and they were asked to distribute the survey questionnaire to their ERP users. All respondents were required to have used ERP system in their daily work. A number of 293 questionnaires were properly filled by respondents and used for the purpose of analysis.

The constructs of purposed model are ERP usefulness, ERP ease of use, and attitude toward ERP use for basic TAM of ERP systems, influence by constructs of external variables. The constructs of external variables are distributed among three second-level constructs which are PCIL, STC, and OPC. PCIL includes factors: experience with computer, computer self-efficiency, personal innovativeness toward IT and computer anxiety. STC includes factors: ERP data quality, ERP system functionality, ERP system performance, and user manuals (help). And OPC includes factors: social influence, fit with business processes, ERP training and education, ERP support and ERP communication. So our model includes 17 first-order factors and three second-order factors. Higher order factors, in our case second-order factors, can thus be created by specifying a latent variable which represents all the manifest variables of the underlying lower order factors.

All the items of factors were measured on a 7-point Likert scale, ranging from "strongly disagree" to "strongly agree" taken from relevant prior researches and adapted to relate to the context of ERP usage. In addition, demographic information has been collected. The instrument was pilot tested with a group of 30 ERP users in 1 organization. The instrument's reliability was evaluated, the Cronbach's alpha values ranged from 0.58 to 0.91, indicating a satisfactory level of reliability exceeding value 0.5 (Hinton et al., 2004). As part of the pretest, comments and suggestions on questionnaire items and items wording have also been taken into account. Based on the results of the pilot testing, revisions and additions were made to the instrument. Pilot participants were included in the main data gathering effort, since they were part of the population of interest. Final scales and items after are listed in Table 6.4.

Covariance-based structural equation modeling (SEM) and component-based SEM, or partial least squares (PLS) approach can be employed to estimate the parameters in a hierarchical model. According to Chin (1998), PLS has several major strengths: it is a predicative technique suitable for situations with less theory development; it places minimal demands on measurement scales; it avoids factor indeterminacy problems and inadmissible solutions; it avoids identification problems of recursive models; it makes no assumptions about the data; it requires no specific distributions for measured variables; it assumes the errors are uncorrelated; it works well with small samples; and it is better suited for analyzing complex relationships and models. Models, which include second-order factors, consist of a higher order factors that are modeled as causally impacting a number of first-order factors (i.e., standard factors with measured indicators, Chin, 1998). Therefore, these second-order factors are not directly connected to any measurement items. PLS allows for the conceptualization of higher order factors by repeated use of manifest variables (Tenenhaus et al., 2005). A higher order factor can thus be created by specifying a latent variable which represents all the manifest variables of the underlying lower order factors. Due to all exposed benefits, we decided to use PLS. The empirical data was analyzed in two stages involving a PLS technique, using Smart PLS 2.0 M3 (Ringle & Will, 2005). In the first stage, all measurement scales have been examined for their psychometric properties, while the second stage focused on hypothesis testing and analysis.

6.4   RESULTS AND ANALYSIS

A number of 293 questionnaires were properly filled by respondents from 44 organizations and used for the purpose of analysis. Survey respondents represented different groups of industries, including IT and telecommunications (44.0%); manufacturing (35.2%); professional, scientific. and technical activities (10.2%); wholesale and retail trade (4.1%); and others (6.5%). Respondents were 51.5% male and 48.5% female. Most of them (67.2%) had high-school education or more. ERP system has been used for 4.73 years in average (min = 0, max = 18, SD = 3.0). A percentage of 53.6% (157) have marked working place as worker (experts and other employees), 31.7% (93) have marked low management (e.g., manager of group or organization unit), 12.6% (37) have marked middle management (e.g., CIO) and 2% (6) have marked corporate government and/or top management. Average total working years is 15.4 years (min = 0, max = 43, SD = 10.2) and average working years on these working place is 7.6 years (min = 0, max = 37, SD = 7.4).

Recall that all the scales were derived from previously developed and validated measures and the instrument's reliability was evaluated through the pilot testing; psychometric properties (measurement model) of these scales were assessed via evaluation of reliability, convergent validity, and discriminant validity each of measurement scale.

For external factors second-order factor procedure has been used. The method of repeated indicators known as the hierarchical component model suggested by Wold (1982) is the easiest to implement. The second-order factor is directly measured by observed variables for all the first-order factors that are measured with reflective indicators. While this approach repeats the number of manifest variables used, the model can be estimated by the standard PLS algorithm (Henseler et al., 2009). Because all of external factors did not meet assessment requirements of the measurement model, we excluded them from further analysis. These external factors are computer self-efficacy and experience with computer from PCIL group, ERP functionality from STC group and ERP support, ERP communications and ERP training and education from OPC group. The final version of model is presented.

We examined two measures of reliability: Cronbach's alpha (α) and composite reliability (CR). As shown in Table 6.4, each of our 11 scales had Cronbach's alpha exceeding 0.70 and CR exceeding 0.8, assuring adequate reliability for our measurement scales.

For convergent validity Fornell and Larcker's assessment criteria has been adopted: all item factor loadings should be significant and exceed 0.7, and the average variance extracted (AVE) for each construct should exceed 0.5 (Fornell & Larcker, 1981). Table 6.4 lists item factor loadings, all of which were significant at $p < 0.01$ and were higher than recommended level of 0.70 and also all values AVE exceeded 0.50. Remainder of our measurement scales shows strong evidence for convergent validity. In Table 6.7 the CR and AVE of the measures in the second-order model are included, which provides evidence for reliable measurements.

Discriminant validity between constructs has been assessed following Fornell and Larcker's recommendation that the square root of AVE for each construct should exceed the bivariate correlations between that construct and all other constructs (Fornell & Larcker, 1981). The inter-construct

TABLE 6.4 Psychometric Properties of the Instrument ($N = 293$).

Construct (source)	Item Mean	Item SD	Load.	α	CR	AVE
Personal innovativeness toward IT (Yi et al., 2006; Thompson et al., 2006)				0.85	0.91	0.77
If I hear about a new IT, I would look for ways to experiment with it	5.40	1.35	0.86			
Among my peers I am usually the first to try out new IT	4.57	1.68	0.86			
I like to experiment with new IT	5.14	1.59	0.92			
Computer anxiety (Venkatesh, 1998; Venkatesh et al., 2003)				0.76	0.87	0.69
Working with a computer makes me nervous[a]	1.57	1.00	0.87			
I get a sinking feeling when I think of trying to use a computer[a]	1.42	0.95	0.87			
I feel comfortable working with a computer	6.37	1.00	0.73			
Data quality (Gattiker & Goodhue, 2005; Venkatesh, 1998; Kositanuri et al., 2006)				0.91	0.93	0.69
The ERP system provides the precise information I need	5.41	1.03	0.82			
The information contents provided by the ERP system meet my needs	5.43	1.25	0.85			
The ERP system provides reports that seem to be exactly what I need	4.90	1.41	0.82			
The ERP system provides sufficient information to my needs	5.41	1.24	0.88			
The ERP system provides complete features I need	4.96	1.44	0.85			
I am satisfied with the speed of interacting with the system	5.21	1.35	0.74			
System performance (Venkatesh et al., 2003; Kositanuri et al., 2006; Liu & Ma, 2006)				0.86	0.90	0.64
It is fast to search data in the ERP system	5.14	1.43	0.79			
The ERP system loads quickly	5.14	1.50	0.72			

TABLE 6.4 *(Continued)*

Construct (source)	Item Mean	Item SD	Load.	α	CR	AVE
I was able to retrieve data quickly	5.45	1.23	0.87			
It is fast to create a new record in this system	5.33	1.35	0.73			
It is fast to use this ERP system	5.23	1.30	0.88			
User manuals (Kelley, 2001)						
The content and index of the user manuals are useful	5.00	1.51	0.88	0.84	0.91	0.76
The user manuals are current (up to date)	5.00	1.59	0.84			
The user manuals are complete	4.54	1.61	0.90			
Business processes fit (Amoako-Gyampah & Salam, 2004; Nah et al., 2004)						
The ERP solution fits well with the business needs of me	5.61	1.27	0.91	0.91	0.94	0.84
The ERP solution fits well with the business need of my department	5.56	1.28	0.89			
All in the ERP system is satisfactory in meeting my needs	5.39	1.29	0.89			
Social influence (Venkatesh et al., 2003, Venkatesh, 1998)						
My supervisor is very supportive of the use of the ERP system for my job	5.94	1.16	0.76	0.84	0.90	0.68
In general the organization has supported the use of the ERP system	6.01	1.03	0.79			
People who influence my behavior think that I should use the ERP system	5.98	1.15	0.88			
People who are important to me think that I should use the ERP system	6.03	1.09	0.87			
ERP usefulness (Davis, 1989)						
Using ERP solution in my job enables me to accomplish tasks more quickly	5.51	1.32	0.91	0.96	0.97	0.89

TABLE 6.4 (Continued)

Construct (source)	Item Mean	Item SD	Load.	α	CR	AVE
Using ERP solution improves my job performance	5.44	1.32	0.96			
Using ERP solution enhances my effectiveness on the job	5.49	1.33	0.97			
Using ERP solution makes it easier to do my job	5.40	1.40	0.93			
ERP ease of use (Davis, 1989)						
My interaction with ERP solution is clear and understandable	5.25	1.34	0.94	0.85	0.93	0.87
I find ERP solution is easy to use	4.90	1.41	0.93			
Attitude toward ERP system (Venkatesh et al., 2003; Nah et al., 2004)						
Using the ERP system is a good idea	5.93	1.18	0.94	0.90	0.95	0.90
I like the idea of using the ERP system to perform my job	5.68	1.32	0.96			

[a]Items have been inverted before processing of statistical data in SmartPLS.

correlation matrix (see Table 6.5) shows that the principal diagonal elements (square root AVE) exceed non-diagonals elements in the same row or columns (bivariate correlations), demonstrating that the discriminate validity of all scales is also adequate.

TABLE 6.5 Intercorrelations of the Latent Variables.

	1	2	3	4	5	6	7	8	9	10	11
1. Personal innovativeness toward IT	0.88										
2. Computer anxiety	0.34	0.83									
3. ERP data quality	0.15	0.19	0.83								
4. ERP system performance	0.10	0.19	0.64	0.80							
5. User manuals (help)	0.11	0.18	0.52	0.47	0.87						
6. Business processes fit	0.16	0.18	0.68	0.57	0.40	0.92					
7. Social influence	0.07	0.16	0.25	0.29	0.26	0.45	0.83				
8. ERP usefulness	0.16	0.17	0.58	0.57	0.35	0.67	0.41	0.94			
9. ERP ease of use	0.19	0.22	0.55	0.58	0.47	0.49	0.22	0.63	0.93		
10. Attitude toward using ERP	0.11	0.14	0.52	0.54	0.33	0.58	0.32	0.74	0.59	0.95	

Overall, these measurement results are satisfactory and suggest that it is appropriate to proceed with the evaluation of the structural model.

The next step in analysis was to examine the path significance and magnitude of each of our hypothesized effect and the overall explanatory power of the proposed model. The hypotheses testing results are based on bootstrapping (with 500 subsamples) to test the statistical significance of each path coefficient using t-tests, as recommended by Chin (1998). Results of this analysis are shown in Table 6.6 and Figure 6.3.

Effect size (f^2) can be viewed as a gauge for whether a predictor latent variable has a weak (0.02), medium (0.15), or large (0.35) effect at the structural level (Henseler et al., 2009). In our model, average f^2 is 0.33 which represent medium effect predictor latent variables at the structural level with none insignificant impact of independent variables on depended variables (see Table 6.6).

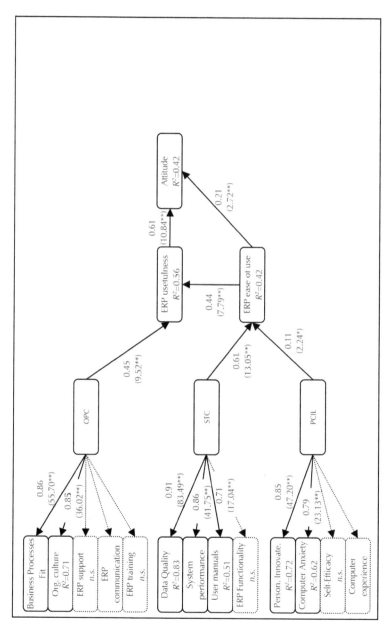

[a]Path significance: ** $p < 0.01$, * $p < 0.05$, n.s. = not significant (shapes are marked dotted).

FIGURE 6.3   Results of structural model analysis.[a]

TABLE 6.6 Parameter Estimation of the PLS Model by the Bootstrap Method and Effect Size.

Dependent variable	Independent variable	Standardized path coefficient	*t*-Values	f^2
Usefulness	OPC	0.45	9.52**	0.38[b]
	Ease of use	0.44	7.79**	0.36[b]
Ease of use	PCIL	0.11	2.24*	0.02[a]
	SCT	0.61	13.05**	0.61[b]
Attitude	Usefulness	0.61	10.84**	0.55[b]
	Ease of use	0.21	3.72**	0.06[a]

$**p < 0.01$; $*p < 0.05$.

[a] A predictor latent variable has a weak (0.02) effect at the structural level (Henseler et al., 2009).

[b] A predictor latent variable has a large (0.35) effect at the structural level (Henseler et al., 2009).

As it can be seen from Table 6.7, the loadings of the first-order factors on the second-order factors exceed 0.7 and second-order factors have significant positive effect on ERP usefulness and on ERP ease of use. PCIL has weak but significant positive effect on ERP ease of use ($\beta = 0.11$, $p < 0.05$), STC has strong positive effect on ERP ease of use ($\beta = 0.61$. $p < 0.01$) and OPC has strong positive effect on ERP usefulness ($\beta = 0.45$, $p < 0.01$). These findings provide empirical support for hypotheses H5–H7.

TABLE 6.7 Path Coefficients External Variables in Second-order Model.

First-order external factors	PCIL $\alpha = 0.79$ $CR = 0.85$ $AVE = 0.49$	STC $\alpha = 0.92$ $CR = 0.93$ $AVE = 0.50$	OPC $\alpha = 0.86$ $CR = 0.89$ $AVE = 0.54$
Personal innovativeness	0.85 ($t = 43.20$)		
Computer anxiety	0.74 ($t = 14.65$)		
Business process fit		0.83 ($t = 37.3$)	
Social influence		0.72 ($t = 15.05$)	
ERP data quality			0.90 ($t = 67.96$)
ERP system performance			0.85 ($t = 38.18$)
ERP user manuals (help)			0.69 ($t = 16.78$)

All *t*-values are bigger then 2.58 and are significant at $p < 0.01$.

The structural model demonstrates predictive power as the variance explained (R^2) in key endogenous constructs and is 0.56 for ERP usefulness, 0.42 for ERP ease of use and 0.58 for attitude toward ERP system. All of R^2 can be described as "moderate" by Chin (1998). The findings show that our model explains large part of variance in the endogenous variables, with an average R^2 of 0.52. Communality and redundancy coefficients are also presented in Table 6.8. They can be used essentially in the same way as the R^2, since they reflect the relative amount of explained variance for latent and manifest variables.

As important part of model evaluation is the examination of fit indexes reflecting the predictive power of estimated inner and outer model relationships, which can be measured by evaluating goodness-of-fit (GoF) coefficient (Tenenhaus et al., 2005). The general criterion for evaluating GoF is to calculate the geometric mean of the average communality and the average R^2 of endogenous variables (Tenenhaus et al., 2005). According to the results in Table 6.8, GoF is 0.56, which can be considered as satisfactory.

Then, the blindfolding approach proposed by Wold (1982) has been followed to calculate the CV-communality and CV-redundancy indexes. The CV-communality index (H^2) measures the quality of the measurement model, where the CV-redundancy index (i.e., Stone–Geisser's Q^2 which Tenenhaus et al. (2005) calls F^2) measures the quality of the structural model. As shown in Table 6.8, the measurement model ($H^2 = 0.50$) shows a little better quality than the structural one ($F^2 = 0.41$).

6.5 DISCUSSION AND CONCLUSION

Most of IT/IS acceptance studies had bring environment, where acceptance of technology was voluntary (Brown et al., 2002; Nah et al., 2004). These environments are different because ERP use is mandatory for their users and each user tasks have impact on tasks of other users. This mandatory context is consistent with all ERP implementation projects in which employee usage is typically compulsory (Hsieh & Wang, 2007; Nah et al., 2004; Yang Chan et al., 2013). Recent studies indicate that evaluating the user's disposition, the nature of the IS, and mandatory usage are all important considerations in evaluating IS usage (Brown et al., 2002; Saeed et al., 2010). IS that entail major reorganization of work practices and prescribed as mandatory are likely to trigger strong user feelings (Saeed et al., 2010).

TABLE 6.8 Explained Variance (R^2), Communality, Redundancy, and Blindfolding Results of CV-communality (H^2) and CV-redundancy (F^2).

	R^2	Communality	Redundancy	CV-communality (H^2)	CV-redundancy (F^2)
OPC		0.54		0.39	0.39
STC		0.49		0.42	0.42
PCIL		0.49		0.28	0.28
ERP usefulness	0.56	0.89	0.33	0.79	0.50
ERP ease of use	0.42	0.87	0.35	0.51	0.34
Attitude toward using ERP	0.58	0.90	0.48	0.58	0.50
Average	0.52	0.70 (0.59[a])	0.39	0.50	0.41

[a]Computed as a weighted average of the different communalities with the weights being the number of manifest variables per each construct.

The role of users' attitude in a mandated environment is important and should not be overlooked (Nah et al., 2004). As noted, Brown et al. (2002) excluding the attitude construct would not provide accurate representation of users' acceptance of IT in the mandated contexts.

Empirical researchers have found support for original relationships of TAM in ERP settings in routine stage (e.g., see Hsieh & Wang, 2007; Lee et al., 2010; Shih & Huang, 2009; Sun et al., 2009), but they are not all support basic hypothesis of TAM. According to Davis (1989) and Davis et al. (1989), PEOU influences PU, PU and PEOU influence attitude toward using the system (AT). Researches of TAM indicate strong empirically support in direction from PEOU toward PU (Davis, 1989; Heijden, 2001). But in context of ERP solutions, empirical researches are in some cases support this relationship (e.g., Amoako-Gyampah & Salam, 2004; Bueno & Salmeron, 2008; Calisir et al., 2009; Hsieh & Wang, 2007; Lee et al., 2010; Yang Chan et al., 2013) and some researches are not support that relationship (Hwang, 2005; Shih & Huang, 2009; Shivers-Blackwell & Charles, 2006).

TAM was originally conceptualized in the context of personal use and ignored the role of organizational work on IT usage or its predictors (Sun et al., 2009). First, organizational users use ERP systems to perform specific tasks and speed organizational work, because of that it is important to examine the role of organizational work in ERP usage. Second, ERP systems are mandatory to use by organizational workers where one user's tasks on the ERP system are tightly integrated with other users' tasks (Nah et al., 2004). In other words, ERP users generally do not have a choice not to use the ERP system, regardless of their attitude. On the other hand, organizations that implement ERP systems want to use their systems to the fullest potential and realize the promised benefits. Because of all that for organizations, it is important that ERP users not just use basic functionality of ERP systems but to use extended functionality them. Hsieh and Wang (2007) have defined extended use as the use behavior that goes beyond typical usage and can lead to better results and returns. Extended use captures the breath and frequency of using different ERP features and functions. If the users find the ERP system to be useful, he or she will be more inclined to fully examine and use its features and functions (Saeed & Abdinnour-Helm, 2008).

In case of ERP implementations, other cognitive considerations beside PU and PEOU may become relevant (Nah et al., 2004). In the ERP context, organizations have to adopt business processes of an implemented ERP system. Although one of the major benefits of ERP systems is that they offer to organizations solution with best business practices it is not necessary that this is the best options for ERP users. Inherent business rules behind the processes gives them little choice but to follow strict business processes of ERP system, unlike the old system, which allowed them many different processes variations (Shanks et al., 2003). So organizations deploy ERP systems to facilitate organizational work rather than to match users' personal preferences or habits. At this presume we view work compatibility strictly as the fit of ERP to organizational work, and not to personal preferences or work habits. Work compatibility, like ERP usefulness and ERP ease of use, is very much a perceptual construct as it is the perception of fit between IT and work that motivates employees to use the system, irrespective of the actual extent of fit (Sun et al., 2009). ERP work compatibility refers to "degree" to which can ERP user do most of their tasks in ERP system. Work compatibility influences ERP usefulness and so it demonstrates the importance of incorporating work compatibility in models of IT usage as exposed, that is, Sun et al. (2009) and Scott and Walczak (2009). In a context of ERP usage, it is expected that relationship between work compatibility and ERP usefulness as the more work compatible ERP system is, the more useful it is for ERP users.

Although that the most important contributions of ERP systems are that they significantly reduce the time to complete business processes and help organizations to share information (Lee et al., 2010) and that organizations usually offer a better work environment for their employees as they are given more efficient system to work with, ERP systems have been plagued with high failure rates and inability to realize promised benefits (Kwahk & Lee, 2008) in routine stage of operation phase. One of the most important reasons seems to be ERP users who do not use ERP system in properly way. Attempt of this research is to improve understanding of how influence of 13 external factors can increase the degree of attitude of ERP users toward ERP system. This work extended previous research by incorporating the groups of external factors: personal innovativeness, computer anxiety, self-efficacy, and computer experience for conceptual factor PCIL; data quality, system performance, user manuals, and ERP functionality for conceptual factor STC; and business processes fit, organizational culture, ERP support, ERP communication and ERP training for conceptual factor OPC. These three conceptual factors have influence on ERP ease of use and ERP usefulness, which have furthermore influence on attitude toward using ERP system. This study also employed SEM (PLS approach) to assess overall model fit to verify the causal relationships between factors. Studying the influence of more external factors on constructs not only contributes to the theory development but also helps in designing interventional programs of organizations.

Empirical researches have found support for original relationships of TAM in ERP settings in routine stage (e.g., see Hsieh & Wang, 2007; Lee et al., 2010; Shih & Huang, 2009; Sun et al., 2009; Yang Chan et al., 2013). Our research confirms their results of influence ERP ease of use and ERP usefulness on attitude toward using ERP system (H2 and H3) and also influence of ERP ease of use on ERP usefulness (H1).

Based on the analytical results, this study found that it can be observed more external factors through second-order factors. In routine stage, external factors personal innovativeness and computer anxiety through second-order factor PCIL have influence on ERP ease of use. While external factors self-efficacy and computer experience did not be significant.

In contrast to most IT implementation research, the fact that ERP implementation research is focused on one technology has enabled the effect of specific technological characteristics to be examined. We have not found a research which has examined system and technology characteristics (SCT) upon the ERP system user acceptance. STC data quality, system performance, and user manuals have strong impact on ERP ease of use, while ERP functionality did not be statistical significant.

Furthermore, business process fit and organizational culture from OPC have strong impact on ERP usefulness. It is important that organization adopt business processes of ERP solutions. Somers and Neslon (2004) pointed out that business process reengineering plays a particular crucial role in the early stages of implementation; it is moderately important in the acceptance stage and tends to be less important once the technology becomes routine stage. But our research shows that business process fit is also important in routine stage. We cannot confirm statement of Lee et al. (2010) that if organization provides sufficient ERP support to organizational workers for their tasks, they are more likely to enjoy their work and improve their performance through usage of the ERP system. Amoako-Gyampah and Salam (2004) discover in their research that ERP user training and education had not only just high impact during implementation phases but also in operation phases especially in routine phase, when training on a continuous basis is required to meet the changing needs of the business and enhance employee skills. Our research shows that ERP users do not think that they need formal or informal training. ERP communication promotes user trust toward the ERP systems and consequently user acceptance and actual usage. ERP communication is viewed as having a high impact from initiation to system acceptance, as it helps to minimize possible user resistance (Somers & Neslon, 2004), but it is not find significant at the routine stage.

The implications for researchers and practitioners, an extended version of TAM through second-order factors was proposed to improve the explanatory power of ERP usage. The new technique (PLS approach) for analysis of model is used. This research has potential for practical application in the degree of ERP system usage. By confirming external factors, organizations should work on their organization culture and business process fit and on the other hand on their ERP system to assure better data quality, system performance, and user manuals to their users and in that way improve degree of attitude toward ERP system.

This study has certain limitations, which may present opportunity for further research. Since the respondents to the survey were limited to enterprises in our country, this study should be extended to other counties. Further research is needed to explore importance of presented external factors in different phases of ERP life cycle and also included some new external factors (e.g., top management support). Because ERP solutions are implemented by different methodologies and approaches importance of external factors by ERP solutions also could be explored. In latest researches of ERP systems, factor work compatibility was presented (see Nah et al., 2004; Sun et al., 2009). Impact of external factors on work compatibility and also impact of work compatibility on TAM should be researched.

KEYWORDS

- TAM-based research
- employee acceptance
- ERP information solutions
- service organizations
- partial least squares (PLS) path modeling

REFERENCES

Adamson, I.; Shine, J. Extending the New Technology Acceptance Model to Measure the End User Information Systems Satisfaction in a Mandatory Environment: A bank's Treasury. *Technol. Anal. Strat. Manage.* **2003**, *15*(4), 441–455.

Agarwal, R.; Prasad, J. Are Individual Differences Germane to the Acceptance of New Information Technologies. *Decis. Sci.* **1999**, *30*(2), 361–391.

Ajzen, I. The Theory of Planned Behaviour. *Organizational Behav. Hum. Decis. Process.* **1991**, *50*, 179–211.

Amoako-Gyampah, K.; Salam, A. F. An Extension of the Technology Acceptance Model in an ERP Implementation Environment. *Inf. Manage.* **2004**, *41*, 731–745.

Boudreau, M. C. Learning to use ERP Technology: A Causal Model. In: The 36th Hawaii International Conference on System Sciences, 2002. Retrieved 7 July 2008, from http://csdl2.computer.org/comp/proceedings/hicss/2003/1874/08/187480235b.pdf.

Bradford, M. *Modern ERP—Select, Implement & Use Today's Advanced Business Systems.* North Carolina State University, College of Management: Raleigh, 2008.

Bradley, J.; Lee, C. C. ERP Training and User Satisfaction: A Case Study. *Int. J. Enterprise Inf. Syst.* **2007**, *3*(4), 33–55.

Brown, S. A.; Massey, A. P.; Montoya-Weiss, M. M.; Burkman, J. R. Do I Really have to? User Acceptance of Mandated Technology. *Eur. J. Inf. Syst.* **2002**, *11*(4), 283–295.

Bueno, S.; Salmeron, J. L. TAM-based Success Modelling in ERP. *Interact. Comput.* **2008**, *20*(6), 515–523.

Calisir, F.; Gumussoy, C. A.; Bayram, A. Predicting the Behavioural Intention to Use Enterprise Resource Planning Systems—An Exploratory Extension of the Technology Acceptance Model. *Manage. Res. News* **2009**, *32*(7), 597–613.

Chin, W. W. Issues and Opinions on Structural Equation Modelling. *MIS Q.* **1998**, *22*, 7–16.

Davis, F. D. Perceived Usefulness, Perceived Ease of Use, and User Acceptance of Information Technology. *MIS Q.* **1989**, *13*(3), 319–340.

Davis, F. D.; Bagozzi, R. P.; Warshaw, P. R. User Acceptance of Computer Technology: A Comparison of Two Theoretical Models. *Manage. Sci.* **1989**, *35*(8), 982–1003.

Fishbein, M.; Ajzen, I. *Belief, Attitude, Intention, and Behaviour: An Introduction to Theory and Research.* Addison-Wesley Reading: Reading, MA, 1975.

Fornell, C.; Larcker, D. F. Evaluating Structural Equation Models with Unobservable Variables and Measurement Errors. *J. Mark. Res.* **1981**, *18*, 39–50.

Garača, Ž. Factors Related to the Intended Use of ERP Systems. *Management* **2011**, *16*(2), 23–42.

Gattiker, T. F.; Goodhue, D. L. What Happens after ERP Implementation: Understanding the Impact of Interdependence and Differentiation on Plant-level Outcomes. *MIS Q.* **2005**, *29*(3), 559–585.

Heijden, H. Factors Influencing the Usage of Websites: The Case of a Generic Portal in the Netherlands. In: e-Everything: e-Commerce, e-Government, e-Household, e-Democracy in 14th Bled Electronic Commerce Conference, Bled, Slovenia, 2001.

Henseler, J.; Ringle, C. M.; Sinkovics, R. The Use of Partial Least Squares Path Modelling in International Marketing. *Adv. Int. Mark.* **2009**, *20*, 277–319.

Hinton, P. R.; Brownlow, C.; McMurray, I.; Cozens, B. *SPSS explained.* Taylor & Francis Group, Routledge: London, 2004.

Hong, K. K.; Kim, Y. G. The Critical Success Factors for ERP Implementation: An Organizational Fit Perspective. *Inf. Manage.* **2002,** *40,* 25–40.

Hsieh, J. J. P. A.; Wang, W. Explaining Employees' Extended Use of Complex Information Systems. *Eur. J. Inf. Syst.* **2007,** *16*(3), 216–227.

Hwang, Y. J. Investigating Enterprise Systems Adoption: Uncertainty Avoidance, Intrinsic Motivation and the Technology Acceptance Model. *Eur. J. Inf. Syst.* **2005,** *14*(2), 150–161.

Iansiti, M. ERP End-users Business Productivity: A Field Study of SAP & Microsoft. In: *Keystone Strategy Research 2007*, 2007. Retrieved 12 June 2010, from http://download. microsoft.com.

Kelley, H. *Attributional Analysis of Computer Self-efficacy* (Unpublished Doctoral Dissertation). Richard Ivey School of Business: London, 2001.

Kositanuri, B.; Nqwenyama, O.; Osei-Bryson, K. M. An Exploration of Factors that Impact Individual Performance in an ERP Environment: An Analysis Using Multiple Analytical Techniques. *Eur. J. Inf. Syst.* **2006,** *15,* 556–568.

Kwahk, K. Y.; Lee, J. N. The Role of Readiness for Change in ERP Implementation: Theoretical Bases and Empirical Validation. *Inf. Manage.* **2008,** *45*(7), 474–481.

Kwak, Y.; Park, J.; Chung, B.; Ghosh, S. Understanding End-Users' Acceptance of Enterprise Resource Planning (ERP) System in Project-Based Sector Engineering Management. *IEEE Trans.* **2012,** *59*(2), 266–277.

Lee, D. H.; Lee, S. M.; Olson, D. L.; Chung, S. H. The Effect of Organizational Support on ERP Implementation. *Ind. Manage. Data Syst.* **2010,** *110*(1–2), 269–283.

Legris, P.; Ingham, J.; Collerette, P. Why do People Use Information Technology? A Critical Review of the Technology Acceptance Model. *Inf. Manage.* **2003,** *40,* 191–204.

Liu, L.; Ma, Q. Perceived System Performance: A Test of an Extended Technology Acceptance Model. *J. Org. End User Comput.* **2006,** *18*(3), 1–24.

Lu, J.; Chun-Sheng, Y.; Liu, C.; Yao, J. E. Technology Acceptance Model for Wireless Internet. *Int. Res.: Electron. Network. Appl. Policy*, **2003,** *13*(3), 206–222.

Markus, L. M.; Tanis, C. *The Enterprise System Experience: From Adoption to success. Framing the Domains of IT Management: Projecting the Future through the Past.* Pinnafex Educational Resources Inc.: Cincinnati, OH, 2000.

Motiwalla, L. F.; Thompson, J. *Enterprise Systems for Management.* Pearson Education, Inc.: Upper Saddle River, NJ, 2009.

Musaji, Y. F. *Integrated Auditing of ERP Systems.* John Wiley & Sons, Inc.: New York, 2002.

Nah, F. F.; Tan, X.; Teh, S. H. An Empirical Investigation on End-users' Acceptance of Enterprise Systems. *Inf. Resour. Manage. J.* **2004,** *17*(3), 32–53.

Pozzebon, M. Combinig a Structuration Approach with a Behavioral-based Model to Investigate ERP Usage. In: Paper presented at the AMCIS 2000, Long Beach, CA, 2000.

Ringle, C. M.; Will, A. *SmartPLS 2.0 (M3)*, 2005. Retrieved 1 June 2009, from http://www. smartpls.de.

Rogers, E. V. *Diffusion of Innovation* (4th ed.). The Free Press: New York, 2003.

Ross, J. W.; Vitale, M. R. The ERP Revolution: Surviving vs. Thriving. *Inf. Syst. Front.* **2000,** *2*(2), 233–241.

Saeed, K. A.; Abdinnour-Helm, S. Examining the Effects of Information System Charac-teristics and Perceived Usefulness on Post Adoption Usage of Information Systems. *Inf. Manage.* **2008**, *45*, 376–386.

Saeed, K. A.; Abdinnour, S.; Lengnick-Hall, M. L.; Lengnick-Hall, C. A. Examining the Impact of Pre-implementation Expectations on Post-implementation Use of Enterprise Systems: A Longitudinal Study. *Decis. Sci.* **2010**, *41*(4), 659–688.

Schwarz, A. *Defining Information Technology Acceptance: A Human-centred, Manage-ment-oriented Perspective* (Unpublished doctoral dissertation), University of Houston: Houston, TX, 2003.

Scott, J. E.; Walczak, S. Cognitive Engagement with a Multimedia ERP Training Tool: Assessing Computer Self-efficacy and Technology Acceptance. *Inf. Manage.* **2009**, *46*(4), 221–232.

Shanks, G.; Seddon, P. B.; Willcocks, L. P. *Second-wave Enterprise Resource Planning System—Implementing for Effectiveness.* Cambridge University Press: Cambridge, 2003.

Shih, Y. Y.; Huang, S. S. The Actual Usage of ERP Systems: An Extended Technology Acceptance Perspective. *J. Res. Pract. Inf. Technol.* **2009**, *41*(3), 263–276.

Shivers-Blackwell, S. L.; Charles, A. C. Ready, Set, Go: Examining Student Readiness to Use ERP Technology. *J. Manage. Dev.* **2006**, *25*(8), 795–805.

Somers, M. T.; Nelson, K.; Karimi, J. Confirmatory Factor Analysis of the End-user Computing Satisfaction Instrument: Replication within an ERP Domain. *Decis. Sci.* **2003**, *34*(3), 595–621.

Somers, T. M.; Neslon, K. G. A Taxonomy of Players and Activities Across the ERP Project Life Cycle. *Inf. Manage.* **2004**, *41*(3), 257–278.

Sun, Y.; Bhattacherjee, A.; Ma, Q. Extending Technology Usage to Work Settings: The Role of Perceived Work Compatibility in ERP Implementation. *Inf. Manage.* **2009**, *46*, 351–356.

Tenenhaus, M.; Vinzi, V. E.; Chatelin, Y. M.; Lauro, C. PLS Path Modelling. *Comput. Stat. Data Anal.* **2005**, *48*, 159–205.

Thompson, R.; Compeau, D.; Higgins, C. Intentions to Use Information Technologies: An Integrative Model. *J. Org. End User Comp.* **2006**, *18*(3), 25–46.

Umble, E. J.; Haft, R. R.; Umble, M. M. Enterprise Resource Planning: Implementation Procedures and CSF. *Eur. J. Operational Res.* **2002**, *146*(2), 241–257.

Uzoka, F. M. E.; Abiola, R. O.; Nyangeresi, R. Influence of Product and Organizational Constructs on ERP Acquisition using an Extended Technology Acceptance Model. *Int. J. Enterp. Inf. Syst.* **2008**, *4*(2), 67–83.

Venkatesh, V. *User Acceptance of Information Technology: A Unified View* (Unpublished doctoral dissertation), University of Minnesota: Minneapolis, MN, 1998.

Venkatesh, V.; Bala, H. Technology Acceptance Model 3 and a Research Agenda on Inter-ventions. *Decis. Sci.* **2008**, *39*(2), 273–315.

Venkatesh, V.; Davis, F. D. A Theoretical Extension of the Technology Acceptance Model: Four Longitudinal Field Studies. *Manage. Sci.* **2000**, *46*(2), 186–205.

Venkatesh, V.; Morris, M. G.; Davis, G. B.; Davis, F. D. User Acceptance of Information Technology: Toward a Unified View. *MIS Q.* **2003**, *27*(3), 425–479.

Wold, H. Soft Modelling: The Basic Design and Some Extensions. *Syst. Under Indirect Obs.* **1982**, 1–52.

Yang Chan, C.; Kyeong Seok, H.; Ki Young, L. An Empirical Study on Factors Affecting Intention to Use an Interworking Information System in a Mandatory Environment. *Int. J. Adv. Comp. Technol.* **2013,** *5*(13), 613–623.

Yi, Y. M.; Fiedler, K. D.; Park, J. S. Understanding the Role of Individual Innovativeness in the Acceptance of IT-based Innovativeness: Comparative Analyses of Models and Measures. *Decis. Sci.* **2006,** *37*(3), 393–426.

Youngberg, E.; Olsen, D.; Hauser, K. Determinants of Professionally Autonomous End User Acceptance in an Enterprise Resource Planning System Environment. *Int. J. Inf. Manage.* **2009,** *29*, 138–144.

CHAPTER 7

LEADERSHIP EFFECTIVENESS: AN EMERGING CHALLENGE TO TODAY'S EDUCATIONAL ORGANIZATION AND ITS PRACTICES IN PRIVATE UNIVERSITIES OF INDIA

SANJEEV KUMAR JHA[1*] and GIRIJESH KUMAR YADAV[2]

[1]*Educational Survey Division, National Council of Educational Research and Training, New Delhi, India*

[2]*National Institute of Occupational Health, Indian Council of Medical Research, Ministry of Health & Family Welfare, Govt. of India, Ahmedabad, Gujarat, India*

Corresponding author. E-mail:krjhasanjeev03@gmail.com

CONTENTS

ABSTRACT

One of the emerging challenges for the universities across the globe is its effective leadership. It is one of the most important aspects to run any organization successfully, likewise educational organization. However, it seems mechanism for leadership development in the higher education is missing. So, if an organization is at top, it must have effective leader. The present chapter is an attempt to highlight the issue and study leadership effectiveness of heads of departments of private and privately sponsored deemed to be universities. The chapter concludes that head of the departments at both types of private universities (PU) have moderate level of the leadership effectiveness. Furthermore, gender and experience of head of the departments have no effect on leadership effectiveness. The IQAC (Internal Quality Assurance Cell) can be used as instrument to develop leadership in the PU.

7.1 INTRODUCTION

The contemporary universities across the globe are facing several challenges. One of the emerging challenges for the universities across the globe is its effective leadership. Though, the positive effects of leadership development in commercial organizations also apply to the academic context (Braun et al., 2009, p. 195). But rarely, a functional mechanism in the universities is found. Braun et al. (2009) highlighted that to date systematic leadership development is largely lacking for employees in higher education (p. 195).

The trend for leadership effectiveness in higher education is not contrary to the global trend. The chapter discusses the leadership effectiveness in the private universities (PU) in India.

7.1.1 UNIVERSITIES IN INDIA

The word university is derived from the Latin word *universitasnagistrorum et scholarium* which roughly means community of teachers and scholars (Wikipedia.org/wiki/university). The word university refers in general to a number of people associated with one body, a society, company, community, etc. "A university is a place where new ideas germinate, strike roots and grow tall and sturdy. It is a unique space, which covers the entire universe of knowledge. It is a place where creative minds converge, interact with each other and construct visions of new realities. Established notions of truth are challenged in the pursuit of knowledge" (Pal, 2009).

In India, university means an institution established or incorporated by or under Central Act, a Provincial Act or a State Act and includes any such institute as may in consultation with university concerned, be recognized by the UG in accordance with regulations made in this regard under the act (MHRD—mhrd.gov.in/overtest11).

In India, there are several types of university level institutions namely Central University, Institutions of National Importance, State University, Private University as well as Deemed to be University. Further, Deemed to be University are of two types (a) Deemed to be University which gets fund from government and (b) Privately sponsored Deemed to be University.

Central Universities, Institute of National Importance and some deemed to be universities (DU) are managed by Central Government. State Universities are managed by State Government, and PU and some DU are managed by private sectors through some nonprofitable society or trust.

In the study the focus was on the PU and private-sponsored (nongovernment funded/granted) DU.

7.1.1.1 PRIVATE UNIVERSITY IN INDIA

Private University in India is emerging out as a new service industry. It produces human resources for not only the country but also for the world. Private University in India, as an organization, is transforming exponentially in order to keep pace with domestic demands. This transformation requires an effective leadership in these universities in order to keep pace with global trend.

Last decade, an unprecedented quantitative growth has been witnessed in Higher Education System (HES). India has one of the largest HESs in the world. During the last decade, an exponential growth in HES has been observed, especially in private institutions. The number of PU reached to 182 in 2014 from just 14 in 2005 (University Grants Commission Annual Reports). Private university system is emerging as a parallel system to its public counterpart as far as its quantitative growth is concerned. Moreover, Chhapia (2009) reported that:

Every second student in India enrolls in a private college/institution. In 2001, when private unaided institutes made up to 42.6 percent of all higher institutes 32.89 percent of Indian students studied in them. By 2006 the share of private institutes went up to 63.21 percent and their student's share went up to 51.53 percent. (p. 7)

One of main reasons for this expansion is highlighted by Pal (2009) as "in the absence of any significant expansion in different sectors of higher education by the States has created a space for the growth of private providers" (p. 32).

Despite this largest status, there is still a wide gap for a huge expansion in HES. National Knowledge Commission [NKC] (2009) of India highlighted that "the portion of our population, in the relevant age group, that enters the world of higher education is about 7 per cent" (p. 62).

Quantitative growth of higher educational institutions (HEIs) is unprecedented. But as far as quality concern, it has several apprehensions to deal with. The NKC (2009) noted that "a quiet crisis in higher education in India, that runs deep" (p. 66). Furthermore, Thorat (2011) highlighted the fact that "68 per cent of the universities and 91 per cent of the colleges are rated average or below average in terms of quality parameters specified by the National Assessment and Accreditation Council (NAAC)" (p. 16). It may be inferred that most HEIs in India are neither scoring higher international ranking in any international reputed ranking system nor getting high ranking in parameters especially developed for Indian context by NAAC.

Thus, there is quantitative growth in the PU in India but at the front of quality these are not at par. Now it is the time on which the quality of these institutions should be developed.

An effective leadership is an instrument which can lead the transformation between quantitative to qualitative growths. In order to develop a world-class university, an effective leadership is prerequisite in the university.

7.1.2 LEADERSHIP EFFECTIVENESS

Leadership means ability to lead an organization. It is the process by which a leader ingeniously directs guides and influences the work of others in attaining specified goals or objectives. It may be defined as a position of power held by an individual in a group, which provides him/her with an opportunity to exercise interpersonal influence on the group members for activating and directing their efforts toward positive goals.

There are several ways to define leadership. It may be defined as a process, that influences other people to achieve common objectives of organization and guides the organization in a way to make it more coherent and cohesive is called leadership (Harman, 2002). It may also be defined as a process of leading people in the right direction in order to achieve goals.

Leadership is typically defined by the traits, qualities, and behavior of a leader (Horner, 2003). The leadership quality of head of the departments is a crucial factor in the success of any department progressively of the university. Leadership as the skill of a person is to influence an individual or a group for achieving their goal. Thus, the interpersonal skill of a person, to influence others was the only aspect, which was considered as criterion of leadership in ancient time. With the advance of civilization, the role of a leader extended to almost all walks of life ranging from social life to occupational and organizational life, from local level to global level. Consequently, the issue became a subject of profound and profuse thinking. Leaders must have followers; one cannot be a leader unless there are people to be led. Leaders have more power to influence followers than the followers have to influence the leaders. The objective of leadership is to influence followers to achieve goals of the group. It is the process by which a person influences others to accomplish an objective and directs the organization in a way that makes it more cohesive and coherent. Though much research was done to identify the traits, no clear answer was found with regards to what traits consistently were associated with great leadership (Horner, 2003). Yukl (1998) defined the leadership as a process of interaction between the leader and the other staff, influencing them toward workplace goals achievement. Leadership is the ability to provide direction toward chosen future aspirations and aligns the followers toward goals achievement (Kelly-Hiedenthal, 2004). Moreover, leadership is viewed as an active interactive process that involves various aspects like planning, organizing, directing, and controlling the group overall. According to Roussel et al. (2006), the activities of an organized group could be affected by a process in which the endeavors to achieve the objectives and tasks are managed by leadership. Leadership can be defined as inducing individuals or a group to take an action in according to the purpose of the leader. Leadership can be formal or informal. As Sullivan and Decker (2004) explain, it can be formal when a manager demonstrates power and authority within a framework of legal approval by the organization. Leadership can be informal when utilized by a staff member who does not have effective leadership skills, ideas, and roles to promote the performance of the work outcomes.

Effective leadership is central to an organization's success (Braun et al., 2009). Furthermore, it develops both managerial and leadership behaviors and qualities.

Leadership effectiveness involves a group with the leader as the main directive element. A transactional view of leadership effectiveness emphasizes between leader and follower relationship in two aspects: first, it deals with the responsiveness of the group gaining specified goals, and second, securing those goals with greatest possibilities consideration for the individuals comprising the group. It depends upon the things are done to produce desired outcomes of group, as well as the outcomes of themselves. Leadership effectiveness means achieving a productive use of human and material resources beyond potential. It contributes to use of human and material resources, at or beyond potential. The leadership effectiveness is one of the important parameter to judge quality in higher education. Gedney (1999) identifies the characteristic of an effective leader is "someone who motivates a person or a group to accomplish more than they would have otherwise accomplished without that leader's involvement" (p. 1).

Leadership is important in providing direction, motivation, creating confidence, high morale, team spirit, and encouragement of initiatives to the group of staff members. An effective leadership provides effective direction to the organizations and its staff/members. To attain the specific objectives of the organization, there must be the direction of activities. The direction of activities is effected through leadership (Brecken, 2004). Leadership is the motivating power to group efforts. Effective leadership motivates the subordinates for higher productivity (Gary, 1996). Motivation is a goal-oriented characteristic that helps a person to achieve their objectives (Lyer, 2015; Oni, 2013). It pushes an individual to work hard at achieving his or her goals. An executive must have the right leadership traits to influence motivation. However, there is no specific blueprint for motivation. An effective leader must have a thorough knowledge of motivational factors for others. He must understand the basic needs of employees, peers and his superiors. Leadership is used as a means of motivating others (MSG [Management study guide], http://management-studyguide.com/leadership-motivation.htm). Leadership creates confidence in the subordinates by giving proper guidance and advice. Locke (1997) found that transformational leadership can express the trust to subordinates and enhance their confidence.

Good leadership increases the morale of the employees that, in turn, contributes to higher productivity. Ngambi (2011) found in his study the existence of relationship between leadership and morale, and those leadership competencies such as communication, fostering trust, and team building set a clear direction for the colleagues that impact on their morale. Neely's (1999) research explored the relationship between employee morale and productivity. His results revealed a pattern that links the productivity of employees with their level of morale.

Effective leadership promotes team-spirit and team-work which is quite essential for the success of any organization. Teams are an essential component of successful organizations today and building and motivating teams are necessary pursuits to attain that success. Teams require continuous nurturing and interaction to maintain high performance throughout their temporary lives. Leadership must now concentrate on motivating and supporting teams using tools that were not previously considered but have become crucial in a globalizing environment. In order for a team to be attuned to success, a combination of attributes are required that include: clear objectives; shared leadership; clear roles and responsibilities; interdependent members; mutual encouragement; and trust between the leader and the team (Sohmen, 2013).

A progressive, forward, and democratic-minded leader always encourages initiative on the part of the followers. Intellectually stimulating leaders encourage employees to think for themselves, question long-held assumptions, and approach problems in innovative ways (Jennifer & Julian, 2013).

Thus, it may be concluded that the leadership is highly important to the development of organizations. An effective leader can create an excellent work environment evolving a certain direction, work motivation, creating confidence, higher morale, team spirit, and encouragement of initiatives among all its subordinates in order to achieve the goal of the organization. Effective leadership can help to run any service organization successfully including educational organization. So, if an organization is at top. It must have effective leaders. Furthermore, an effective head of the departments at private university may lead the department and achieve the goal of the university, meet the global challenges and can keep the pace with global development. Undoubtedly, leadership is an important notion affecting performance of organizations. The realization that schools must transform is placing greater responsibility on school heads as managers and as leaders (NUEPA, 2014). Viewing the importance of the leadership effectiveness it is very essential to investigate the status of leadership effectiveness among the private and DU in India.

7.2 OBJECTIVES OF THE STUDY

1. To compare the leadership effectiveness of head of the departments of Private and DU.
2. To compare the leadership effectiveness of heads of engineering department of Private and DU.
3. To compare the leadership effectiveness of heads of management department of PU and DU.
4. To compare the leadership effectiveness of head of the departments in relation to their gender.
5. To compare the leadership effectiveness of head of the departments in relation to their work experience.

7.2.1 HYPOTHESES OF THE STUDY

1. There will be found a significant difference of the leadership effectiveness of head of the departments from PU and DU.
2. There will be found a significant difference of the leadership effectiveness of heads of engineering department from PU and DU.
3. There will be found a significant difference of the leadership effectiveness of heads of management department from PU and DU.
4. There will be found a significant difference of the leadership effectiveness of head of the departments in relation to their gender.
5. There will be found a significant difference of the leadership effectiveness of head of the departments as function of their work experience.

7.3 METHODOLOGY

7.3.1 SAMPLE

Population of the study was consisted of all the PU and private-sponsored (nongovernment funded/granted) "DU" from Delhi National Capital Region (NCR).

Target population was consisted of all the PU and private-sponsored (nongovernment funded/granted) DU in Delhi NCR established till the date of 31 March 2009 having engineering and management departments.

The date 31 March 2009 was chosen in order to ensure that university must have completed three academic years (six semesters) so that university have some cumulative investment on teaching-learning resources. The two departments (Engineering & Management) were selected as these are common in most of "PU" and "DU" of Delhi NCR. Head of the department (Engineering & Management) & Faculty working in the university for more than 6 months were target population.

Sampling frame enlisted four private-sponsored (nongovernment funded/granted) DU and three PU of Delhi NCR. Head of the departments of engineering and management departments were comprises sampling frame.

7.3.1.1 SELECTION OF THE UNIVERSITIES

Two PU and two DU were selected with the use of "Simple Random Sampling Technique." The use of simple random sampling technique was well justified as a well maintained list of PU and DU were developed and used for sampling. It discarded the biasness of researcher as "Random method ensures the same probability of all possible sample of fix size n" (Kerlinger, 2011) so, it ensures that there was no control of researcher on sample selection.

7.3.1.2 SELECTION OF HEAD OF THE DEPARTMENTS

As common departments were considered for the selection of head of the departments, so heads were automatically selected; thus, there was no scope of researcher's biasness.

In case university has more than one department for Engineering or Management stream and headed by separate heads then one head of the department who has completed 6 months at present position selected. If there was more than one head of the department who has completed 6 months at current position then one head was selected with the use of simple random sampling technique in order to discard biasness of the researcher.

7.3.2 INSTRUMENT USED: ASSESSMENT OF LEADERSHIP EFFECTIVENESS

Leadership is one of the most important aspects to run any organization successfully likewise for the universities. Assessment of leadership effectiveness is a complex work as "there are no universal traits of leaders, certain requirements for leadership effectiveness do exist across diverse situations" (Taj, 2010). The most commonly used measure of leader effectiveness the extent to which the leader's organizational unit performs its task successfully and attains its goals. In some cases, objective measures of performance are available such as profits, profit margin, increase sales, market share, sales relative to target sales, return on investment, productivity, cost per unit of output, cost in relation to budgeted expenditure, and so on. In some cases, subjective ratings of effectiveness are obtained from the leader's superiors, peers, or subordinates. The attitude of followers toward the leader is another common factor of leader effectiveness. Its indicators are related to satisfying the needs and expectations of followers the respect and admiration of the followers for the leader and commitment of the followers (Dhar & Pethe, 2003). In university system, effective leadership can help in achievement of aims and objectives of the University. So, "a systematic leadership development program needs to be developed to ensure academic leadership effectiveness in research universities" (Shahmandi et al., 2011).

The study used leadership effectiveness scale (LES) by Haseen Taj (2010) for the data collection which has high (intrinsic validity, 0.60, 0.78, and 0.67) validity and (test–retest 0.60 and split half 0.64) reliability.

7.3.3 METHOD

Descriptive survey method was employed for the study.

7.3.4   STATISTICAL TECHNIQUES USED

The obtained data are statistically analyzed herewith the objectives of clarifying and explaining the problems raised and hypotheses formulated.

1. The statistical measures employed in this study were as follows:
2. Mean values and SD were calculated for graphical presentation of the data related to different level/type of variables.
3. t-Test has been employed to explore the extent to difference of leader effectiveness among the heads of various departments from PU and DU with help of SPSS-20.

7.4   EMPIRICAL RESULTS

7.4.1   LEADERSHIP EFFECTIVENESS OF HODS OF PRIVATE AND DEEMED TO BE UNIVERSITIES

As shown in Table 7.1 that the result reveals that the group of HoDs of private university was found to be slight more leadership effectiveness ($M = 271.33$, $N = 40$, SD = 12.25) than the group of HoDs of DU ($M = 270.80$, $N = 40$, SD = 13.81). However, the t-ratio (0.18) was found not significant.

TABLE 7.1   Leadership Effectiveness of Heads of Private and Deemed to be Universities.

S. No.	University	No.	Mean	SD	SE_m	t-Ratio	p-Value	Sig.
1	Private	40	271.33	12.252	1.93	0.18	0.85	NS
2	Deemed to be	40	270.80	13.816	2.18			

Hence, the null hypothesis (H_0), that is, there is no significant difference of the leadership effectiveness of head of the departments of PU and DU is accepted. It means the type of universities have no effect on leadership effectiveness of head of the various departments. So, the head of departments at both types of the universities has moderate level of effectiveness as per norms and guideline manual of LES.

7.4.2 LEADERSHIP EFFECTIVENESS OF HEADS OF ENGINEERING DEPARTMENT

As given in Table 7.2 that the result reveals that the group of heads of engineering department of PU was found to be slight more leadership effectiveness ($M = 271.35$, $N = 20$, SD = 11.6) than the group of heads of engineering department of DU ($M = 270.70$, $N = 20$, SD = 13.14). However, the t-ratio (0.089) was found not significant.

TABLE 7.2   Leadership Effectiveness of Heads of Engineering Department.

S. No.	Engineering Dept.	No.	Mean	SD	SE_m	t-Ratio	p-Value	Sig.
1	Private	20	271.35	11.60	2.5	0.089	0.986	NS
2	Deemed to be	20	270.70	13.14	2.3			

Hence, the null hypothesis (H_0), that is, there is no significant difference of the leadership effectiveness of head of the engineering department of PU and DU is accepted. It means that the engineering departments at both types of the universities have similar leadership effectiveness. The heads of engineering department at both types of the universities have moderate level of effectiveness as per the norms and guideline manual of LES.

7.4.3 LEADERSHIP EFFECTIVENESS OF HEADS OF MANAGEMENT DEPARTMENT

As exhibited in Table 7.3 that the result reveals that the group of heads of management department of PU was found to be slight more leadership effectiveness ($M = 271.30$, $N = 20$, SD = 13.17) than the group of heads of management department of DU ($M = 269.90$, $N = 20$, SD = 14.73). However, the t-ratio (0.317) was found not significant.

Hence, the null hypothesis (H_0), that is, there is no significant difference of the leadership effectiveness of head of the management department of PU and DU is accepted. It means that the management departments at both types of the universities have similar leadership effectiveness. The heads of management department at both types of the universities have moderate level of effectiveness as per the norms and guideline manual of LES.

TABLE 7.3 Leadership Effectiveness of Heads of Management Department.

S. No.	Engineering Depts.	No.	Mean	SD	SE_m	t-Ratio	p-Value	Sig.
1	Private	20	271.30	13.17	2.9	0.317	0.345	NS
2	Deemed to be	20	269.90	14.73	3.2			

7.4.4 LEADERSHIP EFFECTIVENESS OF HEAD OF THE DEPARTMENTS IN RELATION TO THEIR GENDER

As exhibited in Table 7.4 that the result reveals that the group of male heads of various department of PU and DU was found to be slight more leadership effectiveness ($M = 272.23$, $N = 39$, SD = 11.84) than the effectiveness among the group of female heads of various department of PU and DU ($M = 269.95$, $N = 41$, SD = 14.02). However, the t-ratio (0.78) was found not significant.

TABLE 7.4 Leadership Effectiveness of Head of the Departments in Relation to their Gender.

S. No.	Gender	N	Mean	SD	SE_m	t-Value	p-Value	Sig.
1	Male	39	272.23	11.84	1.896			
2	Female	41	269.95	14.02	2.191	0.78	0.43	NS

Hence, the null hypothesis (H_0), that is, there is no significant difference of the leadership effectiveness of male and female head of departments of PU and DU is accepted. It means that gender has no effect on leadership effectiveness. The finding is in the consonance with Gedney (1999) and Paustian-Underdahl et al. (2014). The male and female heads of various departments at both types of the universities have moderate level of leadership effectiveness as per the norms and guideline manual of LES.

7.4.5 LEADERSHIP EFFECTIVENESS IN RELATION TO THEIR EXPERIENCE

As shown in Table 7.5 that the result reveals that the group of HoDs of PU and DU having experience below than 5 years and above 5 years experienced was found to be slight different on the score of their leadership effectiveness. Such as, least experienced (below 5 years) were found to be more leadership effectiveness ($M = 271.87$, $N = 60$, SD $= 12.71$) than the effectiveness among the group having more experienced (above 5 years) ($M = 268.65$, $N = 20$, SD $= 13.78$). However, the t-ratio (0.783) was found not significant.

TABLE 7.5   Leadership Effectiveness in Relation to their Experience.

S. No.	Experience	N	Mean	SD	SE_m	t-Value	p-Value	Sig.
1	Up to 5 years	60	271.87	12.71	1.64	0.783	0.43	NS
2	More than 5 years	20	268.65	13.78	3.08			

Hence, the null hypothesis (H_0), that is, there is no significant difference of the leadership effectiveness of HoDs of various departments of PU and PU having experience below 5 years and HoDs having experience more than 5 years is accepted. It means that the work experience has no effect on leadership effectiveness. The finding is in dissonance of the common sense. Possible reasons for such trends in the observed pattern of results may be speculated in terms the individual characteristics and organism factors. All the HoDs of various departments from PU and DU have moderate level of leadership effectiveness as per the norms and guideline manual of LES.

7.5   CONCLUSION

It has been observed that an exponential growth in PU in India. In order to improve their quality, effective leadership is prerequisite. It is the foremost important construct for achieving organizational goal. An effective leader can transform a university in a quality educational institution. But, Braun et al. (2009) highlighted that to date systematic leadership development is largely lacking for employees in higher education (p. 195). Moreover, the chapter concludes that head of the departments at both types of PU have moderate level of the effectiveness. One of the important reason for this explained by Braun et al. (2009) that "faculty are appointed to a senior rank based upon their deep subject knowledge, experience, and scientific accomplishment (e.g., number of publications in international journals), not based on leadership skills. Subsequently, senior faculty members hold leadership positions without adequate preparation" (p. 196). Thus, there is a need to formulate leadership development program for students.

Moreover, if the quality of the university is to enhance, then there is need to develop its leadership capacity with proper leadership development programs. The chapter further concludes that gender and experience of HoDs have no effect on leadership effectiveness.

Further, Bhatkar and Bhatkar (2012) conducted a research on the topic "IQAC (Internal Quality Assurance Cell)" a tool to evolve functional Leadership in the Institutions concluded that Leaders are not born with time and maturity they emerge. The mechanism and procedure of IQAC will be instrumental in developing the leadership traits in institution. Those leasers with these virtues and qualities will then be equipped to handle the challenges in academic administrative and financial reforms. It is the assiduity and perseverance that yields the desired results. The management of the PU should consider IQAC as an instrument to transform its leader as effective leader.

KEYWORDS

- **leadership effectiveness**
- **education**
- **private universities**
- **organizations**
- **quality assurance**

REFERENCES

Bhatkar, V. B.; Bhatkar, N. V. IQAC (Internal Quality Assurance Cell)" a Tool to Evolve Functional Leadership in the Institutions. *Int. J. Educ. Admin.* **2012,** *4*(1), 39–42.

Braun, S.; Nazlic, T.; Weisweiler, S.; Pawlowska, B.; Peus, C.; Frey, D. Effective Leadership Development in Higher Education: Individual and Group Level Approaches. *J. Leadersh. Educ.* **2009,** *8*(1), 195–206.

Brecken, D. Leadership Vision and Strategic Direction. *Qual. Manage. Forum* **2004,** *30*(1).

Chhapia, H. Every 2nd Student in India Enrolls in Pvt. College, 21 June 2009, *Times of India.* http://timesofindia.indiatimes.com/india/Every-2nd-student-in-India-enrols-in-pvt-college/articleshow/4682368.cms?.

Dhar, U.; Pethe, S. *Manual for Dhar & Pethe Leadership Effectiveness Scale (DLES)*. Vedanta Publication: Lucknow, Uttar Pradesh, India, 2003.

Gary, J. *Theories of Work Motivation, Leadership, Organizational Behaviour: Understanding and Managing Life at Work*. Harper Collins College, Concordia University, Publishers: New York, 1996.

Gedney, C. R. *Leadership Effectiveness and Gender*. In: Unpublished Graduation Dissertation at Air Command and Staff College, Air University, Alabama, 1999.

Harman, K. Merging Divergent Campus Cultures into Coherent Educational Communities: Challenges for Higher Education Leaders. *Higher Educ.* **2002,** *44*(1), 91–114.

Horner, M.; Bennett, N., Crawford, M., Cartwright, M., Eds.; *Effective Educational Leadership*. Paul Chapman Publisher (A Sage Publication): London, 2003.

Jennifer, L. R.; Julian, B. Greening Organizations through Leaders' Influence on Employees' Pro-environmental Behaviors. *J. Organ. Behav.* **2013,** *34*, 176–194.

Kelly-Hiedenthal, P. *Essentials of Nursing Leadership & Management*. Delmer Learning: USA, 2004.

Kerlinger, F. N. *Foundations of Behavioural Research*, 2nd ed. Surjeet Publications: New Delhi, 2011.

Locke, E. A. The Motivation to Work: What We Know. *Adv. Motiv. Achieve.* **1997,** *10*, 375–412.

Lyer, R. G. *Essence of Leadership*. Notion Press: Chennai, 2015.

McNaron, M. Using Transformational Learning Principles to Change Behavior in the Operating Theatre. *Aron. J.* **2009,** *89*(5), 859–860.

Neely, G. *The Relationship between Employee Morale and Employee Productivity*. National Fire Academy as Part of the Executive Fire Officer Program: Tulsa, OK, 1999.

Ngambi, H. C. The Relationship between Leadership and Employee Morale in Higher Education. *Afr. J. Bus. Manage.* **2011,** *5*(3), 762–776.

NKC. *National Knowledge Commission (NKC) under Chairmanship of Mr. Sam Pitroda*. Government of India: New Delhi, 2009. Accessed from http://www.knowledgecommission.gov.in/downloads/report2009/eng/report09.pdf (retrieved 30 March 2010).

NUEPA (National Programme Design and Curriculum Framework), 2014. Retrieved from http://mhrd.gov.in/sites/upload_files/mhrd/files/upload_document/SLDP_Framework_NCSL_NUEPA.pdf.

Oni, S. *Challenges and Prospects in African Education Systems*. Trafford Publishing, North America and International: United States of America, 2013.

Pal, Y. *The Committee to Advise on Renovation and Rejuvenation of Higher Education*. Ministry of Human Resource Development, Government of India: New Delhi, 2009.

Paustian-Underdahl, S. C.; Walker, L. S.; Woehr, D. J. Gender and Perceptions of Leadership Effectiveness: A Meta-analysis of Contextual Moderators. *J. Appl. Psychol.* **2014,** *99*(6), 1129–1145.

Roussel, L.; Swansburg, R. *Management and Leadership for Nurses Administrators*, 5th ed., Jones and Bartlett: USA, 2009.

Roussel, L.; Swansburg, R. C.; Swansburg, R. J. *Management and Leadership for Nurses Administrators*, 4th ed. Jones and Bartlett: USA, 2006.

Shahmandi, E.; et al. *Competencies, Roles and Effective Academic Leadership in World Class University*, 2011. Accessed from: www.sciedu.ca/journal/index.php/ijba/article/download/77/52 (retrieved 29 March 2010).

Sohmen, V. S. Leadership and Teamwork: Two Sides of the Same Coin. *J. IT Econ. Dev.* **2013,** *4*(2), 1–18.

Sullivan, E.; Decker, P. *Effective Leadership & Management Nursing,* 6th ed. Pearson Education: New Jersey, USA, 2004.

Taj, H. *Manual for Leadership Effectiveness Scale (LES-TH).* National Psychological Corporation: Kacheri Ghat, Agra, Uttar Pradesh, India, 2010.

Thorat, S. *Higher Education in India. Strategies and Schemes during Eleventh Plan Period (2007–2012) for Universities and Colleges.* University Grants Commission: India, 2011.

Yukl, G. *Leadership in Organizations,* 4th ed. Prentice Hall: Upper Saddle River, NJ, 1998.

CHAPTER 8

EMPLOYEE ENGAGEMENT FOR BETTER TALENT MANAGEMENT

SASMITA MISRA*

Narsee Monji Institute of Management Studies, Hyderabad, India

E-mail: sasmita.misra@nmims.edu

CONTENTS

ABSTRACT

In an era where knowledge dominates the spectrum of industries, the acute shortage of talent to rule the roost has been visible everywhere. It appears as if the corporate world has awakened from the deep slumber of a manufacturing mind set to developing the talent capability of the individuals within the organizations. Talent is the engine of the modern organization, and employee engagement is the mystery ingredient that can transform the engine's output. The organizations which can boast of extraordinary performance are the organizations that have utilized and engaged the high powered individuals better. To quote Branson, "We embarked on consciously building Virgin into a brand which stood for quality, value, fun, and a sense of challenge. We also developed these ideas in the belief that our first priority should be the people who work for the companies, then the customers, then the shareholders. Because if the staff is motivated then the customers will be happy, and the shareholders will then benefit through the company's success." The chapter is about understanding the best practices in engaging high powered employees better and managing the talent shortage in a strategic way.

8.1 INTRODUCTION

Talent management is increasingly seen as a critical element of helping organizations to achieve competitive advantage. The growing economic importance of knowledge-based services and products is driving the need for higher level skills in diverse as construction, pharmaceuticals, defense and high technology, which are already experiencing serious shortfalls of available global talent, with even greater shortages predicted in years to come. Similarly, skilled workers expect to carry on developing their skills, if organizations are to successfully attract new talent; they have to provide growth opportunities and see that the employees associate meaningfulness in the work and work place. The most comprehensive report we have seen was the ISR Employee Engagement Report in 2006. It reported that organizations with highly engaged workforces performed up to 50% better than those with low engagement. Forty-six percent of the employees are reactively or actively disengaged have misguided loyalty or are actually there to mostly impact negatively to the organization. In fact, every manager starting from CEOs to managers should be concerned about the waste of time, money, and resources within their organization. The reason is simple—if people are not engaged, they don't see meaningfulness in the work they do and it further leads to sheer wastage of time and effort. It is imperative therefore to know that engaging staff is the "how" of any strategy and to drive it successfully, the organization should be clear as to what aspects of employee engagement. For example, if an organization wishes to change the culture of its organization, it will depend on the raising on the standard of management and leadership, which in turn will require capable managers to be able and willing to do what is needed to deliver success. Therefore, the key focus is spotting the kind of people who will drive this change with enthusiasm and vigor.

The whole of the effort of the organization is to find out ways and means of creating this vigor and enthusiasm not only in the selected few but also among the rest of the organization. The role of the manager then is to create conditions in which change can occur. Managers need to pay attention to their environments—today if they are not able to create the opportunities and design the jobs as per the competencies of their employees, the day is not far when employees flock to the competing organization to seek meaningfulness for their work. Managers can create a fluid and adaptive organization where there is increasing information flows, improving processes, changing structures, and enabling people to develop the skills required to work in them. It is important for the organization to know that they should act as enablers in future to get the things done to meet high end results altogether. Unfortunately, most companies readily talk the talk when it comes to encouraging employee learning and development, but few know how to convert these concepts into reality. In his book "Learning in Action," David A. Garvin outlines a set of criteria that can be used as a litmus test: like—identifying whether the organization is a learning organization or not; is the organization open to discordant information or not; is the organization able to avoid repetitive mistakes; and can the organization prevent leaking of vital information if key people leave the organization. This few things were critical in knowing the nature of the organization and the depth of involvement of the people who are the assets for that organization. Organization can definitely look forward to work for such a type of environment where meaningfulness is established and the connectivity between the employer and the employee (inclusive of the talented ones) is set—so that it makes the wheel of the organization move at a faster pace.

Companies that are prominent in the "new economy" such as Dell, Microsoft, and Cisco naturally embrace learning and development as part of their culture. But many companies with much longer histories are setting the pace as well. One example is General Electric, where former CEO Jack Welch created an innovative experiment in experiential learning called the change acceleration process. Growing out of the conviction, the only thing certain in the future is uncertainty and the need to manage continuous learning more efficiently is the order of the day, this process draws together participants in small teams and presents them with a real problem to solve.

The concept of employee engagement describes an employee who is committed and fascinated about his/her work. In his book, Getting Engaged: *The New Work place Loyalty*, author Tim Rutledge explains that truly engaged employees are attracted to, inspired by, their work. Engaged employees care about the future of the company and are willing to invest the discretionary effort—exceeding duty's call—to see that the organization succeeds. Engaged employees are those who are emotionally connected to the organization and cognitively vigilant. The concept of engagement was not introduced to management literature until the late 20th century by Bill Kahn, who defined it as "the simultaneous employment and expression of a person's "preferred self" in task behaviors that promote connections to work and to others, personal presence (physical, cognitive, and emotional), and active full role performances" (1990). Put in everyday terms, engagement is a measure of the degree to which people express their identity at work—not only who they are but also who they would like to be. An article in the journal titled *Journal of Happiness Studies* (Schaufeli et al., 2002) identified the physical, intellectual, and emotional components of high engagement, respectively, as "vigor," absorption, and dedication:

- *Vigor*: Highly engaged people are physically active and emotionally resilient. They are able to sustain any amount of hardships in the workplace.
- *Absorption*: Highly engaged employees find their work engrossing and throw themselves into it without thought of time or surroundings or how they look to others.
- *Dedication*: Engaged employees are dedicated to their work because they find it meaningful and fulfilling.

Such behaviors are easily observable and it can be inferred that if an employee brings the enthusiasm to the work place, it creates a learning environment in the workplace. Gallup management journal publishes a semiannual Employment Engagement Index. The most recent US results indicate that:

- Only 29% of employees are actively engaged in their jobs. These employees work with passion and feel a profound connection to their company.
- Fifty-four percent of the employees are not engaged. These employees have essentially "checked out," sleepwalking through their workday and putting time—but not passion into their work.
- Seventeen percent of employees are actively disengaged. These employees are busy acting out their unhappiness, undermining what their engaged coworkers are trying to accomplish.

It has been observed that these highly engaged employees had a significant contribution to usurp change in the organization. Eighty-four percent of the highly engaged workforce believe that they can positively impact the quality of their organization's products, compared with only 31% of the disengaged. Seventy-two percent of the highly engaged employees believe that they can positively affect customer service versus 27% of the disengaged.

Given these data, it is not difficult to understand that companies that do a better job of engaging their employees do outperform their competitors.

In 2011, Towers Perrin's Global workforce study showed that barely one in five employees (21%) are engaged on the job, 8% are fully disengaged, and the remaining 71% fall into categories: enrolled (partially engaged) and disenchanted (partially engaged).

There is a clear association between engagement and retention. The more engaged the workforce the greater the percentage of employees that plan on staying with their current employer.

8.2   CONCEPT OF EMPLOYEE ENGAGEMENT

Before we look at how to improve employee engagement, let's take a look at some definitions:

The extent to which employees feel committed because of a big idea and a big purpose.

Employees make a commitment to stay and engage if "promises made are kept."

Engagement is the extent to which employees enjoy and believe in what they do, feel valued for it and are willing to spend their intellectual effort to make the organization successful.

According to Aon Hewitt's definition of employee engagement employees are engaged they "say, stay and strive." They speak positively about the organization to others; are committed to remaining with their current employer; and are motivated by their organization's leaders, managers, culture, and values to go above and beyond to contribute to business success.

We can therefore summarize most of the definitions of employee engagement to the fact that it can be observed through the commitment levels of the employees and their contribution to the betterment of the organization.

But there is a striking difference between employee engagement and employee satisfaction; employee satisfaction is a measure of an employee's happiness with a company, their particular job or their coworkers among other factors. While an employee's happiness or satisfaction is important and can contribute to their engagement, it's not the same as engagement. Engagement refers to the emotional attachment one has to their job or organization, their enthusiasm for their work or the degree to which they act to further the organizations interests.

Engagement is both about the work environment and the work experience. According to HCI, employees who are engaged:

- use their talents every day;
- provide a consistent level of performance;
- build connections and professional networks;
- are committed to the company and values;
- have high energy;
- broaden what they do and build it; and
- helps make the organization more profitable.

While a moderately engaged employees displays the following behavior patterns:

- is meeting their performance goals; and
- is satisfied with their position in the organization.

In short they're doing what's required. They're just not going the extra mile.

Whereas a disengaged employee often displays the following behaviors:

- often doesn't know what is expected of them—and doesn't ask for clarification;
- may not feel that they have resources they need to do their work effectively;
- does not feel committed to the organization or their coworkers; and
- is actually costing the organization approximately $3400 for every $10,000 in paid salary.

Engagement research also shows that an employee's mindset or attitude about their ability to impact quality, costs, and customer service varies with their engagement levels. The more engaged an employee is, the more they believe they can make a difference and the more they try.

8.3   EMPLOYEE ENGAGEMENT AND BUSINESS OUTCOMES

The three types of outcomes which that matter to employers: increased job performance, increased citizenship behavior and decreased withdrawal behavior (absenteeism and turnover). Each of these is potentially valuable to the organization. Clearly, job performance matters because organizational performance is largely the sum of the performance of individual employees. Organizational citizenship behavior represents things that an employee does that are necessary but not formally a part of the job. All firms depend on employees who take initiative, are proactive, go above and beyond their formal role, and conscientiously do what needs to be done. Withdrawal behavior (absenteeism and especially turnover) is far more expensive than is commonly realized. An incident of turnover typically costs 0.5–2 times annual salary depending on employee skills.

Organizations always want greater job performance and greater citizenship behavior, but they do not always want less turnover. Turnover rates vary inversely with unemployment rates; when employees have few job opportunities, they stay put. At such times, organizations do not need to reduce voluntary turnover and they may welcome it when they need to reduce staffing.

How strong is the relationship between engagement and business outcomes? The contemporary practitioner literature filled with claims that employee engagement is a key to business results. A sampler of such reports:

1. Gallup consulting (2008) claims that a disengaged employee can cost the US economy over \$300 billion in lost productivity and that the difference between organization in the top quartile of engagement on its 12 item survey measure show 18% greater productivity, 12% greater profitability, 27% less absenteeism, 51% less turnover, and 62% fewer safety incidents than organizations in the bottom quartile on engagement.
2. A corporate leadership council study surveyed over 50,000 employees in 39 companies worldwide and concluded that increased engagement can lead to a 57% improvement in discretionary effort, a 20% individual performance improvement and an 87% reduction in turnover intentions.

Overall, there is strong evidence that employee involvement practices foster employee engagement, relatively good evidence shows that employee engagement promotes job performance and strong evidence that employee engagement decrease withdrawal and increases citizenship.

8.4   THE RELATION BETWEEN TALENT MANAGEMENT AND EMPLOYEE ENGAGEMENT

For better employee engagement as we know that we need to identify their competencies and understand their interest areas whereby they can be placed in better fitted jobs. Talent management is nothing but designing and creating systems where the talented individuals are identified and allowed to showcase their talents and the productivity of the employees improves. By doing so, not only the talented individuals but also the average performers get the motivation to come back to workplace. Businesses now aim to give more attention to both talent management and employee engagement. That attention needs to be well directed; those actions need to be well developed.

The first and foremost benefit which happens though better employee engagement program is that the talented individuals who join the organization are oriented in a better fashion about the organization culture rather than putting posters about organizational culture. In the initial days employees come with lots of doubts and hopes; this can be met through an engaging onboarding exercise by which with lot of ease they get to know the organizational culture.

Second important aspect of employee engagement program which has a significant role in managing talent in the organization is engaging employees productively is also developing them and increasing their career enhancement. Employees take up various courses and try to skill themselves in their related fields and this is an engaging exercise. Especially in Asian countries, it has been observed through Gallup survey that employees tend to stay in the organization where they could see career advancement.

Third most critical factor which is taken care by talent management has been performance management. Job involvement comes with better employee engagement practices. Talent joins the company appreciating the company and its product. As talent engages more fully in company's operations, assignments, projects, that appreciation for the company grows. An employee—especially a "talented employee"—who has the opportunity to perform in ways in which he/she sees as valuable consistently seeks to improve the performance.

Fourth, customers naturally prefer to experience quality in service that is provided to them. Research says it is the people with whom customers interact that determine the customer's opinion of that quality. Talent management looks for those candidates who can generate that quality. Employee engagement turns up that quality. With better customer engagement what is observed is better customer retention.

Which company doesn't want satisfied customers? Everyone does right. Therefore, it is imperative for talent management to focus on having more engaging employees in the workplace.

Lastly, it reflects on *employee turnover and increased retention risk.* If intense effort is made to hire talent, equally intense effort should be expended to retain talent. Employee engagement is a specific element of talent management in so far as it boosts a company's ability to hold on to talented employees. People stay with companies they value. The more an employee is allowed and encouraged to engage in job, team and company efforts, the more she sees the value. People stay with companies they value. The more an employee is allowed and encouraged to engage in job, team and company efforts, the more she sees the value. People stay with managers they trust. The more managers and employees engage in continuous communication about expectations, the more trust develops in their relationships. People stay with companies that offer them with personal and professional growth. In short, they should see meaningfulness in the work.

Simply stated, talent management acquires and supports higher levels of skills and knowledge. Employee engagement increases the application of the skills and knowledge and thereby increasing the value for the company as well as for the individual.

8.5 MAKING ENGAGEMENT HAPPEN IN THE ORGANIZATION

A Towers Perrin study in 2005 concluded after several years of surveys showing the link between engagement and performance, that a 5% increase in total employee engagement correlated to a 0.7% increase in operating margin. These employees cost the organization way too much, in terms of what the organization invests on them.

The question therefore is to how to engage cumulatively the head, heart, and the hands of the employees? It is better therefore to draw a model which explains the thought process of engaging a talented group.

8.5.1 THE PROCESS OF ENGAGEMENT

First step: The process of employee engagement starts with the employee himself or herself taking the initiative. He/She has to start thinking about their career goals, leverage their energies, and experience well-being at the work place.

The second step for better engagement has to be led through various organizational processes like revisiting the roles, connecting with the employees about their expectation and trying to understand their value proposition and timely recognition of the efforts put by the employees and engage them with meaningful work.

The first two steps would automatically lead to maximizing the performance of the employee and finally achieving the desired output for the organization. The processes in the organization should aim at connecting the dots—that is, the people. Employees of different strata should connect as to what is the expectation of the management and what are the deliverables at each stage. This sets the momentum for action. Employee engagement should spell authenticity at all levels, which means the processes should take care to remove all barriers to performance. The organization should plan for a learning environment and design to facilitate the employees to achieve better results every day. Potent employee engagement requires powerful recognition system in place within the organization. Each employee who has significantly contributed to the job and brought about the difference should be recognized doing the right thing in the first place itself, so that he continues the habit of doing it right every time. The Kaizen principle of continuous and consistent effort goes on to the process. Engagement is not a onetime survey measure or a steady state. To engage is to fully experience and contribute to the dynamic elements of the work. Work has to be redesigned to utilize the competencies of the people at work so that they can leverage the energies of each individual in the work place. Powerful engagement involves mastery of physical, mental, emotional, and spiritual and organizational energy. Ultimately, work should contribute to employee's well-being. An organization's results are dependent upon the health and productivity of individual employees.

The process of employee engagement can be practiced as follows:

1. Outline the organization's overall objective.
2. Define the scope of work for each particular segment of the organization.
3. Enable the people to understand it thoroughly.
4. Define their roles and purposes.
5. Enable the people to think on a sustainability project.
6. Evaluate at successive steps as to what went right and what went wrong.

The outcome should be well-being of the employees, where the employee sees that there is information flow at all ends. This enables him to see the system as fair and transparent. Employee starts feeling trusted and empowered as an individual and tries to leverage his energies for the organization's cause. The organization in return becomes more productive and efficient and responds to the customer needs and aligns with the business objective better.

8.5.2 EMPLOYEES' TAKING THE FIRST STEP TO MAKE THEIR WORK SPEAK UP

Whose job is engagement? Why will any organization do something for employee engagement? Many employees feel that if they could have a better manager who could understand them and their work they would have contributed more effectively. Yet, if individuals think someone else is responsible for their own engagement, it leaves them feeling more help-less—and even hopeless—in challenging work situations.

1. Therefore, *the first initiative* should be taken by the employees themselves. They should be able to craft their own goals for a particular year to keep them on their toes. When individuals take personal responsibility for their own engagement they are less likely to feel a victim at work place. The external factors should not be governing the employees like a bad boss and toxic work-place to improve their own productivity. Because the more they ponder on those aspects, they tend to get the feeling of being a victim at their work place and work will never measure up.

2. Second, the most important thing for the employee themselves is to have a feeling of partnering with the organization. They should follow the following mantra—contribute—support and challenge the obvious at the workplace.

3. The employees who chart out and plan on the important things where their contribution is of maximum importance and dedi-cate a sufficient amount of time to those important matters have been found to be more productive throughout the day. It's like the 80–20 rule which they should engage themselves with. Eighty percent of their time they should devote to 20% of their important tasks. This saves a lot of man hours and the employees slowly get a sense of achievement and belongingness towards the work place.

4. Fourth, trigger good work on cue. Willpower frequently fades or dissipates over time. Employees need to use structures and trig-gers to cue engaging actions. One can use time to trigger engage-ment such as how well one has distributed oneself throughout the day. The employees can have object or image also to connect with others.

5. Employees need to foster friends and coopt with allies without being skeptical of them. On the surface all employees tend to feel that it is superficial friendship that they develop at the workplace but it has been observed that having at least one friend at the workplace enables you to come to workplace without feeling forced to report to the workplace.

6. Lastly, employees should keep in mind that workplace should be visited daily with a sportive spirit. The ups and downs of the workplace should not affect their sense of self-esteem and meaning that they can contribute to the work. Let's say if one was playing the actual game of snakes and ladders, playing with loaded dice is cheating. But at work one needs to load the dice in favor of progress fused with recognition and celebration of progress. Know that the bad days are inevitable, but bad months or years are intolerable, so ensure ones setbacks are unsustainable.

8.5.3 WHAT ORGANIZATIONS NEED TO DO ABOUT ENGAGEMENT?

In their global Workforce study, Towers Perrin found that four out of five employees are not contributing up to their potential. And four out of 10 workers are disenchanted or disengaged.

Towers Perrin list in the top 10 global drivers of employee engagement as

1. senior management sincerely interested in employee well-being;
2. improved the skills and capabilities of the employees over the last year;
3. organizations reputation for social responsibility;
4. input into the decision making areas of the employees job related field;
5. organizations keenness in resolving customer concerns;
6. setting high personal standards for employees so that they feel challenged;
7. have excellent career advancement opportunities;
8. good relationship of employee with their supervisors;
9. presenting challenging work environment to the workers; and
10. organizations promoting innovative thinking.

Performance management is in many ways the centerpiece of the talent management process. Therefore, for better engagement Mangers should be helped in crafting SMART goals for themselves. Specific, measurable, achievable, realistic, and time frame for reaching the goals. An effective manager uses the performance management process to give employees a framework from which to do their work every day by providing them clear expectations, ensuring they have the skill/knowledge/expertise to accomplish their goals as well as the time, resources, autonomy, and authority.

Providing timely feedback has also been an important step in engaging the employees in productive work. The supervisors or senior managers should talk to the employees to focus on their strengths and also guiding them how to correct the faults typically creates a rewarding relationship between the employee and his/her senior manager. Getting to know the employees through regular, ongoing, two-way dialogue creates a sense of extended family with which employees can easily identify with.

8.6   THE PROCESS OF DEVELOPING AN ENGAGEMENT STRATEGY

After evaluating the US models of employee engagement, it can be brought out that to ensure that employee engagement strategy is linked to business objective there has to be some strategic linkage between the two.

1. Every organization should develop a priority matrix for the employee engagement survey;
2. the matrix developed then has to be understood in terms of organizational vision, mission and values;
3. outline the organization business objectives;
4. define the scope of employee engagement activities;
5. set specific measures of employee engagement strategy;
6. describe a clear path for carrying out these engagement activities;
7. identify the persons responsible for the specific engagement activities; and
8. measure the deliverables.

With this process of employee engagement, it is established by Gallup survey that the four components of employee engagement—well-being, information, fairness, and involvement in all the business process could be achieved.

In their employee engagement research, The Corporate Leadership Council found that up to 765 of employees are "up for grabs" in terms of engagement and could become engagement and could became engaged or disengaged. The potential financial impact of this could be substantial for your organization.

So what an organization can practically do to improve employee engagement. Among other things, organization can

1. provide training and coaching for managers to enhance their ability to motivate and inspire employees which in turn will help improve engagement scores (this is critical since managers have such a direct impact on employee engagement).
2. Organization should ensure support employee development at all levels within the organization. Providing learning and mentoring opportunities and a culture where regular ongoing discussion about career paths takes place.
3. Organizations should ensure that employees get the regular coaching and constructive feedback, they need to direct their work and improve performance. Organizations should make sure that employees have the skills to do that effectively.
4. Organizations should put a succession planning program in place to give high performing, high potential employees' opportunities for development and career progression.

8.7   TIPS FOR HR MANAGERS IN ENGAGING EMPLOYEES BETTER

Enabling managers to have the confidence and capability to be great engagers is the only way HR can truly build and sustain a high performance culture.

HR often owns the employee communication remit and that activity goes a long way to creating "savers" savers are usually hardworking, diligent employees who are committed to the company, while "investors" are people who are willing to invest more of themselves in an organization.

HR needs to give the processes and infrastructure which will enable the employees to develop their confidence and enable them to see career advancement opportunities.

For example, a strong performance management system that doesn't get tangled up in process and complexity is crucial. It's also important to ensure that any performance measurement really does have the performance of people at its heart.

The message will be loud and clear: We measure people performance because the performance of our employees is the most significant measure of business success.

To create investors, HR professionals have the responsibility to ensure the right level of development for every line manager—not just senior leader. It is the line managers who are the game changers on a daily basis when it comes to driving engagement, but they need proper support and development to do so.

HR should develop a culture which motivates the employee to be engaged then disengaged. This leads to mean that the competencies of the employees are identified and the career advancement opportunities for the same should be crafted thoroughly.

8.8   BUSINESS WORLD'S USE OF EMPLOYEE ENGAGEMENT: INDIA

1.   Infosys Technologies Ltd. announced a new social commerce and enhanced employee engagement platform as a part Infosys iEngage. All platforms of Infosys iEngage are now available on Smartphone-enabled devices.

     The enhanced Infosys iEngage employee engagement platform is built on top of industry leading Jive Social Business Software. The platform includes modules such policy manager, ideation manager, and knowledge manager to help foster innovation and increase workforce productivity. The employee engagement platform is powered by the iEngage context engine which creates the social map of an employee in six different dimensions. It provides content, community, and expert recommendations to employees using a robust rules engine. The new mobility module renders all the five Infosys iEngage platforms on smart hand held devices driving increased consumer and employee connect. It also enables users to interact both in a browser-based interface and through smart phone applications. According to Samson David, Vice President and Global Head—Business Platforms, Infosys Technologies, "Building Tomorrow's Enterprise requires organizations to engage with consumers and monetize digital demand. Multiple global companies across industries are successfully leveraging Infosys iEngage for their digital engagement journey. With our Social Commerce and Employee Engagement platforms, companies can successfully drive their sales, marketing, and employee engagement strategies and deliver measurable business value."

2. *Wipro* aims at holistic engagement of its employees by focusing on the initiatives to be targeted in each area of the different stake holders. Planned initiatives in each of its focus areas enable the employees to divert their energies to those significant deliverables. They also emphasize on phased implementation of each project so that they can help the employees to monitor their progress. Wipro induced the company to launch "Josh," a program to create bonding of the employee with the company and for employee engagement. Many companies conduct employee engagement programs to improve employee loyalty.

3. At TCS, the employee involvement is sought through an annual survey called as *PULSE* which is done through an online portal called as Ultimatix opinion polls and proactive employee engagement program called as *PEEP*. It is a mentoring initiative in which employees connect one-on-one with TCS' senior management. *Speak Up*, is a platform for employees to discuss pertinent matters with the company's seniors and to bring themselves closer to the company and its management; thus, developing the channels of communication between employees and the company's top executives. *PROPEL*, Propel provides a forum to discuss issues and ideas to resolve various issues at the inception itself. Camps and confluences are conducted for problem solving, discovery of new ideas, dialogue, reflection, and fun. Through these sessions, the employees are provided a platform for learning, interacting, and problem solving.

The dictionary of engagement differs from organization to organization. For instance, most of the Indian organization has viewed engagement as an initiative for improving the happiness quotient of the employees at the work place. For example, at Google India, employees meet every Friday to have discussions combined with fun and frolic in an informal setup. This acts as a stress buster as well as sets the environment for communicating with the people both horizontally as well as vertically. Though this model caters to the well-being of the employee but the context as to how it can be synergized with the business goals of the organization is yet to be verified.

8.9 CONCLUSION

Best employee engagement practices can be started early in the career of the new entrants in the organization. When it costs millions for the organization to have a group of highly disengaged employees or actively disengaged employees—it is always an added advantage for the employee as well as the organization to have a structured employee engagement procedure in which the employees feel highly motivated and charged to come to the workplace. The organization should start thinking in terms of fun quotient or in terms of happiness quotient to exactly motivate the human capital to come to the organization with the same enthusiasm each and every day. The organizations that are going to execute this in a priority list in the agenda are going to gain a competitive advantage over their competitors. In the long run of the organization, employee engagement is going to foster sustainability and sustainability is to be measured in terms of involvement of the employees in the work place.

The modern organization though has already crossed several milestones in giving the best place to work for physically, for example, like good ambience, favorable atmosphere, but it still has to go a long way in terms of establishing the connect psychologically with its employees. Employees should feel they are valued and engaged and they contribute significantly to the business goals of the organization.

KEYWORDS

- **talent management**
- **employee engagement**
- **strategy**
- **business**
- **organization**

REFERENCES

Erickson, T.; Grafton, L. What it Means to Work Here. *Harv. Bus. Rev.* March, **2007**.

Kahn, W. A. Psychological Conditions of Personal Engagement and Disengagement at Work. *Acad. Manage. J.* **1990,** *33*(40), 700–701.

Leary-Joyce, J. *Becoming the Employer of Choice.* Charted Institute of Personnel Development: London, 2004.

Schaufeli, W. B.; Salanova, M.; Gonzalez-Romá, V.; Bakker, A. B. The Measurement of Engagement and Burnout: A Confirmative Analytic Approach. *J. Happiness Stud.* **2002,** *3*, 71–92.

Spitzer, D. R. *Super Motivation: A Blueprint for Energising Your Organization.* American Management Association: New York, 1995.

RESEARCH STUDIES QUOTED

Gallup Consulting. *Employee Engagement: Whats Your Engagement Ratio?* Gallup: Washington, DC, 2008.

Towers Perrin: www.towersperrin.com.

PART III
Spirituality and Psychological Capital

CHAPTER 9

PSYCHOLOGICAL CAPITAL: A NEW PERSPECTIVE TO UNDERSTANDING TODAY'S EXECUTIVE

ARVIND K. BIRDIE*

Department of Post Graduate Studies, IIMT School of Management, Vedatya Institute, Gurgaon, Haryana, India

E-mail: arvindgagan@gmail.com

CONTENTS

ABSTRACT

In globalization and technology advances, human capital is the most important capital for any organization. Previously, organizations were familiar with the physical capital, money, labor, machinery, etc. After physical capital, organizations assessed human capital which consists of knowledge, skills, and experience of staff and finally, social capital introduced as a network of relationships with other people and now emerging relevance and emphasize on psychological capital is on rise and has roots in positive-oriented psychology movement. Psychological capital is a form of positive organizational behavior which represents the emergence of new trends in the field of human resources management and a positive work environment. This chapter reviews this new concept called "psychological concept" that recently entered into the field of organizational behavior to understand today's executive.

9.1 INTRODUCTION

Since the "formulation" of positive psychology at the onset of the new millennium, researchers have tried to apply positive psychology to organizational settings. For example, Adam Grant promotes pro-social behavior in business settings, Amy Wrzesniewski looks at how employees can foster (perceived) meaning via job-crafting, and Jane Dutton investigates the impact of high-quality connections.

    Positive organizational behavior (POB) defined as POB is defined by Luthans (2002) as "the study and application of positively-oriented human resource strengths and psychological capacities that can be measured, developed, and effectively managed for performance improvement in today's workplace." At the beginning of 21st century, a team of psychologists led by Martin Seligman, decided to examine the results of their research on the past five decades which that called disease-oriented Psychology School. They found significant achievements in finding effective treatments for mental and inefficient behaviors illnesses, but they have little attention to the growth, development, and self-actualization of healthy people. As a result, Seligman and colleagues tried on redirection of psychological research set two new missions:

- Help healthy personality people, for enjoy the happiness and productivity in life.
- Assist to developing human capabilities. Thus school of positive-oriented psychology was created (Luthans et al., 2007a). Psychology, which at the beginning examines the mental illness instead of studying mental health, for a long time ignored the human potential to grow and perfection. But in recent years, a growing number of psychologists believe in the capability of perfection and transformation of human characters. Positive-oriented psychologists' image have shown human nature to be optimistic and promising, because they believe in the capability to develop, growth, prosperity, and perfection of man and become what kind of person he can be. Csikszentmihalyi and Seligman indicated that positive-oriented psychology is mental positive marketing, positive individual traits, and positive organizations oriented. Luthans, positive-oriented organizational behavior leader, saw applications of Positivism in the workplace in respect and strengthen the ability to manage staff not in the weakness (Nelson & Cooper, 2007). Positive-oriented organizational behavior not saying like positive psychology oriented that a new approach is associated with positivism but deal with the need to focus more on theory, research, and effective use of cases, positive attributes, and behaviors of employees in the workplace (Bakker & Schaufeli, 2008). Research indicates changing subject from organizational pure science to considering positive features could make effectiveness of management policies and procedures (Luthans et al., 2009a,b) and improve the physical and mental health of staff (Wright et al., 2009). With avoiding avoidance from constant preoccupation to individual abuse performance and weaknesses by leaders and colleagues, we can pay attention to the strengths and desired qualities and increased confidence, optimism, hope, and confidence among staff (Luthans et al., 2004).

9.2 POSITIVE PSYCHOLOGY AND PSYCHOLOGICAL CAPITAL

There is increased support for the positive outcomes that can result from the focus on positive psychology (Luthans, 2002a,b; Luthans et al., 2007b; Peterson & Seligman, 2004; Sheldon & King, 2001), which shifts the traditional focus on what is wrong with people (i.e., neuroses, deviant behavior, etc.) to what is right with people (positive flourishing, virtues, optimism, hope, etc.) (for a review, see Roberts, 2006 and/or Luthans & Youssef, 2007). Drawing from theory in positive psychology and applying it to the workplace, POB has predominantly focused on advancing knowledge of state-like capacities (Luthans et al., 2007b; Wright, 2003), such as hope (Snyder, 2000, 2002; Snyder et al., 1996), resilience (Masten, 2001), optimism (Seligman, 1998), and self-efficacy (Bandura, 1997).

Luthans et al. (2004a) have developed the idea of positive psychological capital (PsyCap) which is seen as a valuable extension to the concepts of economic, human, and social capital. Current research in the area of POB has evolved into examining a higher order factor composed of these four components into PsyCap, which is defined as "an individual's positive psychological state of development that is characterized by (1) having confidence (self-efficacy) to take on and put in the necessary effort to succeed at challenging tasks; (2) making a positive reference (optimism) about succeeding now and in the future; (3) persevering toward goals and, when necessary, redirecting paths to goals (hope) in order to succeed; and (4) when beset by problems and adversity, sustaining, and bouncing back and even beyond (resilience) to attain success" (Luthans et al., 2007c, p. 3). Drawing from Hobfoll's (2002) psychological resource theory, PsyCap has been shown to be both theoretically (Luthans et al., 2007c) and empirically (Luthans et al., 2007b) supported as a higher order factor, whereas each of the four components are best understood as indicators of a single latent factor. Existing research has demonstrated a strong relationship between PsyCap and multiple employee outcomes, such as manager-rated performance and job satisfaction (Luthans et al., 2007b), trust (Norman et al., 2010), work engagement (Avey et al., 2008), commitment (Luthans et al, 2008a), and absenteeism (Avey et al., 2006).

Given the support for the importance of individual PsyCap in the workplace and the emerging research recognizing the influence of leader PsyCap (e.g., Norman et al., 2010), a logical next step is to understand how these two related sources of PsyCap influence each other in the workplace. Given that the present investigation considers the degree of similarity in PsyCap between two targets of a social relationship (e.g., between a leader and follower), there are at least two theories which may explain the phenomena and are the basis of this study. These theoretical frameworks include person–organization fit, person–supervisor fit.

Very simply, psychological capital is composed of four psychological resources, which are hope, optimism, self-efficacy, and resilience. It can be viewed as "who you are" and "what you can become in terms of posi-tive development" and is differentiated from human capital ("what you know"), social capital ("whom you know"), and financial capital ("what you have") (Avolio & Luthans, 2006; Luthans et al., 2004b). Psycholog-ical capital has been proposed as a viable source of competitive advantage for organizational success (Adler & Kwon, 2002; O'Leary et al., 2002).

9.3 DEFINING POSITIVE PSYCHOLOGICAL CAPITAL

The term psychological capital has its origins in the emerging POB litera-ture (Luthans & Youssef, 2007; Peterson & Seligman, 2004; Seligman & Csikszentmihalyi, 2000; Sheldon & King, 2001). Led by research psychol-ogist Martin Seligman, this movement challenged the field to focus on building positive qualities and traits within individuals and organizations as opposed to focusing on what is wrong and dysfunctional with them. The emphasis of this approach is a call for research that shows the applicability and effectiveness of positive psychological capacities in the workplace (Luthans & Jensen, 2005).

To be included as a POB capacity, the construct must meet the following criteria: (1) theory and research-based; (2) positive and strength-based; (3) valid measurement; (4) state-like (as opposed to trait-like), and hence open to development for performance improvement (Luthans, 2002; Luthans & Youssef, 2007; Luthans et al., 2007c). The relevance of these criteria is linked to the goal of improving workplace performance by learning, enhancing, and developing through self-development programs and training, or through on-the-job applications (Luthans & Jensen, 2005).

Although a number of positive constructs have been examined (see, e.g., Nelson & Cooper, 2007; Turner et al., 2002), to date the four that have been determined to meet the inclusion criteria so far are hope, opti-mism, resilience, and self-efficacy, and when combined, make up the core construct of what has been termed psychological capital (Luthans, 2002a,b; Luthans et al., 2004a,b, 2007c; Luthans & Youssef, 2004, 2007). Psycho-logical capital is defined as "an individual's positive psychological state of development that is characterized by (1) having confidence (self-efficacy) to take on and put in the necessary effort to succeed at challenging tasks; (2) making a positive attribution (optimism) about succeeding now and in the future; (3) persevering toward goals and, when necessary, redirecting paths to goals (hope) in order to succeed; and (4) when beset by problems and adversity, sustaining and bouncing back and beyond (resiliency) to attain success" (Luthans et al., 2007c, p. 3).

Results of studies in field of positive-oriented organizational behavior suggest that psychological capacities such as hope, optimism, resilience, and self-efficacy, next to each, make the factor that is called psychological capital. In other words, some of the psychological variables, such as hope, resiliency, optimism and self-efficacy, make a new latent factor that reflected from all of these variables. In fact, Luthans and colleagues on the development of a positive-oriented organizational behavior in organizations have proposed that positive-oriented psychology capital as a source of competitive advantage for organizations. They believe that psychological capital with emphasis on the positivism psychological variables such as hope, optimism, resiliency, etc. can promote the value of human capital (knowledge and skills of people) and social capital (network of relationships among them) in organization Therefore, psychological capital, made of the positivism psychological variables that can be measured to develop management practices on them. As previously mentioned, these variables are as follows:

- self-efficacy
- hope
- optimism
- resiliency.

In the following, each of these items are explained (Page & Donohue, 2004):

9.3.1 SELF-EFFICACY

The origin of the emergence of self-efficacy is the Bandura Social Cognitive Theory (1997) (we can simply call it confidence). So that defined: Belief in one's abilities to achieve success in a given task by motivation themselves, cognitive resource supply, and taking the necessary actions. People with high self-efficacy have five following features:

(1) Set high goals for themselves and take difficult task to achieve their goals.

(2) Accept challenges and trying to deal with difficult challenges.
(3) They have high motive.
(4) They do anything to achieve their goals.
(5) In dealing with obstacles are consistent.

Five points to regarding self-efficacy are as follows:

(1) Self-efficiency in every field of activity is limited to the same realm; it cannot be extended to other areas of life.
(2) Make self-efficiency in any activity based on training and learn mastery on that activity.
(3) In any activity, although the person has self-efficiency, there is always the possibility of improving their self-efficiency.
(4) Others believe affect the person self-efficiency.
(5) Self-efficacy under the influence of factors, such as knowledge and skills of the person, physical and mental health, external conditions, including the person (Luthans et al., 2007a).

9.3.2 RESILIENCY

Resiliency is a state which can be obtained by the person's development to be able to deal with failures, tragedies, and conflicts of life and even positive events, developments, and more responsibility continue to greater efforts and achieve greater success. Research suggests that some resilient people after dealing with life's difficult situations return to their usual level of performance again. However, the performance of some of these people, after dealing with failures, tragedies, and difficulties, promote.

In general, resiliency people have these features:

(1) Readily accept the reality of life.
(2) Believe that life is meaningful.
(3) They have remarkable ability to adapt quickly to big changes (Luthans et al., 2006b).

To achieve resiliency, a number of skills and attitudes are needed. They are said to be in terms of hardiness. **Hardiness** helps people to convert challenges to lucrative opportunities in dealing with stressful changes. The attitudes of hardiness, including: *commitment, control,* and *challenge-seeking*. *Commitment* is defined as preference to maintain its presence instead of give up in dealing with difficult circumstances and help another on that situation. The meaning of *control* is that the person has belief in his ability to influence the outcome of events. *Challenging* means that individual dealing with challenges rather than incriminate fate seeks to create opportunities for self and others. Also in the formation and incidence hardiness behavior, person should interact with others and use problem solving skills (Maddi & Khoshaba, 2005). In fact, the resiliency is the phenomenon that achieved from human natural adaptive response and in spite of facing the person with serious threat, he is capable of achieving success and overcoming threats. Compared to the other constituent variables of psychological capital, in the literature, little research has been done in terms of resiliency in the workplace (Danahyo, 2004).

9.3.3   OPTIMISM

Seligman (1986), the father of positive-oriented psychology, believes that the optimistic people in dealing with failures and achievements and have these following features:

(1) Rely on general attributions.
(2) Their attributions are stable.
(3) They attribute their success to their innate abilities.
(4) In dealing with failures, they attribute their failure to external, especially factors and unsustainable.

Another explanation for optimism proposed by Carver and Scheier (2003) that *optimistic* people always expect a pleasant happenings occurring in their lives. It is important that psychological capital is realistic optimism, because unrealistic optimism leads to the negative consequences (Luthans, 2006). In fact, in realistic optimism, individual assessment of what can be achieved in the face of what they could not make deals. Therefore, realistic optimism has important role in promotion individual self-efficacy (Luthans, 2007). According to the results of research conducted by Adams (2002), among the organizations participating in his study, those who had higher levels of optimism in employees are more successful than other organizations in organizational performance.

9.3.4   HOPE

Snyder (1991) defined hope as *Positive motivational state of feeling successful in*

(1) *agency (energy toward the target) and*
(2) *planning to achieve the purpose.*

So, hope composed of two components: agency and plans for achieving the goal. The purpose of agency is to achieve the desired result. Thus, hopes need an agency or energy to pursue goals. In addition, other elements of hope are the planning to achieve the goal which not only involves identifying the goals, but it also covers different ways to achieve goals. In other words, hope requires a willingness to succeed and identify, clarify, and pursue ways to achieve success.

Whole psychological capital is greater than the sum of its items (Luthans et al., 2007). Walsh and colleagues have shown in the past few decades in organizational newsletters. Words with a negative approach such as winning or losing, job dissatisfaction, absenteeism and use more than positive word, such as compassion, righteousness, etc. (Baker & Shafly, 2008). So it is obvious that in organization and management area, positive word have been neglected than negative. The application of positive-oriented psychology in organization and management led to create and develop two new approaches in this area:

(1) **Positive-oriented organizational behavior.** That most emphasis on the micro-aspects of organizational behavior and those positive cases of human personality that can develop and grow.
(2) **Positive organizational scholarship.** That most emphasis on the macro-aspects of organizational behavior. Positive-oriented organizational behavior is defined as the study and application of positive psychological capabilities and the strengths of human resources that can be developed and measured and can manage them effectively to improve performance.

It should be mentioned that any psychological variables cannot be considered in the field of positive-oriented organizational behavior. In other words, by taking the following five criteria could define a variable in field of positive-oriented organizational behavior.

The variable that based on theory or research—(1) might valid be measurable, (2) be unique in the field of organizational behavior, (3) possible to educate develop and change, and (4) it has a positive impact on business performance and individual satisfaction.

9.4   PSYCHOLOGICAL CAPITAL THEORY

Psychological capital is conceptualized as an individual's positive psychological state of development characterized by (1) having confidence (self-efficacy) to take on and put in the necessary effort to succeed at challenging tasks; (2) making a positive attribution (optimism) about succeeding now and in the future; (3) persevering towards goals and, when necessary, redirecting paths to goals (hope) in order to succeed; and (4) when beset by problems and adversity, sustaining and bouncing back, and even beyond (resiliency) to attain success (Luthans et al., 2007c, p. 3).

Psychological capital is seen as a resource that goes beyond human capital (experience, knowledge, skills and abilities) and social capital (relationships, networks). It deals with "who you are here and now," and "who you can become" in the proximal future if your psychological resources are developed and nurtured in the workplace (Luthans et al., 2004b; Luthans & Youssef, 2004). To date, surveys support that the four component resources load on the higher order core construct of psychological capital and indicate convergent and discriminant validity with similar positive constructs, such as core self-evaluations and relevant personality traits, such as conscientiousness (Avey et al., 2009; Luthans et al., 2007b).

9.5   PSYCHOLOGICAL CAPITAL MEASUREMENT

In terms of measurement, a valid and reliable PsyCap questionnaire has been developed (Luthans et al., 2007c) and empirically validated (Luthans et al., 2007b). The items used therein were originally drawn from published validated scales commonly used in positive psychology. These individual scales have also been used in previous studies in the workplace (e.g., Luthans et al., 2005; Peterson & Luthans, 2003; Youssef & Luthans, 2007). Six items in this questionnaire represent each of the four components that make up PsyCap. These items were adapted for the workplace from the following standard scales: (1) hope (Snyder et al., 1996); (2)

resilience (Wagnild & Young, 1993); (3) optimism (Scheier & Carver, 1985), and (4) efficacy (Parker, 1998).

Therefore, PsyCap meets the criteria of valid measurement and openness to development, and a growing number of studies have clearly demonstrated that it has impact on the desired outcomes in the workplace. For example, in one major study PsyCap was shown to be positively related to employee satisfaction (Luthans et al., 2007b). There is also increasing evidence that PsyCap is significantly related to desired employee behaviors (and negatively to undesired behaviors), attitudes (e.g., satisfaction and commitment), and performance (Luthans et al., 2007b). Research studies evidently demonstrate the impact that PsyCap may have on satisfaction and/or commitment (Larson & Luthans, 2006; Luthans et al., 2007b, 2008a; Youssef & Luthans, 2007) and absenteeism (Avey et al., 2006).

Several researches on the subject have already provided some evidence that the relative stability of PsyCap makes the return on investment in its development more sustainable in the workplace (Luthans et al., 2007b, 2008b) (the emphasis on demonstrating objective work-related outcomes of PsyCap best fits Pfeffer's (1997) economic model of behavior).

9.6   RESEARCH ON POSITIVITY AT THE WORKPLACE

The recent wave of negativity stemming from corporate and geopolitical problems draws attention to the need for more positivity in the world and the result has been a reemphasis in the use of a positive lens for organizational behavior theory, research, and practice at science and study institutions in the USA. Although the importance of positivity has gained focus through the years, only recently, it has been proposed as a new lens to focus the study on organizational behavior (Cameron et al., 2003; Luthans, 2002a,b; Luthans et al., 2007c; Luthans & Youssef, 2007; Nelson & Cooper, 2007; Roberts, 2006; Turner et al., 2002; Wright, 2003).

9.7   WHY SHOULD YOU CARE (ESPECIALLY IF YOU ARE A CEO OR HR DIRECTOR)?

Well, you should care if you are interested in having a healthy, engaged, and high-performing workforce. A meta-analysis (a type of study that aggregates the results of prior studies) based on 51 empirical investigations found a wide array of positive consequences for workers displaying high (vs. low) PsyCap. From the study abstract:

The results indicated the expected significant *positive* relationships between PsyCap and [...] job satisfaction, organizational commitment, psychological well-being, desirable employee behaviors (citizenship), and [...] measures of performance (self, supervisor evaluations, and objective). There was also a significant *negative* relationship between PsyCap and undesirable employee attitudes (cynicism, turnover intentions, job stress, and anxiety) and undesirable employee behaviors (deviance).

Right around the turn of the last century, the field of psychology began to place greater importance on investigation as to what was right with people and what contributed to human flourishing and growth potential (Seligman & Csikszentmihalyi, 2000; Sheldon & King, 2001; Snyder & Lopez, 2002). Drawn from the recent positive psychology movement (Peterson, 2006; Peterson & Seligman, 2004; Seligman & Csikszentmihalyi, 2000; Snyder & Lopez, 2002), the positive focus extended to the workplace, by focusing on both the value of micro-oriented positivity in individuals as well as macro-oriented positivity in organizations (Cameron & Caza, 2004; Cameron et al., 2003; Luthans, 2002a,b; Luthans et al., 2007c; Nelson & Cooper, 2007; Wright, 2003; Roberts, 2006; Spreitzer & Sonenshein, 2004).

POB was for the first time defined as "the study and application of positively oriented human resource strengths and psychological capacities that can be measured, developed, and effectively managed for performance improvement in today's workplace" (Luthans, 2002a,b). Positive psychological resources, such as hope or resilience, once considered to be "a quality of gifted individuals" (Garmezy, 1974), now has obtained empirical support that they can be developed (Masten & Reed, 2002; Snyder, 2000), as well as other capacities commonly recognized in the field of organizational behavior, such as efficacy (Bandura, 1997; Stajkovic & Luthans, 1998) and optimism (Seligman, 1998). Moreover, a specific construct of psychological capital was introduced (Luthans, 2007).

9.8 RECENT RESEARCH ON PSYCHOLOGICAL CAPITAL IN SCIENCE AND STUDY INSTITUTIONS IN THE USA

Although PsyCap predominately focuses on positivity at the individual level, expanding research in the science and study institutions in the USA has also demonstrated positive relations between collective PsyCap and team performance (Clapp-Smith et al., 2009; Peterson & Zhang, 2011; Walumbwa et al., 2011). There are at present in excess of 45 published PsyCap papers, and the emergence of the first meta-analysis is further evidence of the growth of PsyCap research (Avey et al., 2011a).

Some empirical research indicates that positive appraisals of life domains besides work (i.e., Relationship PsyCap, and Health PsyCap) impact on employee's overall well-being (Luthans et al., 2010; Luthans & Harms, 2013). Furthermore, work-related positivity is viewed as antecedent not only of proximal work outcomes, but also for overall well-being over time (Luthans et al., 2010). A recent meta-analysis has provided further evidence of significant, positive relationships between PsyCap and job satisfaction, organizational commitment, organizational citizenship behaviors, and job performance and negative relationships with turnover intent, cynicism, job stress, and deviance (Avey et al., 2011b).

9.9 CONCLUSIONS AND IMPLICATIONS

There is an expanding research across science and study institutions in the USA demonstrating the positive relationship between PsyCap and desired employee outcomes, and there is also conceptual and empirical evidence that PsyCap can be developed (Luthans et al., 2006a, 2007c, 2008b). Nonetheless, supplementary research is needed to test further whether PsyCap can be developed via the training model as well as to determine its impact on individual performance (Luthans, 2010; Luthans et al., 2006a, 2008b).

KEYWORDS

- psychological capital
- executive
- hope
- optimism

REFERENCES

Adams, V. H.; Snyder, C. R.; Rand, K. L.; King, E. A.; Sigman, D. R.; Pulvers, K. M. Hope in the Workplace. In: *Workplace Spirituality and Organization Performance*; Giacolone, R., Jurkiewicz, C., Eds.; Sharpe: New York, 2002.

Adler, P. S.; Kwon, S. Social Capital: Prospects for a New Concept. *Acad. Manage. Rev.* **2002,** *27,* 17–40.

Avery, D. R.; McKay, P. F.; Wilson, D. C. Engaging the Aging Workforce: The Relationship between Perceived Age Similarity, Satisfaction with Coworkers, and Employee Engagement. *J. Appl. Psychol.* **2007,** *92*(6), 1542–1556.

Avey, J. B.; Luthans, F.; Youssef, C. M. The Additive Value of Positive *Psychological* Capital in Predicting Work Attitudes and Behaviors. *J. Manage. Online First* **2009,** 1–23.

Avey, J. B.; Patera, J. L.; West, B. J. The Implications of Positive Psychological Capital on Employee Absenteeism. *J. Leadership Org. Stud.* **2006,** *13*(2), 42–60.

Avey, J. B.; Reichard, R. J.; Luthans, F.; Mhatre, K. H. Meta-analysis of the Impact of Positive Psychological Capital on Employee Attitudes, Behaviors, and Performance. *Art. Hum. Resour. Dev. Q.* **2011a,** *22*(2), 127–152.

Avey, J. B.; Wernsing, T. S.; Luthans, F. Can Positive Employees Help Positive Organizational Change? Impact of Psychological Capital and Emotions on Relevant Attitudes and Behaviors. *J. Appl. Behav. Sci.* **2008,** *44*(1), 48–70.

Avey, J. B.; Wernsing, T.; Mhatre, K. H. A Longitudinal Analysis of Positive Psychological Constructs and Emotions on Stress, Anxiety, and Well-being. *J. Leadership Organ. Stud.* **2011b,** *18*(2), 216–228.

Avolio, B. J.; Luthans, F. *The High Impact Leader: Moments Matter in Accelerating Authentic Leadership Development.* McGraw-Hill: New York, 2006.

Bakker, A. B.; Schaufeli, W. B. Positive Organizational Behavior: Engaged Employees in Flourishing Organizations. *J. Organ. Behav.* **2008,** *29,* 147–154.

Bakker, A. B.; Schaufeli, W. B.; Leiter, M. P.; Taris, T. W. Work Engagement: An Emerging Concept in Occupational Health Psychology. *Work Stress* **2008,** *22,* 187–200.

Bandura, A. *Self–Self Efficacy: The Exercise of Control.* Freeman: New York, 1997.

Cameron, K. S.; Caza, A. Contributions to the Discipline of Positive Organizational Scholarship. *Am. Behav. Sci.* **2004,** *47*(6), 731–739.

Cameron, K. S.; Dutton, J.; Quinn, R., Eds. *Positive Organizational Scholarship.* Berrett-Koehler: San Francisco, CA, 2003.

Clapp-Smith, R.; Gretchen, V.; James, A. *Authentic Leadership and Positive Psychological Capital: The Mediating Role of Trust at the Group Level of Analysis.* Management Department Faculty Publications. Paper 23, 2009. http://digitalcommons.unl.edu/managementfacpub/23.

Garmezy, N. The Study of Competence in Children at Risk for Severe Psychopathology. In: *The Child in His Family: Children at Psychiatric Risk: III*; Anthony, E. J., Koupernik, C., Eds.; Wiley: New York: 1974; p 547.

Hobfoll, S. Social and Psychological Resources and Adaptation. *Rev. Gen. Psychol.* **2002,** *6,* 307–324.

Larson, M.; Luthans, F. Potential Added Value of Psychological Capital in Predicting Work Attitudes. *J. Leadersh. Org. Stud.* **2006,** *13*(1), 45–62.

Luthans F. The Need for and Meaning of Positive Organizational Behavior. *J. Organ. Behav.* **2002a,** *23,* 695–706.

Luthans, F.; Jensen, S. M. Hope: A New Positive Strength for Human Resource Development. *Hum. Resour. Dev. Rev.* **2002,** *1,* 304–322.

Luthans, F.; Youssef, C. M. Emerging Positive Organizational Behavior. *J. Manage.* **2007,** *33,* 321–349.

Luthans, F. Positive Organizational Behavior: Developing and Managing Psychological Strengths. *Acad. Manage. Execut.* **2002,** *16,* 57–72.

Luthans, F. Positive Organizational Behavior: Developing and Managing Psychological Strengths. *Acad. Manage. Exec.* **2002a,** *16,* 57–72.

Luthans, F. Positive Organizational Behavior: Developing and Managing Psychological Strengths. *Acad. Manage. Exec.* **2002b,** *16,* 57–72.

Luthans, F. The Need for and Meaning of Positive Organizational Behavior. *J. Organ. Behav.* **2002b,** *23,* 695–706.

Luthans, F.; Youssef, C. M. Emerging Positive Organizational Behavior. *J. Manage.* **2007,** *33,* 321–349.

Luthans, F.; Avey J. B.; Avolio, B. J. Peterson, S. J. *The Development and Resulting Performance Impact of Positive,* 2010.

Luthans, F.; Avey, J. B.; Avolio, B. J.; Peterson, S. J. The Development and Resulting Performance Impact of Positive Psychological Capital. *Hum. Resour. Dev. Q* **2010,** *21,* 41–67. DOI:10.1002/hrdq.20034.

Luthans, F.; Avolio, B. J.; Avey, J. B.; Norman, S. M. Positive Psychological Capital: Measurement and Relationship with Performance and Satisfaction. *Pers. Psychol.* **2007,** *60,* 541–572.

Luthans, F.; Avolio, B. J.; Walumbwa, F. O.; Li, W. The Psychological Capital of Chinese Workers: Exploring the Relationship with Performance. *Manage. Organ. Rev.* **2005,** *1,* 247–269.

Luthans, F.; Youssef, C.; Avolio, B. *Psychological Capital: Developing the Human Competitive Edge.* Oxford University Press: Oxford, 2006.

Luthans, F.; Youssef, C. M.; Sweetman, D. S.; Harms, P. D. Meeting the Leadership Challenge of Employee Well-being through Relationship PsyCap and Health PsyCap. *J. Leadersh. Organ. Stud.* **2013,** *20*(1), 118–133.

Luthans, F.; Zhu, W.; Avolio, B. J. The Impact of Efficacy on Work Attitudes across Cultures. *J. World Bus.* **2006,** *41,* 121–132.

Luthans, F.; Avey, J. B.; Avolio, B. J.; Norman, S. M.; Combs, G. M. Psychological Capital Development: Toward a Micro-intervention. *J. Org. Behav.* **2006a,** *27,* 387–393.

Luthans, F.; Avey, J. B.; Avolio, B. J.; Peterson, S. J. The Development and Resulting Performance Impact of Positive Psychological Capital. *Art. Hum. Res. Dev. Q.* **2010,** *21*(1), 4–67.

Luthans, F.; Avey, J. B.; Patera, J. L. Experimental Analysis of a Web-based Training Intervention to Develop Positive Psychological Capital. *Acad. Manage. Learn. Educ.* **2008b,** *7*(2), 209–221.

Luthans, F.; Avey, J.; Avolio, B.; Crossley, C. Psychological Ownership: Theoretical Extensions, Measurement and Relation to Work Outcomes. *J. Organ. Behav.* **2009b,** *30*(2), 173–191.

Luthans, F.; Avey, J.; Jensen, S. Psychological Capital: A Positive Resource for Combating Employee Stress and Turnover. *Hum. Resour. Manage.* **2009a,** *48*(5), 677–693.

Luthans, F.; Avey, J.; Smith, R.; Palmer, N. Impact of Positive Psychological Capital on Employee Well-being over Time. *J. Occup. Health Psychol.* **2010,** *15*, 17–28.

Luthans, F.; Avolio, B. J.; Avey, J. B.; Norman, S. M. Positive Psychological Capital: Measurement and Relationship with Performance and Satisfaction. *Pers. Psychol.* **2007b,** *60*, 541–572.

Luthans, F.; Avolio, B.; Youssef, C. *Psychological Capital*; Oxford University Press: Oxford, UK, 2007a; p 246.

Luthans, F.; Norman, S. M.; Avolio, B. J.; Avey, J. B. The Mediating Role of Psychological Capital in the Supportive Organizational Climate—Employee Performance Relationship. *J. Org. Behav.* **2008a,** *29*, 219–238.

Luthans, F.; Vogelsesang, G. R.; Lester, P. B. Developing the Psychological Capital of Resilience. *Hum. Res. Dev. Rev.* **2006b,** *5*(1), 25–44.

Luthans, F.; Youssef, C. Positive Organizational Behavior in the Workplace: The Impact of Hope, Optimism, and Resilience. *J. Manage.* **2007,** *33*, 774–800.

Luthans, K. W.; Jensen, S. M. The Linkage between Psychological Capital and Commitment to Organizational Mission: A Study of Nurses. *J. Nurs. Adm.* **2005,** *35*(6), 304–310.

Maddi, S. R.; Khoshaba, D. M. *Resilience at Work*. Amacom: New York, 2005.

Masten, A. S. Ordinary Magic: Resilience Process in Development. *Am. Psychol.* **2001,** *56*, 227–239.

Masten, A. S.; Reed, M. G. J. Resilience in Development. In: *Handbook of Positive Psychology*; Snyder, C. R., Lopez, S. J., Eds.; Oxford University Press: Oxford, UK, 2002; pp 74–88.

Nelson, D. L.; Cooper, C. L., Eds. *Positive Organizational Behavior*. Sage Publications: London, 2007.

Norman, S. M.; Avey, J. B.; Nimnicht, J. L.; Pigeon, N. G. The Interactive Effects of Psychological Capital and Organizational Identity on Employee Organizational Citizenship and Deviance Behaviors. *J. Leadersh. Org. Stud.* **2010,** *17*(4), 380–391.

O'Leary, B.; Lou Lindholm, M.; Whitford, R.; Freeman, S. Selecting the Best and the Brightest: Leveraging Human Capital. *Hum. Resour. Manage.* **2002,** *41*(3), 325–340.

Page, L. F.; Donohue, R. Positive Psychological Capital: A Preliminary Exploration of the Construct. Working Paper, Monash University, Department of Management, 2004.

Parker, S. Enhancing Role-breadth Self-efficacy: The Roles of Job Enrichment and Other Organizational Interventions. *J. Appl. Psychol.* **1998,** *83*, 835–852.

Parker, S. Enhancing Role Breadth Self-efficacy: The Roles of Job Enrichment and Other Organizational Interventions. *J. Appl. Psychol.* **1998,** *6*, 835–852.

Peterson, C. *A Primer in Positive Psychology*. Oxford University Press: New York, 2006.

Peterson, C.; Seligman, M. E. *Character Strengths and Virtues*. Oxford University Press: Oxford, UK, 2004.

Peterson, C.; Seligman, M. E. P. *Character Strengths and Virtues: A Handbook and Classification*. Oxford University Press: New York, NY, 2004.

Peterson, S. J.; Luthans, F. The Positive Impact and Development of Hopeful Leaders. *Leadersh. Organ. Dev. J.* **2003,** *24*(1), 26–31.

Roberts, L. M. Shifting the Lens on Organizational Life: The Added Value of Positive scholarship. *Acad. Manage. Rev.* **2006,** *31*(2), 292–305.

Scheier, M. F.; Carver, C. S. Optimism, Coping, and Health: Assessment and Implications of Generalized Outcome Expectancies. *Health Psychol.* **1985,** *4*, 219–247.

Scheier, M.; Carver, C. Optimism, Coping, and Health: Assessment and Implications of Generalized Outcome Expectancies. *Health Psychol.* **1985,** *4*, 219–247.

Seligman, M. E. P.; Csikszentmihalyi, M. Positive Psychology: An Introduction. *Am. Psychol.* **2000,** 55, 5–14.

Sheldon, K. M.; King, L. Why Positive Psychology Is Necessary. *Am. Psychol.* **2001,** *56*, 216–217.

Peterson, S. J.; Luthans, F.; Avolio, B. J.; Walumbwa, F. O.; Zhang, Z. Psychological Capital and Employee Performance: A Latent Growth Modeling Approach. *Pers. Psychol.* **2011,** *64*(2), 427–450.

Snyder, C. R. *Handbook of Hope.* Academic Press: San Diego, CA, 2000.

Snyder, C. R.; Lopez, S. J. *Handbook of Positive Psychology.* Oxford University Press: New York, NY, 2002.

Snyder, C. R.; Irving, L.; Anderson, J. Hope and Health: Measuring the Will and the Ways. In: *Handbook of Social and Clinical Psychology*; Snyder, C. R., Forsyth, D. R., Eds.; Pergamon: Elmsford, NY, 1991; pp 285–305.

Snyder, C. R.; Lopez, S. *Handbook of Positive Psychology.* Oxford University Press: Oxford, UK, 2002.

Spreitzer, G. M.; Sonenshein, S. Toward the Construct Definition of Positive Deviance. *Am. Behav. Sci.* **2004,** 47, 828–847.

Stajkovic, A. D.; Luthans, F. Self-efficacy and Work-related Performance: A Meta-analysis. *Psychol. Bull.* **1998a,** *124*, 240–261.

Turner, N.; Barling, J.; Zaharatos, A. Positive Psychology at Work. In: *Handbook of Positive Psychology.* Oxford, UK: Oxford University Press, 2002.

Wagnild, G.; Young, H. Development and Psychometric Evaluation of the Resiliency Scale. *J. Nurs. Measure.* **1993,** *1*(2), 165–178.

Wright, T. A. A Look at Two Methodological Challenges for Scholars Interesting in Positive Organizational Behavior. In: *Positive Organizational Behavior*; Nelson, D. L., Cooper, C. L., Eds.; Sage: Thousand Oaks, CA, 2007/09; pp 177–190.

Wright, T. A.; Cropanzano, R.; Bonett, D. G.; Diamond, W. J. The Role of Employee Psychological Well Being in Cardiovascular Health: When the Twain Shall Meet. *J. Organ. Behav.* **2009,** *30*, 193–208.

Youssef, C. M.; Luthans, F. Positive Organizational Behavior in the Workplace: The Impact of Hope, Optimism, and Resilience. *J. Manage.* **2007,** *33*, 7.

CHAPTER 10

WORKPLACE SPIRITUALITY AND EMPLOYEE ENGAGEMENT: A STUDY OF THE INDIAN SERVICE INDUSTRY

RABINDRA KUMAR PRADHAN* and LALATENDU KESARI JENA

Department of Humanities and Social Sciences, IIT Kharagpur, Kharagpur 721302, West Bengal, India

Corresponding author. E-mail: rkpradhan@hss.iitkgp.ac.in

CONTENTS

ABSTRACT

In order to prosper and thrive in today's turbulent business environment, organizations are continuously in search of engaged employees. Earlier research findings have brought out the fact that highly engaged employees go above and beyond the core responsibilities outlined in their job descriptions, innovating and thinking out of box to move their organizations forward. Engaged employees carries an incessant will power to give their time and energy for supporting the cause about which they are truly passionate. This kind of ideologies gives birth to the concept such as workplace spirituality. The factor of spirituality encompasses the assumptions that each person has his/her own inner motivations, truths, and desires to be engaged in professional activities that give greater meaning to his/her life and the lives of others. The present study is designed to explore the linkage between workplace spirituality and employee engagement. Data were collected from 172 executives employed with Indian public and private banks & insurance industries. A set of standard tools on workplace spirituality and employee engagement was administered to all the participants. The results revealed that there exists a positively significant relationship between workplace spirituality and employee engagement. At the same time, there are not much of statistical differences between genders on perceiving their organizations differently in regard to dimensions of workplace spirituality and employee engagement. Both theoretical and practical implications of the study are discussed in the light of empirical findings.

10.1 INTRODUCTION

People spend their considerable life span at workplace and make their most significant impact to the society. Knowledge workers employed in industries across the globe are very much reflective as they question themselves about their work profile and are continuously in search of deriving purpose and meaning of their job (Karakas, 2010; Miller, 1997). In the 21st century, establishing employees' psychological connection with their job profile at workplace is a fundamental prerequisite for having a healthy competition among corporate entities. In this context, Harter et al. (2002) has outlined employee engagement as a psychological connection materializing through a meaningful work assignment, incessant involvement in its profile requirements and deriving satisfaction as well as maintaining enthusiasm throughout one's professional tenure. Literature on talent management has emphasized for upholding the ideology of engaging "heads, hands, and hearts" together (Donahue, 2001) with heart (i.e., instilling passion—person's intrinsic motivation) as one of the quintessence for employee engagement. In the context of interpersonal role, engagement is experienced as a state of emotional and intellectual commitment (Falcone, 2006) to an organization or group since an engaged employee is stated to be a person who is fully involved in, and enthusiastic about, his or her work and to be engaged meaning to be emotionally and intellectually committed to one's organization and its job profile (Bhatnagar, 2007; Osman, 2013).

Therefore, engagement is predominantly an energetic state of mind as an engaged employee is vigorous and willing to invest his discretionary effort at work (Schaufeli & Bakker, 2010). Petchsawange and Duchan (2012) have aided engagement with meaningfulness and joy at work, compassion, trust, well-being of employees through fostering *workplace spirituality*. However, there is a dearth of empirical investigation whether spirituality influences human productivity as asserted by academic researchers and professional consultants (Milliman & Ferguson, 2003) and more particularly will it aid nurturing engaged employees in an organizational context. Hence, it can be summed up as there is not much of research has been done amalgamating both the areas so far, there is a need to explore this virgin area to get insight into and its impact on job behavior. Thus, in our chapter, we have pursued to introduce empirically grounded and theoretically defensible research findings on both the constructs. The objective of this research work is to build theoretical and empirical support between workplace spirituality and employee engagement.

10.2   RELEVANCE OF WORKPLACE SPIRITUALITY

"Work" earns a new meaning and significance when it is realized as a calling, a sacred duty, a service opportunity or a higher purpose (Paloutzian et al., 2003). Reave (2005) has said that when work is seen as a calling, it becomes more meaningful and there is a favorable chance for increase in productivity, commitment, and sustenance of well-being. Moore and Casper (2006) has defined work-related feelings as an "internal substance" which defines "value, belief, attitude, or emotion" for one's job profile, toward one's professional colleagues and organization as a whole. Therefore, if spirituality is a countenance of an individual's beliefs and attitudes, then it is assumed that spirituality will differ because of individual differences in an organizational set up. To realize the organizational vision such as engagement-spirituality need to transcend individual differences such as beliefs, race, age, sex, and other similar characteristics.

Spirituality is based on an individual's personal value and philosophy wherein it is expected that an individual need to view themselves as spiritual being whose soul need proper sustenance at work, experiencing a sense of purpose and meaning in work, and a sense of connectedness to one another and toward their workplace and community at large (Ashmos & Duchon, 2000; Mitroff & Denton, 1999). Within the broad definition Mitroff and Denton (1999) has stated in terms of addressing individual differences and organizational productivity that "if a single word best captures the meaning of spirituality and the vital role that it plays in people's lives, that word is interconnectedness with a sense of common purpose." However, Pfeffer (2003) has abridged the spiritual goals proposed by Ashmos and Duchon (2000) and Mitroff and Denton (1999) in their research findings saying that a contemporary professional use to seek four essential constituents in one's workplace:

- An interesting work profile allows for learning and development, while providing a sense of competence and mastery at each level of succession.
- Meaningful work use to instill a feeling of purpose and existence.
- Fosters a sense of connection and positive social relations with colleagues and employees.
- And above all, it provides the ability to live an integrated life, so that the work does not clash with the essential nature of the worker and one's desire to live as a human being.

Milliman et al. (2003) has refined the existing variables through their exploratory empirical assessment between "workplace spirituality and employee work attitude," while proposing three important dimensions of workplace spirituality. The first dimension as *meaningful work*—signifies the degree to which people experience a deep sense of meaning and purpose at work. The *sense of community*—as the second dimension means that people see themselves as connected to each other and that there is some type of relationship between one's inner self and the inner self of other people. The third and the most important dimension is *value alignment*—which measures whether or not individuals experience a strong sense of alignment between their personal values and the organization's mission and purpose.

Workplace spirituality is treated as a kind of cure for the "ills of modern management" (Brown, 2003) and is assumed to be an ideal technique for inducing the trust between employer and employee. With the advent of massive industrialization this has been "allegedly lost with the alienation generated by the dehumanized practices that accompanied the massive processes of downsizing, the abuses of workers and other actions that breached psychological contracts" (Jurkiewicz & Giacalone, 2004).

Workplace spirituality and its consequences can be viewed through the lens of the concept of person–organization fit (P–O fit), a perceptual construct which refers to "judgments of congruence between an employee's personal values and an organization's culture" (Cable & DeRue, 2002). It suggests that when a worker creates a strong bond at his/her workplace by a defined value and belief pattern then better work outcomes may potentially result. Shared person–organization values in this context indicates strong P–O fit, which has been found to positively affecting work attitudes (Balazas, 1990; Posner et al., 1985), job satisfaction and turnover (O'Reilly et al., 1991), and operating unit performance (Enz & Schwenk, 1991). It is presumed that when there is a strong match between worker values and their perceptions of the organization's spiritual values, more positive attitudinal outcomes will result by making an individual to feel a stronger attachment and have better attitudes about, one's organizations and work. For example, we expect that when workers desire working for an organization that practices the spiritual values as openness, connection, truth, personal development and growth, serving and sharing, and finding meaning and purpose through one's work, they will more closely identify themselves with their organizations.

Organizational identification and engagement can be viewed as a worker's perception of congruence or "oneness" with his or her organization (Ashforth & Mael, 1989). Workers who strongly identify with their organizations typically are found to be more supportive (Ashforth & Mael, 1989), make decisions consistent with objectives set by their organizations (Simon, 1997; Smidts et al., 2001), and feel more involved with the mission of their organizations (Cable & DeRue, 2002). This kind of atmosphere creates a positive relationship between worker perceptions of organizational spiritual values and identification with their organizations. Eisenberger et al. (1986) has suggested that when employees perceive organizational support, they in turn contribute to their full potential toward achieving the organizational objectives. One of the explanations for such positive employee responses to perceived organizational support comes from the concept of the norm of reciprocity (e.g., Armeli et al., 1998). The norm of reciprocity is associated with the belief that one should help or benefit the benefactor (Gouldner, 1960). Employee compliance with the norm of reciprocity and the resulting efforts to benefit the organization, in reciprocation of the perceived organizational support, may reflect, to some extent, an employee's transcendence of self-interests. This is supported by the empirical findings such as the observed relationship between employees' perceived organizational support and their lower absenteeism lowering the attrition level and fostering innovative behaviors (Eisenberger et al., 1986). Research performed by University of Southern California's Marshall Graduate School of Business Professor Ian Mitroff (Mitroff & Denton, 1999) has indicated that organizations which identify themselves as spiritual have employees who (1) are less fearful of their organizations; (2) are less likely to compromise their basic beliefs and values in the workplace; (3) perceive their organizations as being significantly more profitable; and (4) report that they can bring significantly more of their complete selves to work, especially their creativity and intelligence.

In a business context, the question arises is, how does spirituality is related to the bottom line of a business and its potential outcomes? Publications and studies in this context have focused on the relationship between workplace spirituality and organizational performance (Giacalone & Jurkiewicz, 2003). A Harvard Business School study has examined 10 companies with strong corporate culture and 10 with weak corporate culture, drawn from a list of 200 leading companies. Researchers in this study has not only explored a dramatic correlation between an organization's spiritual culture and its profitability, but also have found that, in some cases, the more spiritual companies have outperformed others by 400–500% in terms of net earnings, return on investment, and shareholder value (Giacalone & Jurkiewicz, 2003).

Companies that excel at engaging the hearts and minds of their people not only have values, they live them, thereby providing an element of spirituality in their everyday working environment. Therefore, companies that focus on processes that include the spiritual element, such as bringing together employees for motivation at work and encouraging employees to find meaning in work, often increase employee retention, which has a decided impact on profitability. McLaughlin (1998) emphasizes the relationship between spirituality and profitability by asserting—a growing movement across the country is promoting spiritual values in the workplace and pointing to many insights and examples of increased productivity and profitability (p. 11). According to McLaughlin (1998), organizations that want to survive in the 21st century will have to offer a greater sense of meaning and purpose—key elements of spirituality, to their workforce. We have agreed to the point that in today's highly competitive environment, the best talent will seek out organizations that reflect their inner values providing an opportunity for personal development and community service, not just bigger salaries.

Spirituality is a growing torrent of research and its underlying construct is struggling to be empirically proven. There is still a deficiency of depth in terms of its impact, relevance, and significance in a working environment especially in the context of HR effectiveness. At the same time, the "precise means of measurements, using validated instruments, which can help organizations to understand the utility of workplace spirituality" is negligent till date (Giacalone & Jurkiewicz, 2003). The goal of HR department is to attract and maintain highly productive employees, thus it is imperative to understand the most sought after dimension of workplace spirituality and its underlying construct.

10.3   ESSENCE OF EMPLOYEE ENGAGEMENT

Employee engagement is a fairly new but enormously widespread concept in the field of human resource management. Kahn (1990) was the first scholar to study engagement and tried to define it in terms of "personal engagement" as "harnessing of organization members selves into their work roles wherein they are employing and expressing themselves physically, cognitively, and emotionally during role performance." There is a growing cognizance that employee engagement is fundamental to successful business performance, as engaged employees are the "backbone of good working environments because they are industrious, ethical and accountable" (Levinson, 2007; Cleland et al., 2008).

The idea of engagement has claimed to envisage both individual employee outcomes and organizational level success and financial performance (Macey & Schneider, 2008; Shuck & Reio, 2011). This is because, engaged employees are found to be highly energized and resilient professionals in performing their job; putting their heart into their jobs with persistence and willingness to invest effort; exhibiting strong work involvement along with experiencing feelings of significance, enthusiasm, passion, inspiration, pride, excitement, and challenge from their work; and fully concentrating and immersing themselves in their work without noticing how the time passes away (Schaufeli & Bakker, 2004). Macey and Schneider (2008) has defined engagement as "discretionary effort or a form of in role or extra role effort or behaviour," encompassing innovative and adaptive performance and going "beyond preserving the status quo, and instead focus on initiating or fostering change in the sense of doing something more and/or different." Schaufeli et al. (2002) have brought out the perspective of employee engagement stating it as "a persistent positive affective state of fulfillment in employees, characterized by vigor, dedication and absorption.' According to them: *Vigor* is characterized by high levels of energy and mental resilience while working and denotes employees' inclination toward investing energies into their job, and especially the high levels of energy, endurance, and perseverance in the face of difficulties. *Dedication* states about employees' resiliency to their work, their feelings of enthusiasm and significance for the activities, inspiration, pride, and challenge on the activities they do, whereas *absorption* occurs when the employee is pleasantly occupied with work through investing vigor and dedication to his/her work profile.

There is a growing body of business oriented literature that has defined how engaged employees contribute to the overall success of an organization. Kahn (1990) has made an extensive research to identify the psychological conditions necessary to explain moments of personal engagement and personal disengagement among individuals across different situations at work. His research is based on qualitative study of personal engagement and has used observation and interview methods with 16 camp counselors and 16 architectural firm members. He has found that people draw upon themselves to varying degrees while performing work tasks and they can commit themselves physically, cognitively, and emotionally in the various roles they perform or they may choose to withdraw and disengage from their work roles and work tasks. Results of Kahn's study suggest that there are three psychological conditions that shape how people perform their roles—meaningfulness, safety, and availability.

Kahn's identification of these three psychological conditions now serves as a framework for the study of employee engagement. He has defined the state of meaningfulness as one in which workers feel worthwhile/valuable, and that they are making a difference and are appreciated for the work they do. Safety is described as an environment in which people feel it as an ability to act what would be normal for the individual without fear of negative consequences. Safety is found in situations in which workers trust that they will not suffer because of their engagement to their work and where they perceive the climate to be one of openness and supportiveness. Therefore, it can be summed up from his findings that people have dimensions of themselves that they prefer to use and express in the course of role performance. If they can match their preferred actions with the psychological conditions existent in their work environment and work roles, then they will engage with the job.

May et al. (2004) conducted field study in a large Midwestern insurance agency. Using a survey format they explored why some individuals fully engage in their work profiles, while others become alienated or completely disengaged. Results of this study confirmed that engagement differs from simple job satisfaction. They agreed that engagement actually entails the active use of emotions and behaviors in addition to cognitions. Overall, study results have supported Kahn's earlier work where in psychological meaningfulness and safety was found to be positively linked with employee investment in work roles. Additionally, job enrichment and role fit were positively related to psychological meaningfulness. Having supportive supervisor and good relations with coworkers were related to feelings of psychological safety on the job.

Saks (2006) has surveyed 102 employees from a group of manufacturing and service industries to test antecedents and consequences of job and organizational engagement. He has differentiated job engagement from organization engagement and concluded that organizational engagement is a person's attitude and attachment to his/her company, whereas employee engagement is the degree to which an employee is actually absorbed in the performance of his/her own individual job role. Saks drew from Kahn's (1990) work to propose a model of employee engagement in which the antecedents of engagement are identified as (a) job characteristics, (b) perceived organization support, (c) perceived supervisor support,

(d) rewards and recognition, (e) procedural justice, and (f) distributive justice. In this same model, the consequences of employee engagement are identified as (a) job satisfaction, (b) organizational commitment, (c) intention to stay on the job, and (d) organizational citizenship behavior. Results of the survey have shown that psychological conditions leading to organization and job engagement, as well as the consequences of each variable, are different. The study results have confirmed that perceived organization support predicts job and organization engagement. He has concluded that procedural justice predicts organization engagement and that job and organization engagement are both related to employee attitudes, intentions, and behaviors. In particular, job and organization engagement predicts job satisfaction, commitment to the organization, and intention to quit. Overall, the result of the study has suggested that workers who perceived higher organizational support were more likely to reciprocate with greater levels of engagement to their individual job roles.

Though empirical findings and conceptual works on the said variables of engagement have surfaced in the literature over the past two decades and subsidized to much of our understanding of the construct (Macey & Schneider, 2008; Shaufeli et al., 2002), engagement has gained attention of HR scholars and organizational practitioners only in recent years (Kim et al., 2012; Rurkkhum & Bartlett, 2012; Soane et al., 2012; Wollard & Shuck, 2011). The research available till date supporting the possible importance of employee engagement is flawless; yet, research about creating employee engagement and what might be the aftermaths of doing so is remarkably undeveloped. Therefore, there needs rigorous studies in employee engagement to unearth its theoretical underpinning (Saks, 2006) and practical application with other equivalent concept like workplace spirituality along with its antecedents and consequences.

10.4 RESEARCH QUESTIONS AND SUGGESTED INSTRUMENTS FOR STUDY

The literature review presented in this chapter is seeking to examine the hypothesized relationship between the variables of employee engagement and workplace spirituality. In alignment with the purposes for this chapter, these following research questions are enumerated below along with its corresponding instrumentation useful in helping to measure them.

(1) To ascertain the extent to which employee engagement relates to workplace spirituality?

(2) To find out the differences in the engagement and commitment levels of employees among genders?

10.4.1 WORKPLACE SPIRITUALITY SCALE

The scale was established by Ashmos and Duchon (2000) which consists of 21 items. Each item has seven elective responses, which need to be recorded on a 5-point Likert scale fluctuating from strongly disagree to strongly agree. The components like meaningful work scale contain 6 items and its alpha reliability estimate is found to be 0.88. Sense of community scale contains 7 items with its alpha reliability estimate is 0.91. Alignment with organizational values scale contains 8 items and alpha reliability coefficient is 0.94.

10.4.2 EMPLOYEE ENGAGEMENT SCALE

The Utrecht Work Engagement Scale developed by Schaufeli et al. (2001) is consisting of 17 items on the three underlying dimensions, that is, of vigor, dedication, and absorption having "vigor" is measured with six items (items 1, 4, 8, 12, 15, 17), "dedication" with five items (items 2, 5, 7, 10, 13), and "absorption" with six items (items 3, 6, 9, 11, 14, 16). Items are rated on a 7-point scale ranging from 0 (never) to 6 (every day). The scale includes statements such as "my job inspires me" and "I feel happy when I am working intensely." The internal consistencies (Cronbach's alpha) of the UWES-17 are ranged between 0.75 and 0.83 for vigor, between 0.86 and 0.90 for dedication, and between 0.82 and 0.88 for absorption.

10.5 DATA COLLECTION AND ANALYSIS

We have administering the questionnaire to the sample respondents and instructing them that they have been given 30 min for completion. During the instructions, the author researchers has stressed on the fact to the participating subjects about the purpose, potential benefits of the study and their consent. Confidentiality of responses and privacy of the subjects was ensured. Scorings for the entire test were carried out as per the instructions and the raw scores were then tabulated and subjected to various statistical analysis. The collected data have been analyzed using SPSS 20.0. In exploratory analysis measures of central tendency such as mean and standard deviation was carried out to study the nature and distribution of scores on variables of workplace spirituality and employee engagement. In order to respond to the research objective, correlational (two-tailed) analyses and reliability tests were employed. To test whether there are variances between genders, we split the data into two subfiles (male and female) and obtain the desired correlation from each subfile.

We have found that regression analysis is a powerful tool which can be used in a number of ways as it is used to describe the population characteristics or to make predictions about subjects/variables administered for the population or even to test casual hypothesis (Aron & Aron, 1994). We have assumed that, when paired with assumptions in the form of a statistical model, regression can be used for prediction, inference, hypothesis testing, and above all give necessary direction on modeling of casual relationship.

10.5.1   PILOT STUDY

An initial study was made to establish the appropriateness and rationality on the items of both the research instruments used. The instruments were tested using a predetermined sample of 30 respondents drawn from public sector manufacturing organizations other than those organizations used for the actual study. The reliability values on 21 item workplace spirituality scale are found to be 0.82 and 17 item work-engagement scale is 0.76. As per the recommendation of Nunnally (1978), all the reliability values of the instruments met the required threshold.

10.5.1.1   SAMPLING AND DATA COLLECTION

The sample frame or population and the size are drawn by convenience sampling through visiting the public and private banking and insurance establishments in eastern India and approaching the executives through their respective departmental heads for undertaking the survey. A number of 211 questionnaires were distributed to respondents who are working full time in their present organization and key demographic variables (gender and age) were reviewed for further characterizing the sample.

A total of 172 questionnaires were returned, with a response rate of 81.13%, and all of which are deployed for further data analysis. The high response rate can be attributed to the effective administration of these surveys. The sample respondent has found consisting of 32.55% female executives and 29.65% of the total respondents had worked for more than 5 years in their present establishments. The mean age of the participants was 32.76 years (SD = 4.64) with the majority of them (68.60%) were married and regarding the educational level, 68.60% of them had professional post graduate level studies leading to CA, ICWAI, CFA, and MBA level qualifications, and 31.79% had graduate level studies.

10.6 RESULTS AND DISCUSSIONS

The findings have been divided into three sections exploratory researches, testing of hypothesis, inter-correlation matrix and regression analysis. Table 10.1 has revealed that mean and standard deviation of the total sample for workplace spirituality was found to be 4.69 and 0.13, respectively, whereas the mean and standard deviation of employee engagement was found to be 4.52 and 0.16.

TABLE 10.1 Descriptive Statistics for Study Variables (n = 172).

Variables	M	SD
Meaningful work	4.68	0.24
Sense of community	4.65	0.20
Alignment of values	4.74	0.16
Workplace spirituality	4.69	0.13
Vigor	4.49	0.33
Dedication	4.56	0.19
Absorption	4.50	0.23
Employee engagement	4.52	0.16

The correlation table (Table 10.2) has shown that the total variables of workplace spirituality are positively correlated with the dimensions of

TABLE 10.2 Correlation Matrix of Variables of Workplace Spirituality and Employee Engagement ($n = 172$).

Variables	1	2	3	4	5	6	7	8
Meaningful work (1)	1							
Sense of community (2)	−0.045	1						
Alignment of values (3)	0.213**	0.376**	1					
Workplace spirituality (4)	0.663**	0.623**	0.722**	1				
Vigor (5)	0.514**	0.452**	0.448**	0.714**	1			
Dedication (6)	0.066	0.342**	0.558**	0.436**	0.099	1		
Absorption (7)	−0.157*	0.349**	0.298**	0.201**	0.004	0.236**	1	
Employee engagement (8)	0.299**	0.611**	0.672**	0.756**	0.719**	0.587**	0.576**	1

*Correlation is significant at the 0.05 level (2-tailed); **correlation is significant at the 0.01 level (2-tailed).

employee engagement ($r = 0.756$, $p < 0.05$). The findings have revealed that the Indian banking industry is influenced with 57.2% (R square = 0.572, Table 10.3) of spirituality at workplace on employee engagement. Further, the beta coefficient of the simple regression of workplace spirituality on employee engagement (Table 10.4) has revealed the significance (beta = 0.756, $t = 15.078$, $p = 0.000$). The inter-item correlations were found to be significant which has supported the findings of earlier studies. However, the itemized correlation between meaningful work and employee engagement was also found to be in a positive direction ($r = 0.29$, $p < 0.05$) though not significant. We apprehend that the relationship can be expected to be significant with a larger sample size. This supports the findings of Cropanzano and Mitchell (2005) that when individuals receive economic and meaningful work assignment from their organization, they will feel obliged to get themselves engaged whole heartedly and responds for repaying their organization.

TABLE 10.3 Model Summary.

Model	R	R square	Adjusted R square	Std. error of the estimate	Change statistics				
					R square change	F change	df1	df2	Sig. F change
1	0.756a	0.572	0.570	0.09018	.572	227.358	1	170	0.000

aPredictors: Constant, workplace spirituality.

TABLE 10.4 Coefficients.a

Model	Unstandardized coefficients		Standardized coefficients	t	Sig.
	B	Std. error	Beta		
1 (Constant)	1.822	0.191		9.552	0.000
Workplace Spirituality	0.635	0.042	0.756	15.078	0.000

aDependent variable: employee engagement.

In this context, Rothbard (2001) has defined engagement as a psychological presence stating that it involves two critical components: attention and absorption in a meaningful work assignment. Attention over here refers to "cognitive availability and the amount of time one spends thinking about a role" while absorption means "being engrossed in a role and refers to the intensity of one's focus on a role." Work and alignment of personal values with organizational vision are deemed to be a central life activity irrespective of gender deployed in organizational set up. In this connection, the engagement and spirituality dimension has revealed that engaged employees feel personally connected to their work, productive and are less likely to seek alternative employment. In the present study, though there is considerably a small sample size of females ($n = 56$) versus males ($n = 116$), however there is not much of statistically significant differences (female: $r = 0.837$, $p < 0.01$, whereas male: $r = 0.710$, $p < 0.01$, Table 10.5) were found, between genders on perceiving their organizations differently in regard to dimensions of workplace spirituality and employee engagement.

TABLE 10.5 Gender Wise Descriptive Statistics and Correlation Table for Workplace Spirituality and Employee Engagement.

Gender		N	Mean	SD	Workplace spirituality
Female	Workplace spirituality (1)	56	4.69	0.15	1
	Employee engagement (2)	56	4.50	0.17	0.837*
Male	Workplace spirituality (1)	116	4.69	0.12	1
	Employee engagement (2)	116	4.53	0.16	0.710*

*Correlation is significant at the 0.01 level (2-tailed).

The results of multiple regressions as shown in Table 10.6, where workplace spirituality was treated for both genders separately as predictor or independent variable and employee engagement as outcome or dependent variable, it was found that female sample at R square $= 70.1\%$ and male sample at R square $= 50.4\%$ as a significant predictor of employee engagement. The beta values at Table 10.7 have indicated the unique contribution of the dimensions as beta values for female: workplace spirituality $= 0.929$ and male: workplace spirituality $= 0.879$ in predicting employee engagement. The said findings to an extent have challenged the findings of Schaufeli and Baker (2003) that male populations are slightly higher engaged than their female counterparts, but again the differences are very small and hardly bear any significance.

TABLE 10.6 Model Summary.

Gender		R	R square	Adjusted R square	Std. error	Change statistics				
						R square change	F change	df1	df2	Sig. F change
F	1	0.837[a]	0.701	0.696	0.09427	0.701	126.731	1	54	0.000
M	1	0.710[a]	0.504	0.500	0.11333	0.504	115.891	1	114	0.000

[a]Predictors: (constant), workplace spirituality.

TABLE 10.7 Coefficients.[a]

Gender			Unstandardized coefficients		Standardized coefficients	t	Sig.
			B	Std. error	Beta		
F	1	(Constant)	0.147	0.388		0.379	0.706
		Workplace spirituality	0.929	0.083	0.837	11.257	0.000
M	1	(Constant)	0.400	0.384		1.043	0.299
		Workplace spirituality	0.879	0.082	0.710	10.765	0.000

[a]Dependent variable: employee engagement.

Robinson et al. (2004) in their study has suggested on exploring the driving force of employee engagement has suggested that for diversified organizations catering female work force requires the creation of an organizational environment where positive emotions such as involvement, meaningful job assignment and pride to be encouraged, resulting in improved organizational performance, lower employee turnover, and better work–life balance. Finally, empirical evidence is line with the earlier findings suggesting that a spiritual life (irrespective of gender differences) engaged in a profession is likely to be characterized by positive satisfaction, a greater sense of fulfillment, and a better quality of life (Dierendonck & Mohan, 2006; Mohan, 2001).

10.7 BENEFITS OF WORKPLACE SPIRITUALITY ON EMPLOYEE ENGAGEMENT

Employee engagement has been hypothesized with two dimensions: Cognitive Engagement—the extent to which a professional individual is aware of his mission at work and his role in the organization—and Emotional Engagement or physical engagement—the extent to which the worker empathizes with others at work and connects meaningfully with his or her coworker (Luthans & Peterson, 2002). On the other hand, spirituality at work is a term that pronounces the experience of employees "who are obsessive about and energized by their work, find meaning and purpose in their work, feel that they can express their complete selves at work, and feel connected to those with whom they work" (Kinjerski & Skrypnek, 2004).

Therefore, the meaning of workplace spirituality and engagement both advocates a sense of inclusiveness to each other in their underlying construct. Krishnakumar and Neck (2002) has stated "fostering spirituality will lead to the employees feeling complete when they come to work." The primary drivers of feeling completeness are identification of one self with organization's goals and internalization of the organization's values and mission (Chalofsky & Krishna, 2009). Fairlie's (2011) has studied to compare a number of meaningful work features to other work characteristics as correlates and predictors of employee engagement.

Results have suggested that these meaningful characteristics have the strongest correlations with multiple employee outcomes and that they accounted for a substantive amount of variation in employee engagement. Work characteristics that were categorized as meaningful included: self-actualizing work (referring to a job that enables the individual to fulfill their potential and become a "fully function" person), work that is perceived to have a strong social impact (the extent to which one's job matters in the society; legacy), a job that allows one to fulfill their life goals and values, a job that offers a sense of accomplishment, and work that instigates one's belief in achieving their highest career goals within their organization." Second, Chalofsky and Krishna (2009) has suggested that engagement has its foundations in preindustrial society when work was performed in the same neighborhood where people lived, and that work was tied to the well-being of the individual as well as the community. In the same line of it, workplace spirituality also encompasses a feeling of being connected with one's work as well with coworkers (Milliman et al., 2003).

Kahn (1990) has iterated that people experience psychological meaningfulness and engagement, when they are receiving rewarding interpersonal interactions with their coworkers and clients. He has also found that supportive and trusting interpersonal relationships promoted psychological safety and engagement. Cropanzano and Mitchell (2005) have proposed social exchange theory to identify the reasons behind people's engagement to each other in a work setting, whereby relationships evolved over time leads toward commitment as long as the parties involved in the relationship submitted to certain rules of exchange, in that the actions of one party led to a response or actions of the other party. Third, engagement is principally influenced by the culture of the organization, its leadership, the quality of communication, the styles of management, levels of trust and respect, and the organization's reputation. Masson et al. (2008) has suggested that a key lever for engagement, and ultimately effective performance, is an "employee's emotional commitment to the organization and the job, the extent to which the employee derives enjoyment, meaning, pride and inspiration from something or someone in the organization."

10.8   SCOPE FOR FUTURE RESEARCH

Although workplace spirituality and employee engagement have developed autonomously; however, they clearly share many commonalities in their expositions. Basically, the interconnection between them recommends HR functionary to make it exist concurrently in their organizations' as spirituality in workplace may enable employees to fully engage themselves with their work assignments/roles. The need for spirituality is a recurrent theme in corporations and businesses (Karakas, 2010) and it is expected that they need to embrace a set of humanistic and spiritual values in their engagement process so as to enable human hearts, spirits and souls to grow and flourish.

KEYWORDS

- workplace spirituality
- employee engagement
- vigor
- dedication
- absorption
- meaningful work

REFERENCES

Aron, A.; Aron, E. N. *Statistics for Psychology.* Prentice-Hall International Inc.: New Jersey, 1994.

Armeli, S.; Eisenberger, R.; Fasolo, P.; Lynch, P. Perceived Organizational Support and Police Performance: The Moderating Influence of Socio-emotional Needs. *J. Appl. Psychol.* **1998,** *83*(2), 288–297. DOI:10.1037/0021-9010.83.2.288.

Ashforth, B. E.; Mael. F. Social Identity Theory and the Organization. *Acad. Manage. Rev.* **1989,** *14*(1), 20–39.

Ashmos, D. P.; Duchon, D. Spirituality at Work: A Conceptualization and Measure. *J. Manage. Inq.* **2000,** *9*(2), 134–145.

Balazas, A. L. Value Congruency: The Case of the Socially Responsible Firm. *J. Bus. Res.* **1990,** *20*, 171–181.

Bhatnagar, J. Talent Management Strategy of Employee Engagement in Indian ITES Employees: Key to Retention. *Employee Relat.* **2007,** 29, 640–663.

Brown, R. B. Organizational Spirituality: The Sceptic's Version. *Organization* **2003,** *10*(2), 393–400.

Cable, D. M.; DuRue, D. S. The Convergent and Discriminant Validity of Subjective Fit Perceptions. *J. Appl. Psychol.* **2002,** *87*(5), 875–884.

Chalofsky, N.; Krishna, V. Meaningfulness, Commitment, and Engagement: The Intersection of a Deeper Level of Intrinsic Motivation. *Adv. Dev. Hum. Resour.* **2009,** *11*(2), 189–203.

Cleland, A.; Mitchinson, W.; Townend, A. *Engagement, Assertiveness and Business Performance—A New Perspective.* Ixia Consultancy Ltd, 2008.

Cropanzano, R.; Mitchell, M. S. Social Exchange Theory: An Interdisciplinary Review. *J. Manage.* **2005,** *31*(6), 874–900.

Dierendonck, D.; Mohan, K. Some Thoughts on Spirituality and Eudaimonim Well-being. *Ment. Health, Relig. Cult.* **2006,** *9*(3), 227–238.

Donahue, K. B. *It is Time to Get Serious about Talent Management.* Harvard Business School Press: Boston, MA, 2001.

Eisenberger, R.; Huntington, R.; Hutchison, S.; Sowa, D. Perceived Organizational Support. *J. Appl. Psychol.* **1986,** *71*, 500–507. DOI:10.1037/0021-9010.71.3.500.

Enz, C. A.; Schwenk, C. R. The Performance Edge: Strategic and Value Dissensus, *Employee Responsibilities Rights J.* **1991,** *4*, 75–85.

Fairlie, P. Meaningful Work, Employee Engagement, and Other Key Employee Outcomes: Implications for Human Resource Development. *Adv. Dev. Hum. Resour.* **2011,** *13*(4), 508–525. DOI:10.1177/1523422311431679.

Falcone, P. Preserving Restless Top Performers: Keep Your Top Performers Engaged So They Don't Jump Ship Once Job Opportunities Arise. *HR Magazine*, 2006. Retrieved from http://www.allbusiness.com/humanresources/workforcemanagementhiring/8749791.html (accessed during August 2015).

Giacalone, R. A.; Jurkiewicz, C. L. Toward a Science of Workplace Spirituality. In: *Handbook of Workplace Spirituality and Organizational Performance*; Giacalone, R. A., Jurkiewicz, C. L., Eds.; Sharpe: New York, 2003; pp 3–28.

Gouldner, A. The Norm of Reciprocity. *Am. Sociol. Rev.* **1960,** *25*, 161–178. DOI:10.2307/2092623.Greenline. (1995), Mainstream.

Harter, J. K.; Schmidt, F. L.; Hayes, T. L. Business-unit-level Relationship between Employee Satisfaction, Employee Engagement, and Business Outcomes: A Meta-analysis. *J. Appl. Psychol.* **2002,** *87,* 268–279.

Jurkiewicz, C. L.; Giacalone, R. A. A Values Framework for Measuring the Impact of Workplace Spirituality on Organizational Performance. *J. Bus. Ethics* **2004,** *49*(2), 129–142.

Kahn, W. A. Psychological Conditions of Personal Engagement and Disengagement at Work. *Acad. Manage. J.* **1990,** *33*(4), 692.

Karakas, F. Spirituality and Performance in Organizations: A Literature Review. *J. Bus. Ethics* **2010,** *94*(1), 89–106.

Kim, W.; Kolb, J. A.; Kim, T. The Relationship between Work Engagement and Performance: A Review of Empirical Literature and a Proposed Research Agenda. *Hum. Resour. Dev. Rev.* **2012.** DOI:10.1177/1534484312461635.

Kinjerski, V. M.; Skrypnek, B. J. Defining Spirit at Work: Finding Common Ground. *J. Organ. Change Manage.* **2004,** *34*(2), 26–42.

Krishnakumar, S.; Neck, C. P. The "What", "Why" and "How" of Spirituality in the Workplace. *J. Manage. Psychol.* **2002,** *17*(3), 153–164.

Levinson, E. *Developing High Employee Engagement Makes Good Business Sense.* 2007, Retrieved from www.interactionassociates.com/ideas/2007/05/developing_high_employee_engagement_makes_good_business_sense.php (accessed on July 2015).

Luthans, F.; Peterson, S. J. Employee Engagement and Manager Self-efficacy: Implications for Managerial Effectiveness and Development. *J. Manage. Dev.* **2002,** *21*(5), 376–387.

Masson, R. C.; Royal, M. A.; Agnew, T. G.; Fine, S. Leveraging Employee Engagement: The Practical Implications. *Ind. Organ. Psychol.* **2008,** *1,* 56–59.

Macey, W. H.; Schneider, B. The Meaning of Employee Engagement. *Ind. Organ. Psychol.* **2008,** *1*(1), 3–30. DOI:10.1111/j.1754-9434.2007.0002.x.

May, D. R.; Gilson, R. L.; Harter, L. M. The Psychological Conditions of Meaningfulness, Safety and Availability and the Engagement of the Human Spirit at Work. *J. Occup. Organ. Psychol.* **2004,** *77,* 11–37.

McLaughlin, C. Spirituality at Work. *The Bridging Tree.* **1998,** *1,* 11.

Miller, D. W. *God at Work: The History and Promise of the Faith at Work Movement.* Oxford University Press: New York, 2007.

Milliman, J.; Czaplewski, A. J.; Ferguson, J. Workplace Spirituality and Employee Work Attitudes: An Exploratory Empirical Assessment. *J. Organ. Change Manage.* **2003,** *16*(4), 426–447.

Mitroff, I. I.; Denton, E. A. *A Spiritual Audit of Corporate America,* Jossey-Bass: San Francisco, CA, 1999.

Mohan, K. *Spirituality and Well-being: Overview.* In *Integral Psychology*; Mattijs, C., Ed.; Sri Aurobindo Asharam Press: Pondicherry, India, 2001; pp 203–226.

Moore, T. W.; Casper, W. J. An Examination of Proxy Measures of Workplace Spirituality: A Profile Model of Multidimensional Constructs. *J. Leadersh. Organ. Stud.* **2006,** *12*(4), 109–118.

Nunnally, J. C. *Psychometric Theory* (2nd ed.). McGraw-Hill: New York, 1978.

O'Reilly, C.; Chatman, J.; Caldwell, D. People and Organizational Culture: A Profile Comparison Approach to Assessing Person–Organization Fit. *Acad. Manage. J.* **1991,** *34*(3), 487–516.

Osman, M. K. High-performance Work Practices, Work Social Support and Their Effects on Job Embeddedness and Turnover Intentions. *Int. J. Contemp. Hosp. Manage.* **2013,** *25*(6), 903–921.

Pfeffer, J. *Business and Spirit: Management Practices that Sustain Values.* In: *The Handbook of Workplace Spirituality and Organizational Performance*; Giacalone, R. A., Jurkiewicz, C. L., Eds.; ME Sharpe: Armonk, NY, 2003.

Petchsawanga, P.; Duchon, D. *Workplace Spirituality, Mediation and Work Performance.* Management Department, Faculty Publication, University of Nebraska-Lincoln: Lincoln, 2012.

Paloutzian, R. F.; Emmons, R. A.; Keortge, S. G. Spiritual Well-being, Spiritual Intelligence, and Healthy Workplace Policy. In: *Handbook of Workplace Spirituality and Organizational Performance*; Giacolone, R. A., Jurkiewicz, C. L., Eds.; ME Sharpe: New York, 2003.

Posner, B.; Kouzes, J.; Schmidt, W. Shared Values Make a Difference: An Empirical Test of Corporate Culture. *Resour. Manage.* **1985,** *24*(3), 293–309.

Reave, L. Spiritual Values and Practices Related to Leadership Effectiveness. *Leadersh. Q.* **2005,** *16*(5), 655–687.

Saks, A. M. Antecedents and Consequences of Employee Engagement. *J. Manage. Psychol.* **2006,** *21*(7), 600–619. DOI:10.1108/02683940610690169.

Schaufeli, W. B.; Baker, A. B. *UWES-Utrecht Work Engagement Scale: Test Manual.* Unpublished Manuscript, Department of Psychology, Utrecht University: Utrecht, 2003.

Schaufeli, W. B.; Bakker, A. B. Job Demands, Job Resources, and Their Relationship with Burnout and Engagement: A Multi-sample Study. *J. Organ. Behav.* **2004,** *25*(3), 293–315. DOI:10.1002/job.248.

Schaufeli, W. B.; Bakker, A. B. Defining and Measuring Work Engagement: Bringing Clarity to the Concept. In: *Work Engagement: A Handbook of Essential Theory and Research*; Bakker, A. B., Leiter, M. P., Eds.; Psychology Press: New York, 2010.

Schaufeli, W. B.; Salanova, M.; Gnzalez-Roma, V.; Bakker, A. B. The Measurement of Engagement and Burnout: A Two Sample Confirmatory Factor Analytic Approach. *J. Happiness Stud.* **2002,** *3*, 71–92.

Schaufeli, W. B.; Taris, T. W.; Le Blanc, P.; Peeters, M.; Bakker, A. B.; De Jonge, J. Maakt arbeid gezond? Op zoek naar de bevlogen werknemer" ("Does work make happy? In search of the engaged worker"), *De Psycholoog* **2001,** *36*, 422–428.

Shuck, B.; Reio, T. G. The Employee Engagement Landscape and HRD: How Do We Link Theory and Scholarship to Current Practice? *Adv. Dev. Hum. Resour.* **2011,** *13*(4), 419–428. DOI:10.1177/1523422311431153.

Simon, H. A. *Administrative Behavior: A Study of Decision-making Processes in Administrative Organizations.* Free Press: New York, 1997.

Smidts, A.; Pruyn, A.; Van Riel, C. The Impact of Employee Communication and Perceived External Prestige on Organizational Identification. *Acad. Manage. J.* **2001,** *49*(5), 1051–1062.

Soane, E.; Truss, C.; Alfes, K.; Shantz, A.; Rees, C.; Gatenby, M. Development and Application of a New Measure of Employee Engagement: The ISA Engagement Scale. *Hum. Resour. Dev. Int.* **2012,** *15*(5), 529–547. DOI:10.1080/13678868.2012.726542.

Rothbard, N. P. Enriching or Depleting? The Dynamics of Engagement in Work and Family Roles. *Administr. Sci. Q.* **2001,** *46*, 655–684.

Robinson, D.; Perryman, S.; Hayday, S *The Drivers of Employee Engagement.* Institute for Employment Studies: Brighton, 2004.

Rurkkhum, S.; Bartlett, K. R. The Relationship between Employee Engagement and Organizational Citizenship Behaviour in Thailand. *Hum. Resour. Dev. Int.* **2012,** *15*(2), 157–174. DOI:10.1080/13678868.2012.664693.

Wollard, K. K.; Shuck, B. Antecedents to Employee Engagement: A Structured Review of the Literature. *Adv. Dev. Hum. Resour.* **2011,** *13*(4), 429–446. DOI:10.1177/1523422311431220.

PART IV
Work–Life Balance

CHAPTER 11

IS WORK–LIFE BALANCE A SIGNIFICANT FACTOR IN SELECTING A PROFESSION? A STUDY OF WOMEN WORKING IN THE EDUCATION SECTOR IN INDIA

DURGAMOHAN MUSUNURI*

Department of Management Studies, JSS Academy of Technical Education, Noida 201301, Uttar Pradesh, India

E-mail: durgamohan27@gmail.com

CONTENTS

ABSTRACT

India is an important country in the world due to its sheer size in terms of population and geographical territory and also being a key member country of the BRICS (Brazil, Russia, India, China, and South Africa) group. India's education sector with 1.4 million schools and more than 35,000 higher education institutions is considered as one of the largest higher education systems in the world (IBEF, 2015). Indian women have found the education sector as an avenue of employment and the sector employed 3.8% of Indian women in 2012 (Lahiri, 2012). The growth potential of this sector is enormous as the gross enrolment ration in higher education stood at 20.4% only (All India Survey on Higher Education, 2013). Consequently in the decades to come, this sector will remain an attractive employment sector for Indian women.

This chapter focuses in general on the factors that makes education sector attractive for Indian women to find employment and in particular on work–life balance (WLB) as a key factor. WLB describes "the relationship between your work and the commitments in the rest of your life, and how they impact on one another" (Government of South Australia, 2012). The chapter takes into account the insights given by women teachers both at the school and university levels during focus group discussions. Further analysis is carried out to find out whether WLB is a key factor or are there other factors that influence women to take up teaching.

11.1 INTRODUCTION

The quest for the appropriate job or profession that caters to both the professional needs and personal needs of a human being be it a man or woman is an endless exercise. Various factors influence this pursuit that makes it all the more complex and difficult to unravel the key factors. However, in the recent past, a lot of emphasis is being laid on work–life balance (WLB) by employers so that they can attract the right talent and retain them too and by employees too so that they can balance the professional and personal lives. Is WLB a key factor that influences the selection of a profession is a moot question. Social factors, sociocultural factors, environmental factors, apart from WLB have a considerable influence on the selection of the profession and in turn influence other factors. WLB describes "the relationship between your work and the commitments in the rest of your life, and how they impact on one another" (Government of South Australia, 2012).

This chapter focuses in general on the factors that makes education sector attractive for Indian women to find employment and in particular on WLB as a key factor.

11.1.1 INDIA—A MEMBER STATE OF BRICS

India is an important country in the world polity due to its sheer size in terms of population and geographical territory. It is also an important member country of the BRICS (Brazil, Russia, India, China, and South Africa) group. The concept of BRICS group was first propounded by Jim O'Neill of Goldman Sachs, while referring to the investment opportunities in emerging economies.

The BRICS is the acronym for the group of five major emerging economies comprising Brazil, Russia, India, China, and South Africa. It was originally known as BRIC before the inclusion of South Africa in the year 2010. The importance of BRICS countries can be gauged from these facts—they constitute 42% of the world's population; 26% of the world's geographical territory; and 27% of the world's gross domestic product (GDP) (BRICS in Numbers, 2015). BRICS differs from other informal groupings like G7 and G77, as it can be defined as a global forum for new generation. The basis for BRICS influence on the world economy can be easily understood by the fact that the contribution of these countries to the global economic growth has reached 50% in the last decade, resulting in BRICS becoming a leading power in global economic development. The West has now accepted BRICS as a factor of global significance due to their growing potential. The long-term common interests of the member countries are the basis of the BRICS. These common interests for the foreseeable future are as follows:

- reforming the out-of-date global financial and economic structure, which does not consider the increased economic power of BRICS countries and other emerging economies;
- strengthening the principles and standards of international law, discouraging power politics and preventing infringements on the sovereignty of other countries;

- dealing with a wide range of challenges and problems related to economic and social modernization; and
- supporting the complementarily of many sectors of their economies (BRICS-Analysis, 2015).

Every BRICS country has multiple and different attributes making them unique and have a huge growth potential. Brazil is rich in coffee, soya beans, sugarcane, iron ore, and crude oil. Russia has enormous deposits of crude oil, natural gas, and minerals. India is the service provider to the rest of the world with a growing manufacturing base. China is considered as the manufacturer for the rest the world, with its low cost of production. South Africa is the world's largest producer of platinum and chromium and is the 26th largest economy in the world. The above underscores the importance and the complementarity of these five economies and the dependence of the world economy on them.

11.1.2 INDIA—THE URBAN–RURAL CONGLOMERATE

India is home to more than 1.25 billion people, making it the second most populous country in the world, the most populous democracy in the world, with around 32% of population living in cities, that is, nearly 400 million (Rathi, 2014). According to Census 2011, out of India's total population of 1.21 billion the total female population amounted to 587.5 million against a total male population of 623.1 million

According to the report of National Sample Survey Organization, a Government of India's official body (as quoted by Lahiri, 2012), India had an estimated 112 million female workers in the year 2010. It includes all female workers who described themselves as doing a job for at least 30 of the 365 days. This is equal to a little over one out of three women of working age is working, at least part-time.

Traditionally, women have helped the men in agriculture and allied activities and consequently played a significant role in the Indian Economy. The nine sectors where 90% where Indian women work are agriculture, livestock, textiles and textile products, beverage and tobacco, food products, construction, petty retail trade, education, and research and domestic services (Padma, 2004). These nine sectors also appear as avenues of employment as per the report of the National Sample Survey Organization. Some of the important sectors are the following:

Agriculture: Agriculture is the biggest employer for women. Around 68.5% of women work in farming. A significant feature is that the percentage of women employed in farming has come down by 4.8% between 2005 and 2010.

Manufacturing sector: Around 10.8% of Indian working women are in manufacturing sector, with tobacco and clothing amounting to 4.9%.

Construction: The third largest employer is construction sector, where 5.1% of the working women find employment.

Education: Education sector employs around 3.8% of women that is almost 2.5 million women.

It is foregone conclusion that agriculture and rural areas are synonymous to each other and the rural women folk, who are engaged in various farming and allied activities, form the backbone of the working women in India.

However, with rising urbanization and the migration of rural folk to urban areas in search of employment on one side and on the other the policy initiatives of the Government of India and the state governments in the domain of education, the literacy among women has been increasing since 1947, that is, since India's independence. Consequently more avenues of employment opened for women and the also the growth of the services sector, which contributes around 52% of the GDP of India, which amounts to US$ 783 billion in 2014–15 and growing at 9% CAGR (constant prices) compared to overall GDP growth rate of 6.2% in the last 4 years (http://www.ibef.org/industry/services.aspx, accessed on 25 October 2015). The Government of India recognizes the importance of services sector and provides incentives to sectors like health care, tourism, education, engineering, communication, transportation, information technology, banking and finance, etc.

Further, the Government of India has taken a number of initiatives to help the services sector grow more rapidly. Some of these initiatives are

- The Central Government is considering a two-rate structure for the goods and service tax, under which key services will be taxed at a lower rate compared to the standard rate, which will help to minimize the impact on consumers due to increase in service tax.
- By December 2016, the Government of India plans to take mobile network to nearly 10% of Indian villages that are still unconnected.
- The Government of India has proposed to provide tax benefits for transactions made electronically through credit/debit cards, mobile

wallets, net banking, and other means, as part of broader strategy to reduce use of cash and thereby constrain the parallel economy operating outside legitimate financial system.

- The Reserve Bank of India has allowed third-party white label automated teller machines (ATMs) to accept international cards, including international prepaid cards, and has also allowed white label ATMs to tie up with any commercial bank for cash supply (http://www.ibef.org/industry/services.aspx, accessed on 25 October 2015).
- It can be easily understood that services sector will continue to maintain the growth rate if not increase it in the next 3–5 years time, thereby maintain its lead over other sectors.

11.1.3 THE INDIAN EDUCATION SECTOR

Since India became independent on 15 August 1947 and the sovereign republic in 1950. Consequent upon India becoming independent, the Indian Constitution became the basis for all legislations and policy initiatives of the Government of India and various state governments. The provisions of the Constitution that concern education are as follows:

- Provision of free and compulsory education to all children up to the age of 14 years.
- Education is the concurrent responsibility of the Union and the States. However, the coordination and determination of standards of higher and technical education and institutions declared by Parliament to be institutions of national importance are the responsibility of the Union.

Article 21A of the Constitution of India inserted through the Constitution (86th Amendment) Act, 2002 guarantees to provide free and compulsory education to all children in the age group of 6–14 years as a Fundamental Right. The Right of children to Free and Compulsory Education (RTE) Act, 2009 is the legislation envisaged under Article 21A. The RTE Act came into effect on 1 April 2010.

The Ministry that is responsible for education is the Ministry of Human Resource Development (MHRD) of the Government of India is headed by a cabinet minister and has two departments functioning under it.

1. Department of School Education and Literacy
2. Department of Higher Education.

The National Policy on Education (NPE), 1986, as modified in 1992 and the passing of the All India Council for Technical Education Act, 1987 are important milestones in the development education sector in India. The NPE talks about the essence and role of education as a fundamental reiterated the commitment of the Government of India to education and states that "Education will be used as an agent of basic change of status of women. In order to neutralize the accumulated distortions of the past, there will be a well-conceived edge of women. This will be an act of faith and social engineering." This got translated into various schemes and policy initiatives to increase the literacy rate, amongst which the setting up a "Sarva Shiksha Abhiyan" (SSA) is an important milestone.

SSA is the flagship program for achievement of Universalization of Elementary Education in a time bound manner. It is being implemented in partnership with all the State Governments to take care of the needs of 192 million children in 1.1 million human settlements. Under this program, new schools will be opened and at the same time strengthen the physical infrastructure of the existing schools by providing additional class rooms, toilets, drinking water, etc. as well as augmenting the human resources of the existing schools with additional teachers, enhance the capacity of the existing teachers by training them and providing teaching–learning resources. The special focus of SSA is to educate the girls and children with special needs and bridge the digital divide. This program itself will create multitude of employment opportunities in the education sector (source: http://ssa.nic.in/, accessed on 20 July 2015).

The India's education sector with 1.4 million schools and more than 35,000 higher education institutions is considered as one of the largest higher education systems in the world (IBEF, 2015). Indian women have found the education sector as an avenue of employment and the sector employed 3.8% of Indian women in 2012 (Lahiri, 2012). The growth potential of this sector is enormous as the gross enrolment ration in higher education stood at 20.4% only (All India Survey on Higher Education, 2013). Consequently in the coming years and decades, this sector will remain an attractive employment sector for Indian women.

The Indian education sector will be US $110 billion by the financial year 2015, which grew at a CAGR of 16.5% between financial years 2005 and 2012. The higher education segment stood at 34.04% of the total size in FY2010 and grew by a CAGR of 18.13% between the financial years 2004–2010. The market size in financial year 2010 was US$ 17.02 billion.

The entry of nongovernmental organizations that are also not-for-profit organizations especially in the field of higher education—technical and management education has not only increased the number of seats on offer through multiplicity of educational institutions, but also at the same time resulted in severe competition among these institutions to attract prospective students. These private institutions of higher education are run on a self-supporting basis—they have to generate resources in order to fund the education programs being offered. A major source of funds is the tuition fees charged by these institutions. This is another reason for the competition. Another development that has an impact on the higher education sector in India is the advent of private universities in India. There are 174 universities that are competent award degrees as stipulated by UGC Act with the approval of statutory councils.

11.2 FACTORS THAT INFLUENCE THE SELECTION OF A PROFESSION

Önder et al. (2014) described choosing a career as a multicriteria decision making problem and a crucial decision in one's life. While studying the factors responsible for choosing nursing as a profession from the point of view of students who chose nursing and their parents found that different factors mentioned by parents and students. For parents, "academic staff," "want nursing profession," and "job guarantee," and for the students "the security of nursing school," "income of nursing profession," and "developing profession" were the important factors. The study was carried out in Turkey.

Malaysian school students were investigated with regard to the relative influence of parents and peers in choosing accountancy as a career. It was found that mothers are more influential than peers in career selection as accountants among students of secondary schools. Gender differences were also observed. There is a greater influence of family on girl students compared to boys to choose accountancy as a profession (Hashim & Embog, 2015). In another study carried out in Malaysia by Chen and Liew (2015) on "Factors influencing career decision-making difficulties among graduating students from Malaysian Private Higher Educational

Institutions" considered parental authority and personality as factors that influenced career decision-making difficulties (CCMDs) found that parental authority has positive relationship with CCMDs, whereas personality has a negative relationship. Children do not view parental authority passively in the Asian context and parents do have a strong influence on their ward's career decision making (CDM). The authoritarian parenting style is responsible for the significant influence of parental authority and conscientiousness trait is responsible for the significant influence of personality.

Factors that influence career choice of adolescents can be categorized into factors in the external environment of the adolescent and factors that are internal to the individual decision maker. Internal factors are unique to each individual and are rooted in individual personality. Too many students in Ireland initially make the wrong choice as per the Higher Education Authority. Students dropping out of courses in the first year is 9% for level eight programs and is 22% in case of technology courses. In level 6 and level 7 programs, the rate of drop out is 25% and 26% sometimes rising to 33% also. This change of course choice is a traumatic and costly experience for the students and their families. The education system in Ireland is based on quantitative system of course achievement—higher the points achieved at the second level, greater are the range of courses available to students. Consequently, students make their choices based on points they have scored rather than choosing courses that are more suited to them (Ryan, 2014). With the changes in the workplace, concepts of long-term, stable jobs and careers need to be revamped to stay in tune with these changes. As a result, CDM is increasingly seen as a continuing part of one's own involvement in the world of work. Theory of Circumscription and Compromise, as quoted by Albion & Fogarty, 2015, indicates that career aspirations are circumscribed based on gender-stereotype notions of what careers are appropriate. This happens from early childhood. This circumscription of career possibilities limits the number of options to be considered, which will lead to a faster decision making and the people deciding are also reasonably confident of their choice. The authors hypothesized that people who adhere to gender stereotypes will be less undecided than those who do not (Albion & Fogarty, 2015).

In a study carried out among Business School students to find out the factors that influenced students to take finance as a career choice revealed that academic background, quantitative aptitude, interest in the subject, influence of family and friends, overall personality development, and monetary in centivization as the factors (Jha et al., 2013). The authors stated that parents and teachers have a significant impact on the career-related decisions of students. Compensation (pay) plays a vital role in making career-related decisions for men.

A study carried out at a large Northwestern US university among business students to find out the factors that influence a business student's selection of a field of study and/or career, factors that encourage, or deter a business student's choice to enter the field of MIS (Management Information Systems) and strategies that a MIS department can employ to increase awareness about MIS-related careers among business students by Scott et al. (2009) came to the conclusion that job scope and career diversity, the technical nature of IS (Information Systems) jobs and exposure to information on the profession as the important factors. Among all these job scope was the most frequently mentioned characteristic that students cited when asked about the factors influencing their career choice.

Clayton et al. (2012) studied the under representation of women entering the information and communication technology (ICT) programs in Australia based on an empirical qualitative study carried out in three schools in Australia along with existing literature on factors influencing girls' career choices. The authors evaluated the program called "Technology Takes You Anywhere" an annual event in Brisbane, Australia attended by school girls from around 30 schools in south-east Queensland, with the objective of informing schoolgirls, their parents, and teachers about studying ICT and career options in the field and their by increase the girls' confidence and interest in an ICT career (Interventions for ICT, 2015). Based on literature, the authors identified the following factors that influence girls' career choices. The social factors are family, peers, media, role models, and stereotypes. Family has an important influence on study and career choice during childhood and adolescence and family expectations contribute toward girls' decisions to enter ICT. In Australia, the resident computer expert at home is usually father or older brother (male) and the least involved is usually mother or younger sister (female). Consequently, the ICT attitude of their parent provides inspiration or despair among females, as strong tech-savvy female ICT role models are absent at home. Peers have a powerful influence on child's beliefs and behavioral choices. The value of education, gender roles, and academic choices are influenced by peers. If peers reinforce traditional gender roles, then males and females are likely to engage in different activities and acquire different competencies. Students are likely to choose subjects that are considered as appropriate for their gender roles, so that they can avoid negative attention of the peers. Television, movies, magazines, and other mass media reinforce cultural expectations of gender stereotypes and influence the perceptions about ICT. Home, school, and media are the places where gender roles and stereotypes are learned. They are the primary reason for rejecting a certain occupation. Structural factors are teachers, the ICT curriculum, the way it is taught, quality, availability, and maintenance of ICT-related resources. The school environment influences the access to ICT, lack of resources and technical support affects the amount of access available, and the skills taught in the classroom. Many students have reported negative opinions of ICT subjects and expressed general dissatisfaction with the ICT curriculum. The influence of structural factors can range from extremely positive to strongly negative.

Esters and Bowen (2005) studied the factors that influenced the career choice behaviors of students who graduated from an urban agricultural education program. Former students indicated that their parents and friends are the individuals most influencing the career choices. Career opportunities, high school educational experiences, and work experiences are the events reported by former students who chose a career in agriculture. Other career interests, lack of interest in agriculture and lack of career opportunities are cited by those students who did not choose a career in agriculture.

In an article titled "What influences your career choice?" by Venable (2015), the author listed the influence factors. These factors are skills and abilities, interest and personality type, life roles, previous experiences, culture, gender, social and economic conditions, and childhood fantasies. A study was carried out by Ferry (2006) to explore the factors influencing rural young adults' selection of specific careers in rural Pennsylvania. The focus group discussions yielded the following themes, namely, interdependence of family, school, and community culture; different social and economic contextual factors; "ideal job;" barriers; and out migration. Young adults learn about and explore careers through the interaction of family, school, and community, which ultimately lead to career choice. The community's economic and social circumstances influenced the perceptions of the youth about appropriate career choices. Youth belonging to more affluent communities appeared to have more family and school support in career exploration resulting in a wider range of career options for consideration.

The findings of various studies discussed above have been tabulated as follows:

Country/Region	Type of students/Career/ Profession	Factors that influenced career decision making
Turkey	Nursing profession	Parents' view: academic staff, want nursing profession and job guarantee Students' view: security of nursing school, income of nursing profession, and developing profession
Malaysia	Accountancy profession	Gender differences
Malaysia	Graduating students	Parental authority; personality
Ireland	Adolescents	Points scored in the grading system
Australia	Adolescents & adults	Gender stereotype notions
India	Business school students & finance as a major	Academic background, quantitative aptitude, interest in the subject, influence of family and friends, overall personality development, and monetary incentives
USA	High school passed out students	Knowledge about careers
Australia	Girls and ICT	Family, peers, media, role models and stereotypes
USA	Students of urban agriculture programs	Parents and friends are the individuals
Rural Pennsylvania, USA	Young adults	Interdependence of family, school, and community culture; different social and economic contextual factors; "ideal job;" barriers; and out migration

ICT, information communication technology.

The major factors that have an influence on career choices are parents, friends, gender differences, gender stereotypes, apart from knowledge about careers. This points out that across various countries of the world the family, friends, and gender come out as common influencing factors.

11.3   WORKING WOMEN IN INDIA—ISSUES, OPPORTUNITIES, AND THREATS

Religion does mold an individual's behavior, prescribes the expected norms, value systems, customs and practices, which in turn form the basis of the culture. The family that plays a central role in family members' lives characterizes Indian culture. Patriarchal system is practiced with the eldest male member of the family being the head of the family.

First, culture of a society plays an important in deciding the roles of men and women. Culture can be explained as something that is acquired and distinct, consists of norms, values, and beliefs and influences behavior. The Indian civilization is more than 3000 years old with Hinduism as a religion being practiced by around 80% of the Indian population. Indian society like many other societies is a patriarchal society, with the eldest male member heading the family. Being hierarchical and also collectivist in nature also characterize the society, where the head of a family takes the decisions concerning the education of the children, marriage, and other decisions. These decisions are invariably accepted by the other members of the family, due to the hierarchal nature of Indian society and relatively large power distance. Power distance is the degree to which the less powerful members of a society expect and accept the fact that power is distributed unequally (Hofstede, 2015). The status of women is not the same as of men and women are expected to fulfill their gender specific roles, as prescribed by the society.

Second, a major issue faced by working women in India is that they are expected to raise a family (after marriage), take care of the household chores and at the same time contribute to the household budget by working. This leads to situation where working women need to excel in both personal and professional lives. They have the twin loads as compared to a single load for men. Another major issue is the attitudinal bias that Indian women face in certain profession like police, judiciary, armed forces, etc., and politics, which still remain a bastion of men and consequent under representation of women in the Indian Parliament and state legislatures. Yet, another major issue is the tradition-rich Indian culture, which exerts pressure on even successful professional women and interferes in their career path and career growth, by putting pressure to conform to the tradition roles a women is expected to discharge.

Apart from the above major issues another issue that plagues the Indian women is the very low level of education among women. Thirty-one million girls out of 57 million children all over the world are out of school and two-thirds of illiterate adults are women. School-based violence, early marriage, poverty, pregnancy, and discriminatory gender norms are some of the major obstacles to girls' education. School fees and the perceived benefits of girls' domestic work keep girls out of school (Girls & Women, 2015).

Indian women have found certain of the professions as niche areas and are thriving in these professions. First, they are excelling in jobs entailing administrative work by their integrity of character, efficiency, and perfection. Second, Indian women are performing very well in teaching jobs. They have proved to be best teachers due their love and affection for children. Health care is another area, where a large number of women are employed as doctors and nurses (Singh, 2015).

According to a survey by Gallup, as quoted by Lahiri (2012), traditional expectations of the society lead to college-educated women leaving jobs after marriage or after having children to take care of the family. Therefore, white-collar companies find it hard to recruit or keep female workers. As majority of the female workforce in India have only basic education, they are working in poorly paid jobs with no security or benefits and often below the prescribed minimum wages.

Certain professions have become the strongholds of Indian working women over a period of time. Teaching, healthcare, banking, and IT services are some of the sectors where Indian women find employment. Teaching at school level, college or university level is a preferred profession by women. The reasons for this are not difficult to unravel. Ramachandran (2005) states that women chose teaching as profession due to respectability, security, and less work, as they also can manage their home and house. Teaching was also the preferred choice of their parents and husbands. The author further explained that the responses of the women reflect in gender relations in the society.

In an article titled "Teaching vs. home-making: A fine balance?," Vijayasimha I, pointed out that women get into teaching as it is compatible with their roles and responsibilities. Teachers have a notion that work of teaching is especially suited for women as it allows them to perform their socially approved role as homemakers and caregivers and at the same time do a respectable job that has limited demand in terms of time spent outside the home.

Women often get into teaching because it seems to be a career option that is compatible with their other roles and responsibilities. Of course, this is not the only reason why women choose teaching; they cite other reasons such as the possibility of going deeper into a subject, the opportunity to interact with children and the nobility of the teaching profession. Although many women mention their liking for working with children or youngsters as sources of satisfaction in their work, it doesn't seem to be the main reason for choosing a teaching career. A widely held notion among teachers in all school types is that the work of teaching is especially suited for women because it allows them to perform their socially approved role as homemakers and caregivers while also doing a "respectable job" that places relatively limited demands in terms of time spent outside the home. Even without any overt societal or family pressure women feel that teaching gives them the choice of earning an income and the same time take care of the family.

She further quotes Kirk (2008, p. 66). These three ideas—(i) of teaching being a noble profession, (ii) of the work being in accord with women's natural abilities, and (iii) of being in a career that was aligned with family expectations—contain inherent tensions between them and could be understood as a reflection of the tensions between quite different perceptions of women teachers in the given context. "For the woman teacher, they are all three very viable discursive resources from which she constructs her story of becoming a woman teacher."

A study conducted by the Center for Talent Innovation, New York as quoted by Ullas (2013) revealed that the Indian women are on par with many developed countries, except that their rise in the career is stalled. Also, Indian working women get back to work in the shortest possible time as compared to their US, German, and Japanese counterparts. The "daughterly guilt" is a phenomenon found among Indian working women, 80% of whom leave their jobs for the sake of their parents.

The modern Indian woman is working throughout the country, both in urban India and rural Indian at all levels and in all professions at different positions—as politicians, bankers, top bureaucrats, fashion designers, on one hand and on the other as servants, construction workers, or even as beggars. The tradition-rich Indian culture is putting pressure on successful professional women and cuts their career. The male side is only partly responsible for this social pressure. The author opines that the education must precede the active involvement of women in national movements.

Education, freedom, and acceptance by the male side are necessary (Bauer, 2015).

The federal state of Maharashtra of the Indian Union has 297,480 female teachers working in various educational institutions representing 43% of the total teachers employed in schools and junior colleges. Women are willing to work in educational institutions due to flexible job timings, limited working hours and women can go home by evening and be with their families. Teaching is a safer profession and is also a noble profession that gives immense satisfaction. In the higher education sector, more than 40% of teaching staff are women, but women are denied top posts like principals in spite of having the requisite qualifications and eligibility (Joshi, 2014).

From the above, it can be inferred that working women face very many issues some of which are threats too, but at the same time have opportunities too. WLB as a factor plays a key role in choosing a profession by women in general and choosing teaching as a profession in particular.

11.4   WORK–LIFE BALANCE: A CONCEPT

WLB describes "the relationship between your work and the commitments in the rest of your life, and how they impact on one another" (Government of South Australia, 2012). Any working individual faces the challenge of finding a balance between work and daily life. Combining work, family commitments and personal life successfully is important for the well-being of all members in a household. The amount of time a person spends at work is an important element of WLB. Across OECD countries the percentage of men working very long hours is 17% as compared with 7% for women. Moreover, the more people work, the less time they spend on other activities such as personal care or leisure. Women working fewer hours do not result in greater leisure time as time devoted to leisure is roughly the same for both women and men across 20 OECD countries (Work–Life Balance, 2015).

The six components of WLB are as follows:

1. *Self-management*—is the effective utilization of spaces in our lives and the available resources.
2. *Time management*—involves the optimal usage of the time available and the supporting resources.

3. *Stress management*—becoming adept at maintaining tranquility and working toward getting out of pressure-filled situations.
4. *Change management*—involves making periodic and concerted efforts to ensure that the volume and rate of change at work and at home does not overwhelm or defeat you.
5. *Technology management*—ensures technology serves and not abuses.
6. *Leisure management*—acknowledges the importance of rest and relaxation, so that one does not ignore leisure and taking time-off from the work is an important component of the human experience (The Six Components of Work–Life Balance, 2015).

Discussing about WLB of women employees Lakshmi and Gopinath (2013), whose work is based on teaching professionals at the university level, acknowledged that WLB focuses on achievement and enjoyment—that is, woman should be able to have job satisfaction (enjoyment) and also have career growth (achievement) leading to a positive WLB. WLB means the capacity to schedule the hours of professional and personal life so that a peaceful and healthy life can be led. They further quoted that 53% of the women working for 40–45 h/week struggle to achieve WLB. This struggle is a result of the demands of their organizations versus commitments of their homes. They found among their respondents that the average WLB was found to be 73% based on Work–Life Balance Index (WLBI) and falling in the range of 92% and 58%. This shows that there are women who achieved a high WLB and also poor WLB.

Sturges and Guest (2004) studied a sample of graduates in their early years of professional life to explore the associations between WLB, work/nonwork conflict, hours worked and organizational commitment and found that these graduates do seek WLB, but work long hours as they are concerned about success in career. This resulted in the relationship between home and work becoming increasingly unsatisfactory.

In a comparative study of WLB among chartered accountants, doctors, and teachers carried out by Jain (2013), the author defines WLB as people having a measure of control over when, where and how they work. WLB describes the balance between an individual's work and personal life. It includes prioritizing between work (career and ambition) and life (pleasure, leisure, family and spiritual development). Time balance, involvement balance, and satisfaction balance are the key aspects of WLB. Time balance is about the amount of time given to work and non-work roles;

involvement balance is the level of psychological involvement in or commitment to non-work roles; satisfaction balance is the level of satisfaction with work and non-work roles. Working conditions and family support are the two key factors that help create WLB among the professionals studied.

11.5 METHODOLOGY

In social sciences, one of the methods used to get insights into a topic or an issue or get information is focus group interviews. These are "generally anchored in a qualitative paradigm, reflecting a person-centered, holistic perspective for achieving depth understanding of participant's reality" (Hollis et al., 2002). Focus groups are "designed to obtain perceptions on a defined area of interest in a permissive, non-threatening environment" (Krueger, 1994).

This methodology provides an "an insider's view or emic." Further inductive processes are used to analyze and interpret participants' opinions and views. It is particularly useful for obtaining an insight into the participants' knowledge and experiences. This methodology can be used to explore on one hand what people think and on the other to examine how they think and why they think that way (Kitzinger, 1995). Thus, focus group interviews are a valuable means for distinguishing a range of perceptions regarding the factors that makes education sector attractive for Indian women to find employment. Focus group interviews were selected vis-à-vis individual interviews bearing in mind that "the dynamic interaction afforded by focus groups enables the eliciting of a diversity of views from participants and the immediate clarification of issues that affect this diversity" (Hollis et al., 2002).

Thereby, the study used a qualitative method of conducting focus group interviews which provided the researcher with an occasion to explore the stakeholders' perceptions by giving them a forum to express their opinions without the pressure of arriving at a consensus. The population consists of women practicing teaching profession. Nonrandom sampling method (nonprobability sampling method) of convenient sampling method was used to identify the respondents, who were then invited for the focus group interviews. The names and identities of the respondents were kept anonymous so that they can interact openly and without any inhibitions and hindrances.

Each focus group was conducted within the broad focus group study parameters outlined by Krueger (1998) with the help of question that were developed using protocols and strategies expressed by him. Questions were largely open-ended and probes were used to gather richer details about the participants' experiences. Dichotomous questions and questions which required the participants to make noncontextual judgments were avoided (Hollis et al., 2002).

The focus group interviews were conducted at locations convenient to the participants. In order that the interviews remained focused on the topic, the researcher was the moderator, who also facilitated the group interaction. The proceedings were documented as minutes and analyzed later.

The study parameters outlined by Krueger (1998) were used for conducting each focus group with the help of questions. The questions were developed based on the inputs given by him. Open-ended questions were used and wherever necessary probed further to gather more information about the participants' views. Questions that are dichotomous and require making contextual judgment were not used (Hollis et al., 2002).

11.6 DATA ANALYSIS

Thematic analysis was used to analyze the proceedings of the focus group interviews and to identify the common themes. Thematic analysis is a categorizing strategy for qualitative data. The data are reviewed, notes are made start sorting the data into categories. It helps in moving the analysis from a broad reading of data to discovering patterns and developing themes (Thematic Analysis, 2015). The six phases of thematic analysis as enunciated by Braun and Clarke (2006), which are—familiarizing yourself you're your data; generating initial codes; searching for themes; reviewing themes; defining and naming themes; and producing the report, were used to analyze the data collected during the focused group discussions. The findings of the data analysis are as follows.

11.7 FINDINGS

The findings of the focus group interviews are classified under different themes: The themes are as follows:

1. Factors related to profession
2. Societal influences
3. Parents and in-laws' influences
4. Work–life balance issues.

The following table gives an overview of the comments and remarks of the participants (respondents):

Themes	Comments, remarks, and observations
Factors related to profession	Celebration of success of students; professional satisfaction, talking to young and creative minds; joy of reading; analyzing young minds; life-long learning; constant learning; knowledge sharing; handling children; emancipation of thought and creativity; gives lot of respect; work is well-defined; lot of scope for research and writing
Societal influences	Teaching is best for women; safe profession for women; dignified profession; respect for teaching in society; respectable profession; safe and reputed profession for women
Parents' and parents-in-law's influences	Father's influence, father's choice; parents-in-law permitted to teach only
Work–life balance issues	Demands less time compared to other professions; can easily manage family; easy to maintain work–life balance; healthy work–life balance; does not disturb my family life; can manage family with profession; regular timings; no tours; can easily play the roles of teacher, wife, and mother
Personal factors	Internal satisfaction; self-motivation; to remain young and energetic; to be a student while being a teacher; my father was a teacher (role model)

Apart from the above, one aspect that came to the fore was WLB itself as an important factor. The participants in the focused group interviews were of the opinion that WLB is important/very important as working women have to straddle both the worlds. They also expressed the view that WLB is very important that too for a mother and lack of it leads to stress, anger, and frustration:

Work–life balance	Important; mandatory; lack of WLB leads to lack of coordination between personal and professional life; very important; important as a women I have to straddle both the worlds; work is life and life is work; lack of WLB leads to stress, anger, and frustration; very much important; very important for women and that too as a mother

The findings point out that WLB per se and WLB factors are important for the participants followed by other factors and influences. This is inconformity with what the literature has pointed out, wherein the expected role of a woman by the society and the family is important. This further leads to the fact the Indian society being hierarchal in nature and at the same time relationship-oriented the personal life and the time devoted to fulfill the roles in personal life play an important role in maintaining WLB for working Indian women.

The findings of the focus group discussions highlight the fact that WLB is a key to choosing teaching as a profession. Teaching as a profession by its very nature allows people to take care of their household chores and at the same time discharge the responsibilities of a mother, sister, daughter, etc., as perceived by the society. It does have regular timings, no late hours working and above all a well-respected profession. The teaching profession is also tailor made for women, where the attitudes and personality characteristics of a woman match with the expectations of the society, resulting in a win-win situation.

Here, the educational institutions both at the school-level and college/university level should not take it for granted that factors responsible for having a healthy WLB, but enhance and create a conducive atmosphere for the profession to attract more and women.

11.8 CONCLUSION

The study pointed out that in India, an emerging economy, and part of the group of BRICS (Brazil, Russia, India, China, South Africa) countries, WLB is taking center-stage and becoming a key factor in choosing a job or a profession. The teaching profession is no exception to this except that the perception among academics is that teaching profession inherently offers WLB. Companies which want to attract good talent and retain them should delineate the factors in the education sector and replicate them in their respective companies. This will leads to gender diversity on one hand and on the other reduce the employee turnover considerably, as women are not prone to change jobs frequently.

KEYWORDS

- work–life balance
- working women
- education sector
- BRICS

REFERENCES

Albion, M. J.; Fogarty, G. J. *Factors Influencing Career Decision Making in Adolescents and Adults*, 2015 [online]. Available from: www.core.ac.uk/download/pdf/11036623.pdf (accessed on 12 November 2015).

All India Survey on Higher Education, 2013 [online]. Available from: http://mhrd.gov.in/sites/upload_files/mhrd/files/statistics/AISHE2011-12P_1.pdf (accessed on 27 July 2015).

Bauer, V. *India–Women in the Society*, 2015 [online]. Available from: www.hss.de/fileadmin/india/.../India_-_Women_in_the_society_01.pdf (accessed on 11 November 2015).

Braun, V.; Clarke, V. Using Thematic Analysis in Psychology. *Qualitative Res. Psychol.* **2006,** *3*(2), 77–101.

BRICS in Numbers. *BRICS in Numbers*, 2015 [online]. Available from: http://en.brics2015.ru/ (accessed on 31 December 2015).

BRICS-Analysis. *BRICS-Analysis*, 2015 [online]. Available from: http://www.brics.utoronto.ca/analysis/Lukov-Global-Forum.html (accessed on 31 December 2015).

Chen, L. S.; Liew, S. A. Factors Influencing Career Decision-making Difficulties among Graduating Students from Malaysian Private Higher Educational Institutions. In: Proceedings of 8th Asia-Pacific Business Research Conference, 9–10 February 2015 [online]. Available from: www.vuir.vu.edu.au/19360/1/Siriwan_Ghuangpeng.pdf (accessed on 10 December 2015).

Clayton, K.; Beekhuyzen, J.; Nielsen, S. Now I Know What ICT Can Do for Me! *Info Syst. J.* **2012,** *22*, 375–390.

Esters, L. T.; Bowen, B. E. Factors Influencing Carrier Choices of Urban Agricultural Education Students. *J. Agric. Educ.* **2005,** *46*(2), 24–35.

Ferry, N. M. Factors Influencing Career Choices of Adolescents and Young Adults in Rural Pennsylvania. *J. Extens.* **2006,** *44*(3).

Girls & Women. *Girls & Women,* 2015 [online]. Available from: http://www.right-to-education.org/issue-page/marginalised-groups/girls-women (Accessed on 10 November 2015).

Government of South Australia. *What is Work Life Balance?,* 2012 [online]. Available from: http://www.safework.sa.gov.au/worklifebalance/wlb_show_page.jsp?id=111580 (accessed on 27 July 2015).

Hashim, H. M.; Embog, A. M. Parental and Peer Influences upon Accounting as a Subject and Accountancy as a Career. *J. Econ., Bus. Manage.* **2015,** *3*(2), 252–256.

Hofstede, G. *Dimensions,* 2015 [online]. Available from: http://geert-hofstede.com/national-culture.html (accessed on 5 November 2015].

Hollis, V.; Openshaw, S.; Goble, R. Conducting Focus Groups: Purpose and Practicalities. *Br. J. Occup. Ther.* **2002,** *65*(1), 2–8.

IBEF, 2015 [online]. Available from: http://www.ibef.org/industry/education-sector-india.aspx (accessed on 27 July 2015).

Interventions for ICT, 2015 [online]. Available from: http://www.publications.awe.asn.au/interventions-for-ict (accessed on 15 November 2015).

Jain, P. A Comparative Study of Work Life Balance among CA, Doctors and Teachers. *Int. J. Multidiscip. Res. Soc. Manage. Sci.* **2013,** *1*(4), 58–65.

Jha, R. R.; Priyadarshini, C.; Ponnam, A.; Ganguli, S. Factors Influencing Finance as a Career Choice among Business School Students in India: A Qualitative Study. *IUP J. Soft Skills* **2013,** *7*(4), 48–56.

Joshi, P. *Higher Education Sector Continues to Ignore Women for Top Posts,* 2014 [online]. Available from: http://archive.indianexpress.com/news/higher-education-sector-continues–to-ignore-women-for-top-posts/1214075/ (accessed on 27 July 2015).

Kitzinger, J. Introducing Focus Groups. *BMJ* **1995,** *311*, 299–302.

Kirk, J., Ed. *Women Teaching in South Asia*; Sage, 2008; pp 3–33, 56–86.

Krueger, R. *Focus Groups: A Practical Guide for Applied Research,* 2nd ed. Sage Publication: Thousand Oaks, CA, 1994.

Krueger, R. A. *Developing Questions for Focus Groups.* SAGE: Thousand Oaks, CA, 1998.

Lahiri, T. *By the Numbers: Where Indian Women Work,* 2012 [online]. Available from: http://blogs.wsj.com/indiarealtime/2012/11/14/by-the-numbers-where-indian-women-work/ (accessed on 27 July 2015).

Lakshmi, K. S.; Gopinath, S. S. Work Life Balance of Women Employees—With Reference to Teaching Faculties, *ABHINAV,* **2013,** *II,* 53–62.

Önder, E.; Önder, G.; Kuvat, Ö.; Taş, N. Identifying the Importance Level of Factors Influencing the Selection of Nursing as a Career Choice Using AHP: Survey to Compare the Precedence of Private Vocational High School, Nursing Students and Their Parents. *Proc.—Soc. Behav. Sci.* **2014,** *122*, 398–404.

Padma. *Women Workers in India in the 21st Century—Unemployment and Under Employment,* 2004 [online]. Available from: http://www.cpiml.org/liberation/year_2004/febraury/WomenWorkers.htm (accessed on 27 July 2015).

Ramachandran, V.; Pal, M.; Jain, S.; Shekar, S.; Sharma, J. *Teacher Motivation in India,* 2005 [online]. Available from: www.teindia.nic.in/efa/Vimla.../TeacherMotivation_inIndia_2008.pdf (accessed on 10 November 2015).

Rathi, A. *India's Urban Work Boom Is Leaving Women Behind*, 2014 [online]. Available from: http://www.thehindu.com/news/national/indias-urban-work-boom-is-leaving-women-behind/article5681042.ece (accessed on 27 July 2015).

Ryan, C. H. Factors Influencing Adolescent Career Choice with Particular Emphasis on the Role of Personality, The Future of Education International Conference, 2014.

Singh, R. *Status of Women in Indian Society*, 2015 [online]. Available from: https://www.bu.edu/wcp/Papers/Huma/HumaSing.htm (accessed on 10 November 2015).

Scott, C; Fuller, M. A.; MacIndoe, K. M.; Joshi, K. D. More than a Bumper Sticker: The Factors Influencing Information Systems Career Choices. *Commun. Assoc. Inf. Syst.* **2009**, *24*, 7–26.

Sturges, J.; Guest, D. Working to Live or Living to Work? Work/Life Balance Early in the Career. *Hum. Resour. Manage. J.* **2004**, *14*(4), 5–20.

Thematic Analysis. *Thematic Analysis*, 2015 [online]. Available from: http://isites.harvard.edu/icb/icb.do?keyword=qualitative&pageid=icb.page340897 (accessed on 25 November 2015).

The Six Components of Work–Life Balance. *The Six Components of Work–Life Balance*, 2015 [online]. Available from: http://work-lifebalance.com/ (accessed on 27 July 2015).

Ullas, S. S. *Three Cheers to Indian Working Women*, 2013 [online]. Available from: http://timesofindia.indiatimes.com/city/bengaluru/Three-cheers-to-Indian-working-women/articleshow/19391221.cms (accessed on 27 July 2015).

Venable, M. *What Influences Your Career Choice?*, 2015 [online]. Available from: http://www.onlinecollege.org/2011/05/17/what-influences-your-career-choice/ (accessed on 27 November 2015).

Vijaysimha, I. *Teaching vs. Home-making: A Fine Balance?* 2012 [online]. Available from: http://www.teacherplus.org/cover-story/teaching-vs-home-making-a-fine-balance [accessed on 27 July 2015].

Work–Life Balance. *Work–Life Balance*, 2015 [online]. Available from: http://www.oecd-betterlifeindex.org/topics/work-life-balance/ (accessed on 27 July 2015).

PART V
Managing Stress

CHAPTER 12

MINDFULNESS: A CONCEPT AND ITS ROLE IN THE WORKPLACE

ARVIND K. BIRDIE* and KULDEEP KUMAR

Department of Management, IIMT School of Management, Vedatya Institute, Gurgaon, India

Corresponding author. E-mail: arvindgagan@gmail.com

CONTENTS

ABSTRACT

The present workplace and world is blooming complicated and stressful day by day. The present chapter reviews the role of mindfulness in workplace. Mindfulness is gaining a growing popularity as a practice in daily life, apart from Buddhist insight meditation and its application in psychology. Not that long ago, meditation was seen widely as the preserve of hippies and saffron-clad monks, unsuited for the business world. Nowadays, a growing number of businesses are recognizing what mindfulness has to offer. Being mindful makes it easier to savor the pleasures in life as they occur, helps employees become fully engaged in activities, and creates a greater capacity to deal with adverse events. By focusing on the here and now, many people who practice mindfulness find that they are less likely to get caught up in worries about the future or regrets over the past, are less preoccupied with concerns about success and self-esteem, and are better able to form deep connections with others.

12.1 INTRODUCTION

In today's world of work, employees and leaders face multiple stress-inducing demands and pressures as well as constant connectivity through smart phones, social media, and tablet computers. People during work manage fluctuating priorities, working with increased expectations, balancing competing demands for our personal and professional goals, and handling ongoing conflict and ambiguity in complex environments. Consulting firm AON Hewitt estimates that over 35% of US employers in 2013 offered stress reduction programs to their employees and that estimate is expected to grow (AON Hewitt, 2013). HR and talent management professionals increasingly looking for ways to reduce employee stress and many employers like Google, General mills have found that introducing mindfulness into their workplace not only lowers employee stress, but also improve focus, clarity of thinking, decision making, emotional intelligence and more.

12.1.1 WHAT IS MINDFULNESS?

> …inhabiting this moment, our only moment, with greater awareness shapes the moment that follows, and if we can sustain it, actually shapes the future and the quality of our lives and relationships in ways we often simply do not appreciate…. (Kabat-Zinn, 1990)

Mindfulness has its origin that go back 2500 years and uses an anchor—often breathing—to center attention and to bring awareness to the present moment. According to Frances Weaver in The Week, "the goal of mindfulness practice is to "quiet the mind's constant chattering-thoughts, anxieties and regrets." Mindfulness practitioners learn to focus on the present in everything they do and to accept events in the present moment (Weaver, 2014).

Psychologist Ela Amarie of the Switzerland based consultancy Mindful Brain observes that there are three characteristics of mindfulness: intention, attention, and attitude. The goal of mindfulness is to recognize and accept inner thoughts and feelings.

Mindfulness is a quality of consciousness, more specifically defined as "paying attention in a particular way: on purpose, in the present moment, nonjudgmentally" (Kabat-Zinn, 1994).

> Mindfulness is a state of conscious awareness in which the individual is implicitly aware of the context and content of information. It is a state of openness to novelty in which the individual actively constructs categories and distinctions. In contrast, mindlessness is a state of mind characterized by an overreliance on categories and distinctions drawn in the past and in which the individual is context-dependent and, as such, is oblivious to novel (or simply alternative) aspects of the situation. (Langer, 1992: 289)

Mindfulness entails self-regulation of attention so that attention is concentrated on the present (Bishop et al., 2004). One's attention remains focused on the "unfolding of experience moment by moment" (Kabat-Zinn, 2003). Thoughts, feelings, and bodily sensations are considered to be "objects of observation" (Bishop et al., 2004) but not something on which one should elaborate (i.e., direct attention toward thinking about the thought, feeling, or sensation). Such elaboration would take one out of the present moment, and thus, distract focus from the current experience. It would also require use of resources that could be devoted to attention and present-moment awareness. In addition, elaboration often involves judgment (e.g., this is a "good" event or this is a "bad" experience because of how it is making me think or feel). Mindful awareness is fostered by acting as an "impartial witness" (Kabat-Zinn, 1990) to one's own experience. This means stepping back from one's tendency to categorize and judge one's experiences, a practice which "locks us into mechanical reactions" (Kabat-Zinn, 1990) of which we may not even be aware. The nonjudgmental quality of mindfulness leads to equanimity, as emotional disturbance often comes from our interpretation of the event rather than the event itself. Mindfulness also encourages one to realize that the thoughts, feelings, and sensations that one observes are simply experiences in the mind or body, and not something that one should "over-identify" with (e.g., a thought is a thought, but you are not your thought). Mindfulness also involves one's orientation to experience (Bishop et al., 2004). Mindfulness encourages approaching one's experiences with a "beginner's mind" (Kabat-Zinn, 1990), as if experiencing the event for the first time. With such an approach, one brings to their experience "an orientation that is characterized by curiosity, openness, and acceptance" (Bishop et al., 2004). Acceptance in this sense refers to receptivity to seeing things as they actually are in the present moment (Kabat-Zinn, 1990). Each moment is viewed as unique, and if one brings to the moment preconceived ideas, assumptions, or expectations, one will not be able to experience the moment as it truly is. Fundamental to the idea of being open to and seeing one's experience as it is in reality is the attitude of letting go or nonattachment. One learns to "put aside the tendency to elevate some aspects of our experience and to reject others" (Kabat-Zinn, 1990). This attentiveness to and acceptance of one's full experience allows an individual to respond effectively rather than react habitually to the situation and experience (Bishop et al., 2004; Kabat-Zinn, 1990). The following quote illustrates the philosophy of approaching each experience with a beginner's mind:

> When I say that I 'know' my wife, it is that I have an image about her; but that image is always in the past; that image prevents me from looking at her—she may already be changing.... So the mind must be in a constant state of learning, therefore always in the active present, always fresh; not stale with the accumulated knowledge of yesterday. (Krishnamurti, 1972, p. 8, as cited in McIntosh, 1997)

12.2   CHARACTERISTICS OF MINDLESSNESS

12.2.1   *TRAPPED BY CATEGORIES*

The creation of new categories … is a mindful activity. Mindlessness sets in when we rely too rigidly on categories and distinctions created in the past (masculine/feminine, old/young, success/failure). Once distinctions are created, they take on a life of their own (Langer, 1989).

12.2.2   *AUTOMATIC BEHAVIOR*

Habit, or the tendency to keep on with behavior that has been repeated over time, naturally implies mindlessness. However, …mindless behavior can arise without a long history of repetition, almost instantaneously, in fact. (Langer, 1989: 16)

12.2.3   *ACTING FROM A SINGLE PERSPECTIVE*

So often in our lives, we act as though there were only one set of rules. (Langer, 1989: 16)

To be mindful means to be fully in the "here and now," moment-to-moment. Mindfulness includes awareness of current external stimuli, such as externalevents or objects, as well as of internal processes and states, such as emotions, perceptions, sensations, and cognitions. The observing, witnessing stance of mindfulness is associated with a reduction in mental commentary and judgment (e.g., Weick & Putnam, 2006).

Mindfulness can be contrasted with mindlessness. Being mindless can be defined as neither paying attention to, nor having awareness of, the activities one is engaged in or of the internal states and processes (e.g., emotions) one is experiencing. Modes of being that are characteristic of mindlessness are, for example, performing tasks on autopilot, daydreaming, worrying about the future, or ruminating about the past (Brown & Ryan, 2003).

12.2.4   IS MINDFULNESS A PERSONALITY TRAIT?

Mindfulness might be a personality trait rather than a cognitive ability. It might be useful to consider a well-regarded trait theory of personality and to inquire as to whether mindfulness resembles any of the traits proposed. ... The most popular trait theory today is probably the big-five theory.... Although there are certainly other theories, big-five theory has gained such overwhelming comparative acceptance that I will limit my discussion to this theory alone. Although different investigators sometimes have given the big five different names, they generally have agreed on five key characteristics as a useful way to organize and describe individual differences in personality. The following descriptions represent the five traits:

1. *Neuroticism*—characterized by nervousness, emotional instability, moodiness, tension, irritability, and frequent tendency to worry.
2. *Extraversion*—characterized by sociability, expansiveness, liveliness, an orientation toward having fun, and an interest in other people.
3. *Openness to experience*—characterized by imagination, intelligence, and aesthetic sensitivity, as well as openness to new kinds of experiences.
4. *Agreeableness*—characterized by a pleasant disposition, a charitable nature, empathy toward others, and friendliness.
5. *Conscientiousness*—characterized by reliability, hard work, punctuality, and a concern about doing things right. Mindfulness seems potentially related to openness to experience. There is almost certainly some overlap. Moreover, research suggests that openness to experience itself is correlated with cognitive abilities. So, it would seem potentially fruitful to pursue the relation between the two constructs. Mindfulness also may bear some relation to conscientiousness. Studies are needed that correlate mindfulness with these traits to see if indeed there is a relation (Sternberg, 2000:21).

12.3   MINDFULNESS AND CULTURAL INTELLIGENCE

As a facet of CQ [cultural intelligence], mindfulness (at a highly developed level) means simultaneously:

- Being aware of our own assumptions, ideas, and emotions; and of the selective perception, attribution, and categorization that we and others adopt;
- noticing what is apparent about the other person and tuning in to their assumptions, words, and behavior;
- using all of the senses in perceiving situations, rather than just relying on, for example, hearing the words that the other person speaks;
- viewing the situation from several perspectives, that is, with an open mind;
- attending to the context to help to interpret what is happening;
- creating new mental maps of other peoples' personality and cultural background to assist us to respond appropriately to them;
- creating new categories, and recategorizing others into a more sophisticated category system;
- seeking out fresh information to confirm or disconfirm the mental maps;
- using empathy—the ability to mentally put ourselves in the other person's shoes as a means of understanding the situation and their feelings toward it, from the perspective of their cultural background rather than ours (Langer, 1989; Thomas, 2006: 85).

12.4 EMPIRICAL STUDIES ON MINDFULNESS

Emerging research also suggests that mindfulness affects a person's social relationships. For example, Wachs and Cordova (2007) found that mindfulness increased people's ability to identify and communicate emotional states with their partner as well as regulating their anger expression, which led to an increase in marital quality. Other research suggests that mindfulness increases people's ability to cope with relationship stress (Barnes et al., 2007). These positive effects may be partly due to mindfulness increasing empathic concern toward relationship partners (Block-Lerner et al., 2007). It has also been argued that mindfulness moves people from an adversarial mind-set to a more collaborative mind-set in mixed-motive interactions such as negotiations (Riskin, 2002).

Research suggests that mindfulness is associated with a better ability to function in relationships as it helps people relate to others emotionally

(Wachs & Cordova, 2007), cope with relationship stress (Barnes et al., 2007), and emphasize with relationship partners (Block-Lerner et al., 2007). In addition, research found mindfulness to be related to higher emotional intelligence and self-regulation (e.g., Brown & Ryan, 2003), which implies a better recognition and understanding of others' emotional states as well as a better understanding and regulation of one's own emotions (Arch & Craske, 2006).

Reb et al. (2014) examined the influence of leaders' mindfulness on employees' well-being and performance and found that supervisors' trait mindfulness is positively associated with different facets of employee well-being, such as job satisfaction and need satisfaction, and different dimensions of employee performance, such as in-role performance and OCBs. They also explored whether one measure of employee well-being, psychological need satisfaction, plays a mediating role in the relation between supervisor mindfulness and employee performance. These predictions were tested in two studies using data from both supervisors and their subordinates. Results were consistent with hypotheses. It contributes to understanding of mindfulness by examining its interpersonal effects in a very important domain of human life: the workplace.

Scholars are now asking, how would a mindful approach to work affect one's work life and work outcomes?

Kabat-Zinn (1990) foreshadows the possibilities:

> When you begin to look at work mindfully, whether you work for yourself, for a big institution, or for a little one, whether you work inside a building or outside, whether you love your job or hate it, you are bringing all your inner resources to bear on your working day. In all likelihood, if we saw work as an arena in which we could hone inner strength and wisdom moment by moment, we would make better decisions, communicate more effectively, be more efficient, and perhaps even leave work happier at the end of the day. (pp. 389, 393)

Shapiro and Carlson (2009) have suggested that mindfulness meditation can also serve as a mean of self-care to help combat burnout rates. Future research on not only how therapists' practice of mindfulness meditation helps facilitate trainee development and affects psychotherapy is needed, but the ways in which therapists' own practice of mindfulness meditation can help with burnout rates and other detrimental outcomes of work-related stress.

Narayanan et al. (2011) found that employee trait mindfulness was positively related to task performance, and this relation was partly mediated by the lower emotional exhaustion experienced by more mindful employees.

At this time, little research is available regarding the effects of mindfulness in a work setting. Most studies that do exist continue to demonstrate the positive effect of mindfulness on mental health; in this case, its ability to reduce stress and burnout in the workplace, particularly for health-care professionals (Cohen-Katz et al., 2004, 2005; Galantino et al., 2005; Irving et al., 2009; Klatt et al., 2009; Mackenzie et al., 2006; Pipe et al., 2009). An exploratory qualitative study, however (Hunter & McCormick, 2008), suggests much broader effects. Hunter and McCormick's (2008) results indicate that a mindful approach to work may result in more external awareness at work, more acceptance of one's work situation, increased ability to cope with and remain calm in difficult work situations, increased adaptability, and more positive relationships at work. Anecdotal evidence also hints at the power of mindfulness. For example, National Basketball Association (NBA) coach Phil Jackson, former coach of the championship (1991–1993) Chicago Bulls and current coach of the championship (2000–2002) Los Angeles Lakers, considers the mindfulness sessions he holds for his players to be a competitive secret (Lazenby, 2001) that contributes to his team's success.

Mindfulness has been shown to contribute to the development of empathy (Block-Lerner et al., 2007; Cohen-Katz et al., 2005; Shapiro et al., 1998; Tipsord, 2009) and to be related to positive affectivity (Brown & Ryan, 2003), both of which are positively related to citizenship behavior (Borman et al., 2001; Organ & Ryan, 1995; Settoon & Mossholder, 2002). More proximal processes likely mediate the effect of mindfulness on an individual's performance and citizenship behavior at work. I propose that two intermediary variables, experienced affect and high-quality work relationships, link the independent variable mindfulness and the dependent variables of performance and citizenship behavior. I suggest that mindfulness is associated with one's experienced affect. Mindful individuals tend to be more skilled emotionally, as mindfulness scales have correlated positively with measures of emotion awareness and regulation (i.e., emotional intelligence; Baer et al., 2004, 2006; Brown & Ryan, 2003; Feldman et al., 2007) and negatively with scales measuring difficulty in identifying and describing feelings (Baer et al., 2004, 2006; Wachs& Cordova, 2007).

Because more mindful individuals are expected to be more emotionally aware and more effective at regulating emotions, they should experience more positive affect and less negative affect as compared to less mindful individuals. Such relationship would be consistent with meta-analytic findings that mindfulness has a moderate, positive relationship with trait positive affect and a moderate, negative relationship with trait negative affect (Giluk, 2009). Shiota et al. (2004) argue that positive emotions are critical to the formation and maintenance of social bonds. Those who experience more frequent positive affect are better able to develop social relationships, have more friends, and enjoy a stronger network of support (Lyubomirsky et al., 2005), while the expression of frequent negative affect is likely to negatively impact relationships (Labianca & Brass, 2006).

Mindfulness has been shown to positively affect mental health and psychological well-being, physical health, and quality of intimate relationships (Baer, 2003; Brown & Ryan, 2003; Brown et al., 2007; Grossman et al., 2004). A recent conference paper (Hunter & McCormick, 2008) presented a small exploratory qualitative study in which eight managers and professionals described a broader set of workplace outcomes stemming from their mindfulness practice, including increased external awareness, more acceptance of one's work situation, increased ability to cope and remain calm in difficult work situations, increased adaptability, and more positive relations at work.

12.5   HOW IT BENEFITS BUSINESS AND WORK

12.5.1   EMOTIONAL WELL-BEING

Napoli et al. (2005) reported the results of integrated mindfulness and relaxation work with 225 children with high anxiety, aged between 5 and 8 taking part in the attention academy program in a school context. The intervention constituted 12 sessions of 45 min each. The children showed significant decreases in both test anxiety and ADHD behaviors and also an increase in the ability to pay attention. The study was reasonably strong methodologically, being a randomized control trial with a large sample, and the use of objective measures of attention.

Wall (2005), in a small study, outlined effort program to teach MBSR and Tai Chi in a mainstream school to 11–13-year olds in the United States, which brought perceived benefits such as improved well-being, calmness, relaxation, improved sleep, less reactivity, increased self-care. Self-awareness and a sense of connection with nature.

Schonert-Reichl and Lawlor (2010) investigated a mindfulness-based program, delivered by teachers, involving 10 lessons and three times daily practice of mindfulness meditation. Overall, there was a significant increase in scores on self-report measures of optimism and positive emotions. Teacher reports showed an improvement in social and emotional competence for children in the intervention group and a decrease in aggression and oppositional behavior.

12.6 CONCLUSION

Mindfulness in the workplace is a relatively new area of mindfulness application and research. Nevertheless, initial studies have documented beneficial effects of mindfulness training on conditions related to work stress. Present workplace is a chaotic place to be and definitely if mindfulness enhances focus and attention, increases self-awareness and the awareness of others raises levels of resilience and emotional intelligence, and strengthens cognitive effectiveness it is need of the hour for leaders and executives.

It has been shown that mindfulness training:

- Decreases perceived stress, improves sleep quality, and the heart rhythm coherence ratio of heart rate variability in employees (Wolever et al., 2012).
- Reduces perceived stress and increased mindfulness in working adults (Klatt et al., 2009).
- Improves multitasking-related problems—HR staff showed better memory for tasks, more concentration on a task, and less switching between tasks (Levy et al., 2011).
- Mindfulness has also been shown to contribute directly to the development of cognitive and performance skills in the young. They often become more focused, more able to approach situations from a fresh perspective, use existing knowledge more effectively and pay attention. Mindfulness also helps nurture imagination and improves mental health, according to Manfred Ke DE Vries, INSEAD distinguished professor of leadership and development (Williams, 2014).

ACKNOWLEDGMENTS

All of the authors named in this chapter have made equal contributions to the development of the ideas presented. Consider it a joint effort and want to thank all authors to share in equal credit for this chapter.

ADDITIONAL READING

Spencer-Oatey, H. Mindfulness for Intercultural Interaction. A Compilation of Quotations. *Global PAD Core Concepts*, 2013. Available at Global PAD Open House http://go.warwick.ac.uk/globalpadintercultural.

Davis, D. M.; Hayes, J. A. What are the Benefits of Mindfulness? A Practice Review of Psychotherapy Related Research. *Psychotherapy* **2011,** *48*(2), 198–208.©2011 *Am. Psychol. Assoc.* **2011,** *48*(2), 198–208, 0033-3204/11/$12.00. DOI:10.1037/a0022062; http://www.apa.org/pubs/journals/features/pst-48-2-198.pdf.

KEYWORDS

- **mindfulness**
- **workplace**
- **Buddhist**
- **meditation**
- **behavior**

REFERENCES

AON Hewitt Staff. *AON Hewitt 2013 Health Care Survey*, 2013. AON Hewitt. Retrieved from http://www.aon.com/attachments/human-capital-consulting/2013 Health Care Survey.pdf.

Arch, J. J.; Craske, M. G. Mechanisms of Mindfulness: Emotion Regulation Following a Focused. *Behav Res Ther.* **2006,** *44*(12), 1849–1858.

Baer, R. A. Mindfulness Training as a Clinical Intervention: A Conceptual and Empirical Review. *Clinical Psychol.: Sci. Pract.* **2003,** *10*, 125–143.

Baer, R. A.; Smith, G. T.; Allen, K. B. Assessment of Mindfulness by Self-report: The Kentucky Inventory of Mindfulness Skills. *Assessment* **2004,** *11*, 191–206.

Baer, R. A.; Smith, G. T.; Hopkins, J.; Krietemeyer, J.; Toney, L. Using Self-report Assessment Methods to Explore Facets of Mindfulness. *Assessment* **2006**, *13*, 27–45.

Barnes, S.; Brown, K. W.; Krusemark, E.; Campbell, W. K.; Rogge, R. D. The Role of Mindfulness in Romantic Relationship Satisfaction and Responses to Relationship Stress. *J. Mar. Fam. Ther.* **2007**, *33*(4), 482–500.

Bishop, S. R.; Lau, M.; Shapiro, S.; Carlson, L.; Anderson, N. D.; Carmody, J.; Segal, Z. V.; Block-Lerner, J.; Adair, C.; Plumb, J. C.; Rhatigan, D. L.; Orsillo, S. M. The Case for Mindfulness-based Approaches in the Cultivation of Empathy: Does Nonjudgmental, Present-moment Awareness Increase Capacity for Perspective-taking and Empathic Concern. *J. Mar. Fam. Ther.* **2007**, *33*(4), 501–516.

Borman, W. C.; Penner, L. A.; Allen, T. D.; Motowidlo, S. J. Personality Predictors of Citizenship Performance. *Int. J. Select. Assess.* **2001**, *9*, 52–69.

Brown, K. W.; Ryan, R. M.; Creswell, J. D. Mindfulness: Theoretical Foundations and Evidence for Its Salutary Effects. *Psychol. Inq.* **2007**, *18*, 211–237.

Brown, K. W.; Ryan, R. M. The Benefits of being Present: Mindfulness and its Role in Psychological Well-being. *J. Pers. Soc. Psychol.* **2003**, *84*(4), 822–848.

Cohen-Katz, J.; Wiley, S.; Capuano, T.; Baker, D. M.; Deitrick, L.; Shapiro, S. The Effects of Mindfulness-based Stress Reduction on Nurse Stress and Burnout: A Qualitative and Quantitative Study, part III. *Holist. Nurs. Pract.* **2005**, *19*(2), 78–86.

Cohen-Katz, J.; Wiley, S.; Capuano, T.; Baker, D. M.; Shapiro, S. The Effects of Mindfulness-based Stress Reduction on Nurse Stress and Burnout: A Qualitative and Quantitative Study. *Holist. Nurs. Pract.* **2004**, *18*(6), 302–308.

Feldman, G.; Hayes, A.; Kumar, S.; Greeson, J.; Laurenceau, J. Mindfulness and Emotion Regulation: The Development and Initial Validation of the Cognitive and Affective Mindfulness Scale-Revised (CAMS-R). *J. Psychopathol. Behav. Assess.* **2007**, *29*, 177–190.

Galantino, M. L.; Baime, M.; Maguire, M.; Szapary, P. O.; Farrar, J. T. Association of Psychological and Physiological Measures of Stress in Health-care Professionals During an 8-week Mindfulness Meditation Program: Mindfulness in Practice. *Stress Health* **2005**, *21*, 255–261.

Giluk, T. L. Mindfulness, Big Five Personality, and Affect: A Meta-analysis. *Pers. Individ. Diff.* **2009**, *47*, 805–811.

Grossman, P.; Niemann, L.; Schmidt, S.; Walach, H. Mindfulness-based Stress Reduction and Health Benefits: A Meta-analysis. *J. Psychosomat. Res.* **2004**, *57*, 35–43.

Hunter, J.; McCormick, D. W. Mindfulness in the Workplace: An Exploratory Study. In: *Weickian Ideas*; Newell, S. E.; Ed. (Facilitator); Symposium Conducted at the Annual Meeting of the Academy of Management, Anaheim, CA, 2008.

Irving, J. A.; Dobkin, P. L.; Park, J. Cultivating Mindfulness in Health Care Professionals: A Review of Empirical Studies of Mindfulness-based Stress Reduction (MBSR). *Complement. Ther. Clin. Pract.* **2009**, *15*, 61–66.

Kabat-Zinn, J. *Wherever You Go, there You Are: Mindfulness Meditation in Everyday Life.* Hyperion: New York, 1994.

Kabat-Zinn, J. *Full Catastrophe Living: Using the Wisdom of Your Body and Mind to Face Stress, Pain and Illness.* Delacorte: New York, 1990.

Kabat-Zinn, J. Mindfulness-based Interventions in Context: Past, Present, and Future. *Clin. Psychol.: Sci. Pract.* **2003**, *10*, 144–156.

Klatt, M. D.; Buckworth, J.; Malarkey, W. B. Effects of Low-dose Mindfulness-based Stress Reduction (MBSR-ld) on Working Adults. *Health Educ. Behav.* **2009,** *36,* 601–614.

Krishnamurti, J. *You Are the World.* Harper Collins: New York, 1972.

Labianca, G.; Brass, D. J. Exploring the Social Ledger: Negative Relationships and Negative Asymmetry in Social Networks in Organizations. *Acad. Manage. Rev.* **2006,** *31,* 596–614.

Langer, E. J. *Mindfulness.* Perseus Books: Cambridge, MA, 1989.

Langer, E. J. Matters of Mind: Mindfulness/Mindlessness in Perspective. *Consciousn. Cogn.* **1992,** *1,* 289–305.

Lazenby, R. *Mindgames: Phil Jackson's Long Strange Journey.* Contemporary Books: Chicago, IL, 2001.

Levy, D. M.; Wobbrock, J. O.; Kasznaik, A. W.; Ostergen, M. Initial Results from a Study of the Effects of Meditation on Multitasking Performance. In: Proceedings of the 2011 Annual Conference Extended Abstracts on Human Factors in Computing Systems, 2011, 2011–2016.

Lyubomirsky, S.; King, L.; Diener, E. The Benefits of Frequent Positive Affect: Does Happiness lead to Success? *Psychol. Bull.* **2005,** *131,* 803–855.

Mackenzie, C. S.; Poulin, P. A.; Seidman-Carlson, R. A Brief Mindfulness-based Stress Reduction Intervention for Nurses and Nurse Aides. *Appl. Nurs. Res.* **2006,** *19,* 105–109.

Napoli, M.; Krech, P. R.; Holley, L. C. Mindfulness Training for Elementary School Students. *J. Appl. School Psychol.* **2005,** *21*(1), 99–125.

Narayanan, J.; Chaturvedi, S.; Reb, J.; Srinivas, E. Examining the Role of Trait Mindfulness on Turnover Intentions and Job Performance: The Mediating Role of Emotional Exhaustion. Working Paper, National University of Singapore, 2011.

Organ, D. W.; Ryan, K. A Meta-analytic Review of Attitudinal and Dispositional Predictors of Organizational Citizenship Behavior. *Pers. Psychol.* **1995,** *48,* 775–802.

Pipe, T. B.; Bortz, J. J.; Dueck, A. Nurse Leader Mindfulness Meditation Program for Stress Management: A Randomized Controlled Trial. *J. Nurs. Adm.* **2009,** *39,* 130–137.

Reb, J.; Narayanan, J.; Chaturvedi, S. Leading Mindfully: Two Studies on the Influence of Supervisor Trait Mindfulness on Employee Well-being and Performance. *Mindfulness* **2014,** *5*(1), 36–45. DOI: 10.1007/s12671-012-0144-z.

Riskin, L. The Contemplative Lawyer: On the Potential Contributions of Mindfulness Meditation to Law Students, Lawyers and their Clients. *Harv. Negot. Law Rev.* **2002,** *7,* 1–67.

Schonert-Reichl, K. A.; Lawlor, M. S. The Effects of a Mindfulness-based Education Program on Pre-and Early Adolescents' Well Being and Social and Emotional Competence. *Mindfulness* **2010,** *1*(3), 137–151.

Settoon, R. P.; Mossholder, K. W. Relationship Quality and Relationship Context as Antecedents of Person- and Task-focused Interpersonal Citizenship Behavior. *J. Appl. Psychol.* **2002,** *87,* 255–267.

Shapiro, S. L.; Carlson, L. E. The Art and Science of Mindfulness: Integrating Mindfulness into Psychology and the Helping Professions. American Psychological Association: Washington, DC, 2009.

Shapiro, S. L.; Schwartz, G. E.; Bonner, G. Effects of Mindfulness-based Stress Reduction on Medical and Premedical Students. *J. Bhav. Med.* **1998,** *21,* 581–599.

Shiota, M. N.; Campos, B.; Keltner, D. Hertenstein, M. J. Positive Emotion and the Regulation of Interpersonal Relationships. In: *The Regulation of Emotion*; Philippot, P., Feldman, R. S.; Eds.; Lawrence Erlbaum: Mahwah, NJ, 2004; pp 127–155.

Sternberg, R. J. Images of Mindfulness. *J. Soc. Sci. Issues* **2000**, *56*(1), 11–26.

Thomas, D. Domain and Development of Cultural Intelligence. The Importance of Mindfulness. *Group Organ. Manage.* **2006**, *31*(1), 78–99.

Tipsord, J. The Effects of Mindfulness Training and Individual Differences in Mindfulness on Social Perception and Empathy. Doctoral Dissertation, University of Oregon. Dissertation Abstracts International-B, 70/11, 2009.

Wachs, K.; Cordova, J. V. Mindful Relating: Exploring Mindfulness and Emotion Repertoires in Intimate Relationships. *J. Mar. Fam. Ther.* **2007**, *33*(4), 464–481.

Wall, R. B. Tai Chi and Mindfulness Based Stress Reduction in a Boston Public Middle School. *J. Pediatr. Health Care* **2005**, *19*(4), 230–237.

Weick, K. E.; Putnam, T. Organizing for Mindfulness: Eastern Wisdom and Western Knowledge. *J. Manage. Inq.* **2006**, *15*, 275–287.

Weaver, F. The Mainstreaming of Mindfulness. The Week. April 2014. Retrieved from http:/the week.com/article/index/the mainstreaming-of-mindfulness-mediation.

Williams, R. Why Reflection and "Doing Nothing" Are Critical for Productivity. Financial Post, August 2014. Retrieved from http://business.financialpost.com/2014/08/15/why-reflection-and-doing-nothing-are critical-for-productivity/?.

Wolever, R.; et al. Effective and Viable Mind Body Stress Reduction in the Workplace: A Randomized Controlled Trial. *J. Occup. Health Psychol.* **2012**, *17*(2), 246–258.

WEB REFERENCES

http://ir.uiowa.edu/cgi/viewcontent.cgi?article=1859&context=etd. Retrieved on 25 August 2015.

http://switchandshift.com/how-mindfulness-improves-workplace-performance. Retrieved on 20 August 2015.

http://www.amanet.org/training/articles/Stress-Management-and-Mindfulness-in-the-Workplace.aspx. Retrieved on 27 August 2015.

http://www.mindfulness-works.com/. Retrieved on 22 August 2015.

www.bangor.ac.uk/mindfulness/work.php.en. Retrieved on 20 August 2015.

CHAPTER 13

STRESS MANAGEMENT: CONCEPT

NAVDEEP KAUR KULAR*

Vedatya Institute, Gurgaon, Haryana, India

*E-mail: novikular@rediffmail.com

CONTENTS

ABSTRACT

The share of services measured in terms of gross domestic product is consistently rising in the emerging economies. The rapid growth has brought its own challenges and opportunities for the service organizations. The performance of the organizations is dependent upon the quality and consistency of service delivery by employees. The well-informed customers demand customized, efficient, and valuable service instantly. The co-creation of services in disruptive and hypercompetitive environment in which the organizations operate today makes the role of employees all the more challenging. The employers face an uphill task of recruiting and retaining people with right skills and expertise in an age of diminishing loyalties and changing cultural values. The employees on the other hand are under constant stress of upgrading knowledge and technical skills, working in cross functional teams and adapting to changing organizational structures and systems. This chapter explores the causes of stress for employees and employers and the measures employed to manage the same in the service organizations. The literature has been reviewed to determine the potential causes of stress, its effect on employees and employers, and preventive and remedial measures taken by organizations. Round table and focus group discussions were conducted to arrive at the managerial implications for organizations. The organizations need to be adaptive and innovative with appropriate organizational climate and systems to mitigate the negative stress and enable the employees to be proactive and dynamic in service co-creation in this ever changing world.

13.1 INTRODUCTION

The share of services in the gross domestic product of the emerging economies has been steadily rising. According to the study by Associated Chamber of Commerce of India, the services sector in India contributed 67.3% to the gross domestic product of the country in the year 2013–14. The proportion of people employed in the services sector in the corresponding time-period stood at a little over a quarter of the workforce at 27%. The census data of India show that the growth in employment has been at a rate of 1.4% against the rise of people seeking jobs at 2.23% in the decade 2001–2011. In the years of high growth of the Indian economy spanning between the years 2004–2005 and 2009–2010, five million jobs were lost in India (The Hindu, 2015a).

There is a profound, unpredictable, and turbulent change taking place in the environment in which the organizations operate. The society is ever changing and the rate of change is ever increasing (Lehrer et al., 2007). There is rise in urbanization as the people move from the villages to the towns and cities in search of jobs. With the rising incomes, there are change and lifestyle and dietary habits which are taking toll on the people's health and organizational productivity. Globalization and inter linkages in the economies of different nations have brought with them their own opportunities and challenges. People today experience choices in every sphere of life which were not available to the previous generations. In the complex modern world, everything seems to be in a state of fluidity—be it career, acquaintances, values, or religion. Group cohesiveness and social order are diminishing at a fast pace. There is breakdown of joint families even in the developing nations where strong emphasis has been on the hierarchy over the centuries. The nuclear families are weakened as the teenagers want to ape the west in social norms in the age of internet and social connectivity. Increased individual freedom and ambition are leading to social isolation. Sense of belonging and confidence is reduced as the communal and spiritual world is being replaced with the materialistic and individualistic world (Lehrer et al., 2007).

The service industry professionals are in direct and face-to-face contact with the customers in most of the services. They have to adapt and serve according to the demands of the customers. The performance of the organizations is dependent upon the quality and consistency of service delivery by employees. The well-informed customers demand customized, efficient, and valuable service instantly. The co-creation of services in disruptive and hypercompetitive environment in which the organizations operate today makes the role of employees all the more challenging. The study of stress has never been as pertinent as it is today as the society experiences high level of stress. The subject is extensively explored (Dewe et al., 2010) as there is dramatic proliferation of research in the area with hundreds of articles published each year. The field is immensely popular with the researchers, popular press, and the practitioners. The numerous books published on the topic are informative, intriguing, contradictory, and challenging (Cotton, 2013). Importance of health and well-being of the productive workforce has been highlighted by many researchers. In this myriad of literature, the organizations struggle to identify the stress in employees, its roots causes and the best ways to manage stress in the employees. This chapter builds on the theoretical background of stress and its management in literature and integrates it with the perspective from an emerging economy. The aim is to simplify the meaning of stress for organizations as well as employees and helps the practitioners to manage the stress better.

CHALLENGES UNIQUE TO A DEVELOPING NATION

The economic prosperity and technological progress has made it easier to do certain tasks but there is a socio-cultural change taking place as well. By the year 2020, 45% of the population of India is going to be under the age of 25 (Das, 2012). According to the study by Associated Chambers of Commerce in India, 13 million people are entering the workforce each year (The Hindu, 2015a). This demographic dividend brings its own challenges. The Tower Watsons report says that the starting salaries in India are less than one-fifth of those in Singapore or South Korea (The Hindu, 2015c). In a period of high growth also there has been loss of five million jobs in India. There is rapid urbanization and the public infrastructure is stretched to the limit. The first-time graduates from the villages and towns migrate in the hope of landing a dream job. There is variability in the standard of education across various schools and colleges in the country. This leads to wide skills gap between what the job seekers possess and what the employers look for. The young work-force finds it difficult to respond to the demanding situations with the limited resources they have. The frequent job hopping by the young employees in search of the best job leads to high attrition rates in the organizations. Consequently, inexperienced employees are in the supervisory roles in many organizations. They face stress due to their own inexperience and pass the stress on their subordinates. They fail to guide and mentor the staff under them and just play the role of a taskmaster. The era of instant gratification, consumerism and changing values has led to nuclear families. This has left many without the safety net to cope with in times of distress. Stress has been identified as the foremost lifestyle risk factor ahead of obesity and physical inactivity (Business Line, 2014).

Source: Das (2012), Business Line (2014), The Hindu (2015a,c).

13.1.1   WHAT IS STRESS?

The first nontechnical use of the term stress was made in the 14th century to refer to hardship or adversity (Lumsden, 1981 cited by Lazarus, 2006). Derived from the Latin word "stingere," stress is viewed in different ways (Saiyadain, 2003). Business people may view it as tension and frustration, an air traffic controller may view it as lack of concentration, whereas a medical professional may view it as the physiological impact it causes. Robert Hooke used the term in the late 17th century in engineering sense to describe the ability of man-made structures like bridges to take loads without collapsing. When these ideas were applied to the body and mind, the psychologists used the terms stress stimulus and stress response. The term stress has undergone changes in meaning, and there is no precise definition for it. Originally, it was thought to be the external stimuli faced. The stress was something thought to be external to the person in the environment. Then, the processed or perceived stimulus as viewed by each individual was referred to as stress. This is the viewpoint of researchers contributing to the cognitive–appraisal model of stress. Other researchers have termed the response to the stimulus as stress (Lehrer et al., 2007). Stress is the body and mind's response to the some event. Stress can be the feeling of doubt of not being able to cope. It is the perception of resources being inadequate to match to demands made. "Stress can be chronic or episodic, positive or negative, a problem or a challenge" (Lehrer et al., 2007).

13.2   THEORIES OF STRESS

13.2.1   FIGHT OR FLIGHT RESPONSE

Since the early days mankind has learnt to survive in the harsh environmental conditions through the fight or flight syndrome. When faced with and encounter with the wild animals, the human beings either gathered their strength to fight them or ran away and saved their lives. The normal response is to face the situation or to run away from the situation if perceived as harmful. The body releases hormones such as adrenalin and cortisone as a response to the dangerous situation encountered. The activation of the bodily systems increases our heart rate, blood pressure, tightens the muscles, accumulates energy, and increases metabolism. Such reactions protect us from infections and inflammations detrimental to the body.

The stress is this response to the internal and external demands placed on individuals. Modern day work environments do not pose such extreme situations. The organizational employees do not engage in bodily fights. In most cases, they cannot leave the meeting midway. There can be arguments and discussions which lead to heightened amount of tension. The cause of tension can be a difficult customer, uncooperative employee, or a demanding boss. When such situations come up too frequently and for prolonged durations, the body has to time and again gather the resources to cope with them. The greater the stressor the higher the level of response hormones required to generate the response. The repeated activism of the bodily mechanism can lead to physical and psychological damage. The stress is accumulative in nature.

13.2.2 GENERAL ADAPTATION SYNDROME

Hans Seyle dedicated his life to stress research and identified the three stages in which stress occurs in an individual. The first is the *alarm stage* in which the individual gets shocked and the body's stress mechanisms are activated. The second is the *adaptation stage* in which the individual tries to come back to the normal state after the arousal is taken away. The third is the exhaustion stage in which the burnout occurs because of extended metabolic demands on the body due to a stressor.

Lazarus (2006) has focused on the way the individual appraises the situation. The primary appraisal is to assess whether the event poses harm, threat, or challenge to the individual. This is primarily to determine "what is at stake?" The secondary appraisal consists of assessing the resources/abilities the individual possess to tackle the situation. The individual answers the question "what can I do about it?" Different perceptions of the same external stressor can be there, dependent on the expected events and the reality. Stimulus–Response psychology was the result of positivist approach to research which focused on the behavioral outcomes of the stimulus. Stimulus–Organism–Response model is more prevalent now. The "O" in this refers to the thought processes in-between the stimuli and response and includes the motivations, belief of self and the world, situational intents, and resources among others (Lazarus, 2006). Lazarus has highlighted the role of emotions in how a person appraises environment and copes with stress. His list of 15 emotions comprises of "anger, envy, jealousy, anxiety, fright, guilt, shame, relief, hope, sadness, happiness, pride, love, gratitude, and compassion."

Snyder and Lefcourt (2001) say that stress is the interpretation and response to the perceived external threats and people can dismiss the feeling or manipulate the situation. According to Hart (1999), stress is the result of complex interactions of different interrelated variables. These factors span the work and nonwork experiences and also the personality of the individual. McEwen and Lashly (2002) have added the lifestyle factors of diet, sleep quality and quantity, smoking and alcohol impacting the stress response. Some people are more prone to stress than others. The individual differences are their due to our distinct biological make-up, health behaviors, lifestyle choices, social, spiritual, and financial resources to cope with stress. The reaction to stressors protects us and is important for survival but can cause harm, and even death.

13.3 STRESS AND PERFORMANCE

If the arousal level is very low then the individual may feel lethargic, lazy, and show a low performance level. Some amount of stress is required to motivate the individual to work and utilize his personal powers. Seyle (1974) identified two types of stress. Eustress is the positive stress which can be used in beneficial way. It leads to constructive outcomes and is associated with good health. The pressure of work to an extent can keep the employees motivated, provide learning opportunity according to the resources, and personality traits they possess (World Health Organization, 2015). The positive reappraisal of the situation makes the person aware of his or her personal power and ability and leads to personal growth (Folkman & Moskowitz, 2000).

The second is distress or more commonly referred to as stress that leads to anger, frustration, anxiety, and aggression. It has destructive outcomes. If the person is under distress for a prolonged period, then he experiences fatigue and exhaustion. The malfunction of the body systems causes illness and disease. If not cured, then the individual can face burnout.

There are physiological, psychological, and behavioral symptoms of stress. Stress affects the mental, physical, and emotional well-being of individuals. The violence and vandalism can cause disruption in work and lower productivity (Dollard et al., 2003). Stress can impact the health of the employees and business of the organization (World Health Organization, 2015). High cases of nervous breakdown, chronic fatigue, substance abuse, divorce rates, and low retention rates are reported from the information technology sector in India (Business Line, 2013).

The performance of the stressed employees can be poor due to lack of focus and concentration. They may not be able to perform to the optimum of their ability. This can lead to deadlines for work not adhered to and targets not being met. This can further cause the irritability in the employee and hamper the smooth working relationship between the employees. There is deterioration in the boss–subordinate relationship which becomes the cause of further stress. Researchers have said that stressed employees find difficulty in making decisions (Clarke & Cooper, 2004; Cooper & Cartwright, 1997). Stress has also been linked to increased accidents at the workplace (Clarke & Cooper, 2004). All this results in the poor motivation levels in the workforce and a low morale of the employees. The organizations find it difficult to retain employees and the attrition rate is high.

There is cost associated with stress. The medical and insurance claims by employees are the cost to the company. Additional costs can be due to sick leave, absenteeism, and presenteeism. Presenteeism (Dewe et al., 2010) is the state when an employee comes to work in ill health and has low performance levels. Johnston et al. (2009) estimated that for every 47 cents spent on the treatment of depression, 53 cents are being spent on absenteeism, presenteeism, and disability. So the indirect costs of stress are even higher than the direct costs. The annual loss of $300 billion is incurred each year due to work-related stress and its consequences such as productivity loss, absenteeism, employee turnover, health-care costs, and alcoholism as per the estimates of American Institute of Stress (Galinsky et al., 2001). The stress level exhibited by the employees is positively associated with absenteeism, sickness leave, and health-care costs. The productivity of the organizations is negatively associated with stress. So, to be successful companies need to take care of the stress levels.

13.4 IDENTIFICATION OF STRESSORS/MEASUREMENT OF STRESS

The *Social Readjustment Rating Scale* commonly known as the Holmes–Rahe scale was first developed in 1967 and has been revised and updated by Dr. Rahe number of times. Lazarus and Folkman (1984) pointed out that everyday hassles too can cause stress. The minor irritants such as getting stuck in the traffic jam or dealing with a difficult customer too cause stress. These irritants can annoy a person several times a day and can be measured using the *Daily Hassles Scale*. Other than the major events and the minor irritants, there can be *Central hassles* that represent the ongoing problems in individual's life (Gruen et al., 1988). The *Profile of Moods State* and *Perceived Stress Scale* have been developed for specific purposes such as to measure the work stress, anger, anxiety, Type A behavior, for teenagers, women, elderly, and so on.

13.5   WHY DO WE HAVE STRESS?

The globalization of the business world has increased the competitive intensity and has raised the quality standards and customer expectations at the same time. The spate of mergers and acquisitions has brought in organizational restructuring leading to mass scale lay-offs in many industries. The ones retained have to shoulder the excessive responsibility of supervising more number of subordinates, working longer hours and meeting tight deadlines to counter the increasing competitive pressures. Organizations are struggling hard to keep pace with the technological advancements. The employees are expected to master the new technology at the earliest. The work days have stretched beyond the normal 8 h office time. The all-pervasive technology has invaded the private space and time of the individuals making them accountable 24/7. More and more number of people are taking work home, working on weekends, and traveling more (Choudhary, 2013). The pressure of coping with enhanced responsibilities causes stress. Fear of redundancy is an added cause of tension. People are working more for the same level of compensation. The cost of living is increasing each year with the rising inflation. The increase in population in the productive years in emerging economies competes for the limited job opportunities. According to the report by Indian Staffing Federation, 1.7 million people are employed in the flexi-working industry in the private sector. The number of people in the flexi-working industry is expected to be 9 million in the next 10 years (Sarumathi, 2015). The concept of variable pay brings in an element of uncertainty and increases stress.

Barling et al. (2005) have studied the sources of job stress in different contexts and different individuals. Cooper et al. (2001) have examined the sources and effects of occupational stress. Colligan and Higgins (2006) have identified the factors that contribute to workplace stress. The European Commission's (1999) "Guidance on work related stress" explains the causes and consequences of stress. It also provides useful checklist and questionnaires for determining the workplace stress. Dollard et al. (2003) studied the work and nonwork factors that cause stress in service professions. The causes stress can be subdivided into the following three broad categories:

13.5.1 WORK CHARACTERISTICS

13.5.1.1 WORK CONTENT

The nature of the work itself can be a cause of stress. Those performing routine repetitive jobs which lack variety experience stress. Many job designs are repetitive and monotonous and underutilize the skills of the workforce. The meaninglessness of tasks and under stimulation fails to challenge the employee's abilities and can be a source of stress.

13.5.1.2 WORK OVERLOAD/INTENSIFICATION

When the work demands are greater than the resources at hand, the individual feels stressed. The excessive work coupled with the training, skills, and experience falling short of expectations can lead to stress. Knowledge explosion is happening in every discipline with the passage of time. It has become impossible for any one person to be a specialist in a specific field. Continuous assimilation of knowledge is required to keep pace with the changes (Lehrer et al., 2007). The pressure to upgrade skills can be a cause of stress (McCormack, 2014). The targets are revised every quarter and when unachievable targets are set with tight deadlines the employees feel the pressures of work overload. Accountants and consultants work under tight deadlines. The organization gives incentives to perform but the stressful situations take their toll on the employee's health. Covering for employees who are absent or on leave leads to greater work for those present. Work intensification happens because of the way work is allocated, organized, controlled, and rewarded (Dewe et al., 2010) under situations of inadequate staffing. Excessive work in too little time, just too much work or failing to meet targets are all causes of stress (Choudhary, 2013).

The traveling time gives an opportunity to be out of office but leads to backlog in the work at office. It leaves the employees physically fatigued. Time spent away from family causes stress. Early morning and late night flights or long distant flights lead to jet lags and sleep deprivation. The cumulative effect of all these factors can lead to poor performance at work.

13.5.1.3   WORK SCHEDULES

The long and inflexible hours of work lead to stress. The unsocial hours of work can lead to stress. The employees working in shifts have more stress as compared to the ones working in day time only. The rest times and breaks should be scheduled within the parameters of law to reduce stress. The employees of business process organizations working in night shifts over a long duration of time-period experience burnout. The employees of media and advertising agencies which require creativity have erratic work schedules (Bhatt, 2013).

13.5.1.4   WORK CONTROL

If the employees have no control over work content and work processes that can lead to stress, if the pace of work and working hours are rigidly defined leaving little scope for flexibility, then it is a cause of stress. The nonparticipation in decision making can also cause stress. On the other hand, the complete lack of direction can also be a stressor.

13.5.1.5   PHYSICAL WORK ENVIRONMENT

McCormack (2014) has highlighted environmental causes to be one of the sources of work stress. The design and layout of the work environment either boosts the morale of the workforce or is a cause of stress. If the heating, lighting, ventilation, and humidity levels are too low or too high, the workforce is unable to work at its optimal level and feels stressed. The vibrations or noise levels of the work environment can cause stress. If the workplace design is not ergonomic in nature, then the body gets tired after a short duration of work. The elements of hygiene, privacy, convenience, and safety are also important. If the place to get food and drinks/

refreshments is not available on or near the place of work then it causes stress. Accessibility of the workplace can be a source of stress, inadequate transport and child-care facilities are also stressors.

STRESSORS IN THE INDIAN SCENARIO

The banking industry professional spend long hours at work, take financial risks and deal with demanding customers. All these factors add to the stress level of the employees (Mehra, 2014).

Long working hours, sitting in front of the computer, lack of knowledge and training of technological advancements, and time away from family are the main stressors for the employees in the information technology sector. These are enhanced when the company has a strict policy for break times and dress code (Business Line, 2013). In the information technology industry, the real time, speedy service has to be delivered to clients, traveling to far off places for work and spending time away from family increases stress (Naidu, 2004).

The business process organizations have employees working in shifts. The unnatural working hours, and working on festival days causes stress in the employees. They are the receiving end of many of the customer frustrations and grudges and are supposed to provide quick solutions to the customer complaints with limited resources at hand on a daily basis.

The education services professional are expected to up skills themselves and use the latest multimedia technological tools in pedagogy and assessment. With each passing year the process of codification of knowledge is taking place and the rise of e-learning institutions is threatening the traditional face-to-face teaching in classrooms.

The educators in public institutions have to deliver lessons with bare minimum resources and infrastructure in place in remote areas of the country. They face crunch of resources to implement the latest technology in their institutions. The level of poverty is high in the country and the teachers are under pressure to keep the children from dropping out due to financial needs of the family.

The hospitality industry employees face intense competition in keeping the utilization rate high due to mass expansion in the sector in the boom years. They are expected to personalize and cocreate the service with the customers. The employers in the industry are stressed due to lack of appropriate competencies in the potential employees, rising costs due to inflation, high cost of funds, and high level of competition in the industry.

The disruption the business model of the private car hire services has caused stress to the incumbents in the industry. They have huge sunk costs due to ownership of cars and monthly outflow of the wages and salaries. There is an added pressure to upgrade technology and provide a faster, efficient, and cheaper service to the customers.

The start-ups in the on demand delivery services in India are stressed to make a mark in the industry. The stressors are to get the adequate funds, recruit, and retain key resources and to get a get enough traction in a competitive market.

13.6   ORGANIZATIONAL CONTEXT OF WORK

13.6.1   ORGANIZATIONAL STRUCTURE, CULTURE, AND POLICIES

The periods of change and transition lead to higher levels of stress in the organization. The changes in organizational structure and culture can be due to instable environmental conditions or mergers and acquisitions. These are usually accompanied by retrenchment, redundancies, and outsourcing. All these, directly impact the people working for the organization and are a cause of stress. The bureaucratic culture which requires endless approvals can be a source of stress. The organizational policies determine the level of consultation the employees are engaged in (Butts et al., 2009). The communication policies determine whether free and fair views are acknowledged or not. If there is no mechanism for feedback and redressal, then it is a cause of stress. The stress can be caused by the lack of clarity regarding organizational strategies and objectives or the task at hand. Lack of clarity can be pertaining to input or contribution to plans.

The politics within the organizations thrives on policies and is a source of stress for many. Lack of organizational policies to support work–life balance is a major stressor.

13.6.2 INTERPERSONAL RELATIONSHIPS

Working in cross functional teams comprising of people from different regions, religions with different personality types result in stress. Most of the times the employees do not have the flexibility to choose their team members and that leads to meshing of multicultural individuals each with his own distinct personality and cultural values. The teams have to work together for long durations to accomplish the tasks at hand. Interdepartmental work brings more challenges as each one has its own agenda and way of working embedded in its culture. The disagreements during meetings cause stress. Interpersonal conflicts can be with superiors, subordinates, or colleagues. The supervisor can show a biased behavior toward certain employees leading to stress and frustration in others. Individuals react in different ways to the same stimuli. Interpersonal conflict, social exclusion, and social conflict are all stressors. World Health Organization (2015) report mentions inconsiderate or unsupportive supervision, poor relationships with colleagues, bullying, harassment, or violence as causes of stress. Relationships, sexism, racism, discrimination, and harassment, lack of communication, impersonal treatment are all stressors (Hicks & McSherry, 2006). Employees facing criticism, abusive boss, conflicting values, having no friends in office, their view not being recognized or people going back on their words face stress (Choudhary, 2013). The employees working in customer interface environment in many service industries face a unique situation. The customers have high expectations in an era of internet and demand instant service. The employees have to regularly deal with difficult customers in an emotionally charged environment in many service industries.

13.6.3 ROLE IN ORGANIZATION

Role conflict and role ambiguity occur where an employee is supposed to undertake multiple tasks at a time and has no clarity regarding how to prioritize them. An absolutely vague job where there is no clarity regarding what has to be done is a source of stress. Lack of knowledge regarding how their part fits into the whole, what they are supposed to do, how they are supposed to do, when they are supposed to do, and how it will lead to the overall goal can be a cause of stress. World Health Organization (2015) reports too mentions unclear role, conflicting role, and isolated or solitary work as a cause of stress. Lack of any social contact with fellow employees may lead to alienation and cause stress (Choudhary, 2013). The situation becomes all the more difficult when the responsibility for accomplishment lies with the employee but the requisite authority is not there to channelize the resources necessary for execution. Clarity about the task and to have adequate and reliable tools to accomplish it is important. The stress can be due to the machine is being inefficient or unreliable which hinders the delivery of quality service. Poor work organization and change in job roles can be a cause of stress (McCormack, 2014).

13.6.4   CAREER DEVELOPMENT

The issues related to career development can be a source of stress. Not enough credit or compensation for the job carried out. There can be status and pay issues on which the organization and the employee may not be on the same page. If there is lack of respect and appreciation in addition to the inadequate compensation, then it can cause stress. The contractual jobs with no benefits are a source of stress. The ambiguous performance appraisal systems are seen as unfair by the employees and raise their stress levels. Not getting a promotion can be a cause of stress but someone else getting a promotion may cause stress too (Choudhary, 2013).

13.7   PERSONAL CHARACTERISTICS

13.7.1   PERSONALITY AND LIFESTYLE FACTORS

Employees with certain personality traits experience more stress than others. People who are indecisive and fail to be assertive are more stressed. The ones who take things lightly develop the habit of procrastination experience stress when the deadlines approach. The temperament of the person determines whether he is able to keep his cool or not when debates get heated at workplace. The failure to do so leads to stress.

13.7.2 GENDER DIFFERENCES

Women experience more stress than men in general. They experience a role conflict between the role played at the office and home. At workplace, a woman has to be competitive to meet the targets, whereas at home they play the traditional role of nurturer and caregiver. In addition to this, the gender disparity in compensation is a source of stress.

13.7.3 YOUNG VERSUS EXPERIENCED EMPLOYEES

In general, the young and inexperienced employees face more stress as compared to the experienced ones because of lack of on-the-job training. They do not have the repository of past experiences to guide them to face the difficult situations encountered at work.

STRESS IN WOMEN EMPLOYEES

Worldwide research has shown that women in male-dominated offices face interpersonal and workplace stressors. They come across social isolation, periods of low or high visibility, coworkers doubts about competence, sexual harassment, mobility obstacles, and low level of social support at work (Economic Times, 2015; World Health Organization, 2007). In India, there is lack of awareness about the sexual harassment law in 97% of the private companies (The Hindu, 2015b).

In India, which is at the cross roads of transition to a modern workplace and still holding on to its traditional values poses a great challenge for its women employees. There is a stark contrast between the physical infrastructure inside the office and outside. Many multinationals offer great opportunities and challenges, but the social support and the physical infrastructure have failed to reduce the stress for women. There is dearth of childcare facilities at work and in general, the state of public transport is dismal at most places. There is a rising issue of women safety and security. Nielsen study has said that Indian women are the most stressed in the world. In

India, 87% of women in India felt stressed all the time and 82% found no time to relax (Goyal, 2011). The expectations from women have increased and they contribute to the rising consumerist household aspirations. A great number of them are the first generation of women stepping out to work in their families and are expected to shoulder all the household responsibilities as well. The competitive, assertive women change the role completely at home. The business travels, networking, and late nights contribute to career advancement but are a source of stress for women.

Source: World Health Organization (2007), Goyal (2011), Economic Times (2015), The Hindu (2015b).

Choudhary (2013) has summarized the reasons for stress in simple words, such as—facing criticism, abusive boss, carelessness, conflicting values, no friends in office, my view not being recognized, people going back on their words, saying something foolish, no promotion yet again, failing to be assertive, looking back with deep regrets, conflict with friends, other taking me for granted, not keeping my cool, just too much work, an absolutely vague job, boring job, disobedient subordinates, no end to seeking approvals, can't find ideal solution, failing to meet targets, procrastination, and indecisiveness.

The survey by Regus found that employees in India are most stressed by work pressure, personal finance, commute, and the instable world economy (Business Line, 2012). The Indian employees feel conflicting or unclear job expectations, inadequate staffing (uneven distribution of work in the group and lack of support), and paucity of work–life balance as the three top stressors (Business Line, 2014).

Only 38% of the employers in India feel that stress management should be top priority for health and productivity workshops (Business Line, 2014).

13.8   MANAGEMENT OF STRESS

The major cause of stress for individuals is their work (Choudhary, 2013). Very few people are in the jobs that they like or are passionate about. Finding a job that you are passionate about is a difficult task. It is difficult to imagine that any job in this world would be stress free. It is difficult to find one which harnesses your potential well and matches your abilities and interests. Only once you are in the job that you can experience the challenges the work offers and whether you have the adequate capabilities to manage it. The perks of the jobs such as benefits, status, pay, and pension keep many in their current roles. So, reducing anxiety and coping with stress are required to perform well and maintain your physical and mental well-being. The effective management of stress is a key to happy and successful life nowadays (Mace, 2014). The stress can be managed through organizational, professional, and personal support (Dollard et al., 2003).

The unhappy and unhealthy employees fail to perform and are prone to more sick leaves and absence. Their morale and efficiency are low resulting in low productivity. It is imperative for the company to take care of the physical and mental well-being of the employees.

According to Smith (2002), stress management is the ability of an individual to anticipate, prevent manage and recover from the ill effects of perceived threats and coping deficiencies. The stress can be managed by either change in attitude toward stress or altering the situation that has caused stress. Kompier and Cooper (1999) have described management of stress at four levels depending upon its severity. The first level is the modification of the appraisal of the situation that causes stress. The second stage is to improve the individual and organizational fit to manage stress. The third is primary prevention in which the focus in on elimination of the source of stress. The secondary and tertiary prevention aims at reducing the ill effects of stress. The stress management interventions are the techniques used to appraise the stressful situations and to deal more effectively with the symptoms of stress (Murphy, 1996).

Proactive or task focused coping strategies increase the levels of stress that the individuals can handle. The emotion focused coping strategies such as use of smoking or alcohol to mitigate the effects of depression or anxiety are used by individuals when they are unable to deal with the stressful situations (Clarke & Cooper, 2004). The emotion focused coping strategies are evident in individuals having external locus of control. The individual with internal locus of control suffer fewer stress symptoms (Hurrel and Murphy, 1991 cited in Clarke & Cooper, 2004). The employees can be trained have a positive outlook toward work and life. Making it a habit to think like an optimist can help to reduce the stress levels. The employees can be trained to use proactive or task focused coping strategies in stressful situations. Control coping and avoidance coping are the terms used by Koeske et al. (1993 cited in Clarke & Cooper, 2004) to describe the whether the individual is able to take control and deal with the situation or runs away from it. Time management training should be administered to prepare employees to cope with the additional work load (Cotton, 2013). The assertiveness training should be provided to the employees so that they are better able to manage interpersonal relationships in the organization.

13.8.1 ELIMINATING IRRITANTS OF DAILY LIFE

The irritants of daily life are the most common cause which prevents the individuals from concentrating on their work. Many companies have recognized this fact and provide assistance in elimination or reduction of these irritants. The company transport, childcare, and house-help supports take the burden off the employee's shoulders. Assistance in paying bills and liaison with government offices ensures that productive hours are not wasted on these routine hassles. The support and guidance provided for house hunts and school admissions during relocation of employees take care of the stress due to changes initiated by the company's moves.

13.8.1.1 IMPROVING WORK STRUCTURE AND ENVIRONMENT

The major causes rooted in the work structure and environment can be analyzed and mitigated. Cotton (2013) places emphasis on identification of potential causes of stress and minimization of the same.

13.8.1.2 JOB DESIGN

The design of the job should be such that it provides enough challenge for the employee but should not overburden the employee. The techniques of job rotation and job upgradation are a solution for repetitive and monotonous jobs.

13.8.1.3   ENVIRONMENT

The stressors in the work environment have been highlighted in the previous section. The workplace should be designed to reduce stress and enhance efficiency. It should provide opportunities for employees to take breaks to de-stress themselves (Cotton, 2013).

13.8.1.4   RELATIONSHIPS

The relationships at workplace with superiors, subordinates, and colleagues are another cause of stress. Lehrer et al (2007) propose the use of social skills training for the employees so that they maintain cordial relations at work place.

13.8.1.5   CULTURE

The culture of the organization can be an enabler or disabler in maintaining a low-stress environment in the organization. Every attempt should be made to have an open, flexible culture so as to synergize the efforts of the employees.

13.8.1.6   ORGANIZATIONAL POLICIES

Clear organizational policies reduce the ambiguity and bias. These should be clarified and communicated to all concerned in an open manner so as to reduce stress.

The Regus survey found that three-fourths of the employees in India feel that flexible working hours would reduce their stress (Business Line, 2012).

WORK STRUCTURE TO REDUCE STRESS

The advertising agencies have intense demands and odd working hours. Ogilvy and Mather monitored its employees to see if they were spending practically every weekend in office to complete the projects, then they were advised to distribute the workload with others to avoid burnout. It also allowed the young employees to play music while at work for relaxation and to take care that it does not disturb others.

To cater to clients in the Western world, the employees in India have to work or take calls at odd hours which leads to stress and burnout. To counter this, companies like Google and Sabre are permitting their employees to fix call timings that suit their schedules also.

Wipro rotates one-third of employees on a project after 6 months pave way for change from constant night shifts (Sengupta, 2011). Citibank India employees working long hours can leave office one day early in month. Smartphones have been provided to the employees by many companies including HDFC Life so as to improve flexibility (Bhatt, 2013).

Source: Sengupta (2011), Bhatt (2013).

13.8.2   INCREASING THE STRESS THRESHOLD

A conscious effort should be made to increase the threshold level of stress that an individual can bear without any ill effects. Some of the researchers have named it as stress inoculation training to prevent stress. The following areas require particular attention to increase the stress levels that the employees can take without any adverse effects.

13.8.2.1   PROBLEM SOLVING/COPING

The employees can be trained to keep the focus on the task at hand and gather the resources available to tackle the same. They should be advised to keep away from any maladaptive and irrational thoughts. The focus should be on rational discourse. The complex problems can be broken into manageable wholes to solve one at a time.

13.8.2.2   TIME MANAGEMENT

One of the causes of stress is the paucity of time to accomplish the tasks allocated to the employees. Cotton (2013) has proposed the use of time management sessions to train the employees to better manage their time.

13.8.2.3   ASSERTIVENESS TRAINING

Stress builds up when the employees are not able to put their viewpoint forward at the appropriate time and forum. Cotton (2013) has also proposed the use of assertiveness training as a useful tool to make the employees resilient in competitive environments.

ONLINE COUNSELING TO REDUCE STRESS

There is a social stigma attached to counseling in India. The nature of the IT industry is such that employees have to provide speedy solutions in the real on time basis. The have to travel to far off places on work and sometimes stay there for extended duration of time without their families. Companies have come up with the novel method of providing help to distressed employees through online sessions in addition to the regular face-to-face meetings. The online medium provides anonymity and privacy for the employees where they can have candid chats with the counselors regarding the interpersonal conflicts, harassment at work, loneliness, lack of self-confidence, etc. offered. The guidance and counseling on balancing the personal and professional life is provided. The careful use of words overcomes the lack of audio and visual clues for the counselor. The employers screen the credentials of online counselors before recommending them to the employees. The director of 1to1Help.Net says that the worldwide web is full of the online counselors, but only the local players can empathize and connect with the individuals of a particular region. The ones based in distant countries are not aware of the nuances of the local culture and fail to provide therapy in face-to-face sessions if that is required.

To capture the demand in Tier II and Tier III cities the e-learning initiatives have been launched so as to reduce the cost of delivery. One such organization by the name Ripples based in Kormangala is servicing clients in 22 cities across the length and breadth of India.

Source: Naidu (2004), Roy (2014).

13.8.3 PHYSICAL TECHNIQUES TO MITIGATE STRESS

The physical techniques of stress management focus on the movement of body parts to reduce stress. Following are some of the physical techniques for stress management.

13.8.3.1 WALKING, RUNNING, STRETCHING EXERCISES, WEIGHT TRAINING, AND TAKING PART IN A SPORTING EVENT

Walking, running, stretching exercises, weight training, and taking part in a sporting event can mitigate stress. Walking is the simplest physical technique that helps in maintaining a healthy body and mind. It keeps the weight in check, saves us from many diseases, and builds the immunity of the body. Stretching exercises keeps the body fit and flexible. They improve the agility and response of a person. Weight training builds the muscles and strengthens the body. The toned body is source of confidence and pride for a person. The in-house gym or discounted memberships of nearby gyms can act as a motivation to employees to do some physical exercises. The pool table and table tennis table at the workplace give and opportunity to meet and play with fellow employees.

Through these measures, a person can get rid most of the pains and aches which are the common cause of mental stress and agony. An organization can encourage its employees to incorporate one or more of these routines in their life to lead a healthy and more meaningful life. Many organizations provide facilities such as gym, pool table, table tennis, and so on to bust the workplace stress. Organizational policies such as organizing a sports day, various sporting events between divisions, departments or sites sends the right signals to the employees. Participation in walking competitions and marathons by the employees as a unit encourages team spirit and wards off the ill effects of any organizational stress. It acts as a glue to bring together people from diverse backgrounds into a homogeneous whole in a social setting for the benefit of the organization. A healthy workforce is a happy workforce and is less prone to sick leaves and absenteeism.

13.8.3.2 PROGRESSIVE RELAXATION

It can be used for skeleton muscle relaxation. The ones over which we have voluntary control such as the one which help us in moving, lifting, walking, etc. (Smith, 2002). Progressive relaxation (also called Jacobsonian muscle relaxation) is the tensing up and relaxing the muscle groups (Lehrer et al., 2007).

13.8.4 PSYCHOLOGICAL TECHNIQUES TO MITIGATE STRESS

13.8.4.1 BREATHING

Fast breathing, increased heart rate, and high blood pressure are all related to stress. By exercising conscious control over one's breathing and heart rate variability (Lehrer et al., 2007), one can reduce these ill effects. Focus on breathing takes the mind away from stressful situations and aids in de-stressing.

13.8.4.2 MINDFUL MEDITATION

Mindful meditation starts by focusing on the inhalation and exhalation of the breath. Its helps the individual to become aware of and accept all the thoughts and feelings that occur at that moment without reacting. Not being judgmental about feelings, thoughts, body sensations from moment to moment aids in reducing stress. The mental focus is on the body sensations (Smith, 2007). Meditation workshops have practical applications and have been included in many corporate training sessions (Deshpande, 2012).

13.8.4.3   MANTRAS MEDITATION

Meditating on mantras is influenced by Eastern values and cultures. The repetition of the mantras takes the attention away from the things that bother us and brings inner calm. It emphasizes on feelings of oneness with the universe. The aim is to achieve inner peace.

13.8.4.4   MIND TRAINING

Dalai Lama proposes the use of Mind Training, the Buddhist traditional method of transforming adversity to advantage (Hindustan Times, 2011). The root cause of all stress is negative emotions. The inner peace can be achieved by training the mind to recollect good fortune. The virtues of kindness, compassion, and respect help to accomplish goals without fear and anxiety. The human intelligence and strong determination should be used to solve the problems instead of worrying about them.

SPIRITUAL QUOTIENT OF CORPORATE EMPLOYEES

Asia has always been the center of spirituality. The modern development has brought economic prosperity to the masses but they have lost their peace and tranquility. The work pressures cause fatigue, tiredness, depression, and burnout among even the young workforce. As a preventive and remedial measure the companies are turning toward spirituality to improve the physical and mental well-being of their employees.

The Future group has appointed world's first chief belief officer to take care of the holistic well-being of its employees. The spiritual quotient of the corporates is being enhanced by the training modules such as Apex of the Sri Sri Shankar's Art of Living. Other organizations such as Isha Foundation, Ashwani Dyana Foundation and Swami Sukkha Bodhananda's Prasanna Trust are working with corporates to reduce the stress levels of the employees. The breathing and meditation techniques are being taught to the workforce of many companies. The spiritual gurus claim that there is marked improvement in workplace relationships and productivity as a result of these courses. The spiritual sessions focus on the holistic well-being of mind, body, and soul and have a deeper impact than the offsite training or run off the mill management development programs. Capgemini which had administered the Apex Programme to its women employees in Bangalore, Mumbai, and Kolkata. The Yash Birla Group wants all its employees to undergo it. Indraprastha Gas Limited regularly sponsors its employees for the Isha Foundation's Inner Engineering Programme. Barclays Wealth employees have taken the Art of Living Programme. In companies such as Crompton Greaves, Art of Living has even been administered to the workers of the Nashik plant. GMR hold regular workshops with the Prasanna Trust.

The participants feel that the team bonding after the workshop increases leading to fewer work relationship conflicts. The energy levels are boosted. The employees take fewer cigarette breaks. The Yash Birla Group feels that the old values of loyalty, sincerity, and integrity resurface after these holistic programs. The underlying motto of this intervention is that the organization that prays together, stays together.

Source: Amarnath (2011).

13.8.5   PHYSICAL AND PSYCHOLOGICAL TECHNIQUES

13.8.5.1   YOGA

The Yoga *asanas* range from lying on the ground in a state of complete relaxation to the ones that stretch the body to a limit. The use of breathing techniques integrated with the various body postures brings physical and psychological benefits. Yoga harmonizes the body and mind leading to increased fitness, body strength and flexibility, and inner peace and calm. The use of Yoga leads to mitigation of workplace stress (Deshpande, 2012).

13.8.5.2   QIGONG

Qigong has its origin in Chinese philosophy, medicine, and martial arts. It makes uses of slow bodily movements with deep breathing and calm meditative state. It is useful in reducing stress.

13.8.5.3   AUTOGENIC TECHNIQUE

This Training Method of Schultz acknowledges debt to yoga and Zen as Schultz spent considerable amount of time in Japan (Lehrer, 2007). It is passive repetition of the suggested relaxation exercises (Smith, 2002). Thoughts bring out relaxation. Phrases or images can be used to relax the mind. For example, the thought of ice cream brings saliva, same way the thought generates relaxation. To de-stress, thought of warmth, easy, and slow breathing can be used to generate the same.

CORPORATE YOGA IN A NOVEL WAY FOR WORK–LIFE BALANCE

The centuries old Yoga has been reincarnated to help the stressed professionals. Corporates are holding Yoga classes for work–life-related issues. These classes are held either on the premises of the company or at the service provider's premises. Dell has a dedicated Yoga studio to de-stress its employees. Crompton Greaves plans to open yoga rooms in 22 centers in India and 18 abroad (Amarnath, 2011).

The novel ways to de-stress the employees while on work is being practiced by Total Yoga trainers. Named Desktop Yoga, while administering the trainer sits on the floor while the employees are at their workstations. The exercises are done without taking any scheduled breaks. The yoga provider invested a small amount of one lakh and earned a profit of 50% on the yoga modules (Roy, 2013).

Companies across industries are lapping up the innovative ways to reduce the stress in their employees.

Source: Amarnath (2011), Roy (2013).

The mental imagery and relaxation techniques reduce the arousal that causes stress. The different techniques make some individuals feel relaxed and energized while others feel sleepy, joyful, rested, and peaceful. Smith (2007) has analyzed the relaxation and renewal techniques in detail and has categorized the relaxation states into four groups—"basic relaxation, core mindfulness, positive energy, and transcendence." He proposes that mix of techniques should be selected from the basic approaches like "balanced diet" so as to achieve the desired psychological state. Yoga for the posture, progressive muscle relaxation for the muscles, breathing exercises for control over breathing, autogenic training for body focus, positive self-talk for emotions, and meditation for attention are recommended.

13.8.6 PHYSIOLOGICAL TECHNIQUES

13.8.6.1 BIOFEEDBACK

Biofeedback involves monitoring of the physiological parameters of the body such as heart rate, breathing, and muscle activity. Constant feedback of the same is provided in order to make some improvements. This information along with some change in thinking, emotions, and behavior lead to improvements in the parameters being measured.

13.8.6.2 NEUROFEEDBACK

Neurofeedback is feedback about the brain activity which is measured through the sensors being placed on the scalp. This can be given through the video or sound and the positive activity and negative activity can be differentiated by different colors or sounds. The aim of this kind of feedback is to self-regulate the brain activity.

13.8.6.3 PHARMACOTHERAPY

Pharmacotherapy entails the use of medicines to treat the extreme cases of stressed employees. The prolonged stress can lead to burnout and many serious bodily ailments which require treatment under the guidance and supervision of trained medical practitioners.

13.8.7   PURSUIT OF HAPPINESS

The activities pursued away from work which add meaning to our existence and enrich our lives take us to a whole new platform. Music, yoga, pursuing some hobby (Cotton, 2013), all help in de-stressing. Some of such activities are listed below which help in mitigating the stress caused at workplace.

13.8.7.1   ART

The thing of beauty is a joy forever. When a person immerses in some artistic pursuit then he or she is immune to the noise of the outside work and can totally relax and enjoy the creative process. Pursuing art is a great stress buster. Art-based interventions produce long–term benefits as compared to the conventional interventions (Romanowska et al., 2011).

13.8.7.2   MUSIC

The way exercise takes care of the body, the music heals the soul. The melodies take away the distress and negativity and soothe the listener.

13.8.7.3   HUMOR

Laughter acts as a curative and relaxes the bodily muscles, increases blood circulations, and reduces blood pressure (Sengupta, 2007). Humor helps in reframing of the stressful situation, reduces the social distance, and creates playful and creative work environment (Deshpande, 2012). There are many stand-up comedians nowadays who help in mitigating the stress of corporate clients.

13.8.7.4   AROMA

Aroma has a positive role in reducing the stress levels and improving our well-being. Aromatic oils provide relief and relaxation and can invigorate our body and mind (Sengupta, 2007).

13.8.7.5   THEATER

The theater takes the participants and the audience into a different world. Pursuing theatrical activities encourages team work and acts as a stress buster. The research of Theorell et al. (2013) found that the cultural activity at workplace has a protective effect on the emotional exhaustion of employees.

13.8.7.6   SOCIAL SERVICE OR WORKING WITH NGO

Many corporate employees work and support the social services for the general good of the underprivileged sections of the society. These activities bring a purpose and meaning in life and provide contentment and satisfaction of doing a good deed.

13.8.7.7   SABBATICAL

Sabbatical provides a means to highly stressed employees to step back from the active work life and pursue some life-enriching experiences. It is a great way to give back to the society in which we all work and flourish.

MANAGING STRESS TO IMPROVE PRODUCTIVITY

According to Roy (2013), the corporate professional today is stress, depressed, and well dressed. The increase in stress levels of corporate employees have been addressed by the companies in various manners. There has been a spurt in the institutes offering stress management training. Companies realize that these measures would lead to increase in productivity of the employees. Employees of many companies such as Wipro, HDFC Life, and Titan undergo stress management sessions. The profit margins in this segment are 25–30% and this market is growing at 40% (Roy, 2013). Some have also introduced the anger and behavior management courses. The government employees from various departments too undergo the stress management training.

The companies are trying to weave the benefits of health along with the work schedules. Vodafone provides gymnasium facilities and conducts yoga lessons. It has a health portal to which all employees in the country have access and visits by physician once a week (Mehra, 2014).

Citibank has "Live Well @ Citi" program which focuses on health, diet, lifestyle, and wellness. The company empowers its employees through flexible work hours, concierge service to run errands, and has a corporate holiday program (Mehra, 2014).

Infosys and Oracle have a doctor and psychologist on call. Employees can meet them face to face and discuss their issues. Infosys also has an internal system in place where the experienced employees counsel the juniors and the confidentiality is maintained. Oracle has "bunker cells" where the employees can go for a short duration for relaxation. Many firms encourage group activities outside of the work environment like film, dance clubs and theater to nurture camaraderie among employees. Regular sessions on stress management, counseling, and yoga are arranged (Business Line, 2013).

The program named "Bullet Proof" was adopted by Phillips, where the senior managers were trained to look for signs of stress in subordinates in addition to managerial training (Sengupta, 2011).

The Federal Bank in Kerala encourages its employees to pursue their passions through recreation clubs, take a sabbatical of three years in addition to the engagement programs that it holds. The employees can also commit to social or environmental cause through its Corporate Social Responsibility activities (Mehra, 2014).

The survey by Towers Watsons in the Asia Pacific Edition of staying @ work shows that India leads other Asian countries in managing work-related stress. One out of three employers has administered e some sort of stress or resilience management program in 2013 (Business Line, 2014). The flexible work hours is the top-most solution offered by employers to reduce stress and 50% of employees make use of it. The stress management interventions include education and awareness campaigns, yoga, and tai chi.

Source: Sengupta (2011), Roy (2013), Business Line (2013, 2014), Mehra (2014).

Progressive relaxation, biofeedback, pharmacotherapy originated in the West. These techniques are technological, mechanistic, and physiological in nature. While administering these techniques, less focus on inner peace and more on setting things right based on parametric research (Lehrer et al., 2007). Biofeedback and Respirations can be used for automatic arousal of the body function and systems on which we have no control such as rapid heartbeat, excessive perspiration, shallowness of breath (Smith, 2002).

The health and wellness is a holistic concept and encompasses intellectual, spiritual, and environmental dimensions in addition to the physical, psychological, and social factors (Alters & Schiff, 2009).

The Regus survey found that three fourths of the employees in India feel that flexible working hours would reduce their stress (Business Line, 2012).

13.9   MANAGERIAL IMPLICATIONS

Stress is an area of concern for the organizations. Managers often fail to notice that employees are suffering from chronic stress (McCormack, 2014). The first step should be to raise awareness about stress. Managers should be trained to identify the potential of stress in employees and minimize them (Cotton, 2013; Dewe et al., 2010).

As is evident from many researches that low level of challenge in repeat, repetitive, and monotonous jobs is a cause of stress in employees. The jobs should provide enough challenge for the employees. The organizations should take care of the factors under their control to minimize stress.

13.9.1   JOB DESIGN

Goal setting, personal direction, having a sense of purpose, pursuing an activity with passion may direct the stimulus responses to feeling of fulfillment (McGrady in Lehrer et al., 2007). The job enrichment, job enlargement or job upgradation techniques can be deployed to make the person's skills and abilities better match the job. The job rotation can break the monotony of performing a repetitive work and provide the employee the opportunity for learning and development.

13.9.2 PHYSICAL WORK ENVIRONMENT

The physical work environment should be made hygienic, safe, and stress free to keep the employees motivated. The physical comfort of the work environment should be monitored and kept at the level prescribed by the company's policies. The workplace design should reduce the stress in the employees and enhance their efficiency (Cotton, 2013). Many companies in India are working toward this goal. The ambience of the offices, hotels, and even hospitals is being improved through better design and decorating the interiors with the use of art pieces. Many more organizations need to follow the best practices emerging in this sphere.

13.9.3 INTERPERSONAL RELATIONS

The resources for prompt response to customer queries, problems, and complaints should be at hand with the employees. The companies need to train the employees to show rational and logical response to stress and not enter into fights and arguments with customers and colleagues. They need to take cool-off breaks before coming back to debate on area of conflict.

13.9.4 CHANNELS OF COMMUNICATION

The management should communicate and clarify the vision of the organization with all employees in unambiguous manner. The channels of communication should be clear and open to all employees (Dewe et al., 2010). The difficulties faced by the employees should reach the top management and solution for the same should be provided at the earliest. The grievance resolution mechanism and the open house policy of the organization should be in place. This makes the bond between the organization and the employees strong and keeps the stress at bay.

13.9.5 MENTORSHIP PROGRAM

The experienced employees can guide the new employees on the ways of work and behavior in the organization. They can help the juniors in the times of crisis. Positive coping strategies should be taught to the junior employees by the experienced employees. They should be advised to refrain from maladaptive coping in which irrational appraisal of the situation is carried out. The logical thinking based on facts and reason should be encouraged.

13.9.6 STRESS RESILIENCE TRAINING

The psychological capital which comprises of the positive resources of resilience, efficacy, optimism, and hope determine the individual differences in the levels of stress (Avey et al., 2009). These factors also impact the intentions of the employees to quit or undertake job search. Attempts should be made to maximize the psychological capital of the organization withstand turbulent changes and increased competitive levels faced by the organizations today. Some researchers use the term psychological hardiness to describe the individual's capacity to deal with stress. They propose the strategies to enhance commitment, control, and challenge through which the impact of stress can be minimized.

13.9.7 EMPLOYEE WELLNESS PROGRAMS

The employee wellness programs can reduce the level of stress in the organization (Cotton, 2013). These programs provide the counseling and consulting services for the employees and their families. These can act as preventive as well as remedial measure for managing occupational stress. Arthur (2000) says that the employee assistance programs offer consultation, assessment, referral, and short-term treatment for psychological difficulties. These difficulties arise as a result of the interaction between the individual and organizational factors. There is evidence of success of such programs. The programs have been criticized for focusing at the individual level rather than the organizational level.

13.9.8 HEALTH AND PRODUCTIVITY STRATEGY

The employee health and well-being meant only medical facility and insurance policy of the company 10 years ago in India. Now, the annual assessment of the health through biometric feedback, health check-ups, interventions to manage stress to weight, all form part of the health and productivity strategy. One report says that three out of four employees expects more focus in this sphere in the coming years. Ninety-six percent of companies perceive health and productivity's role to be moderate to essential in their health strategy. Bajaj Allianz has seen a growth of 25% in their health insurance packages over last year (Mehra, 2014). Globally the health and productivity effectiveness in companies meant 34% higher revenue per employee according to the Tower Watsons report. The first step for health and productivity strategy is to have a dedicated health and safety officer in the organization. The next is to have a stress policy in place to counter stress. The stress survey and audits can gauge the stress in the organization and appropriate steps can be taken to mitigate the same.

13.9.9 ORGANIZATIONAL CLIMATE

A conducive organizational climate should be created to nurture talent. The employees should feel free to share their ideas for business and organizational development. The flexible culture of team working and participative management reduces stress levels in the organization. Organizational climate needs to be managed to minimize the stress in the organization (Dewe et al., 2010).

13.9.10 ADAPTIVE AND INNOVATIVE STRUCTURES AND SYSTEMS

Totterdill and Exton (2014) propose that workplace innovation provides learning and development opportunities for the employees and also take care of their health and well-being. The key areas to focus for workplace innovation are sustainable workplaces, meaningful work, harmonizing technological innovation, and human potential. Ackfeldt and Malhotra (2013) in their empirical study found that empowerment of employees helps in mitigating the dysfunctional effects of role ambiguity on commitment. The professional development counters the detrimental effects of role conflict on commitment in organizations. The empowerment of employees in a participatory work structures are negatively related to job stress and have positive impact on employee health and well-being

(Butts et al., 2009). Dewe et al. (2010) also propose the use of adaptive and innovative organizational structures and systems to reduce stress in organizations.

13.9.11　CHANGE MANAGEMENT PROGRAM

Effective change management systems can reduce the stress in organizations (Dewe et al., 2010). Change brings with it the uncertainty and the fear of the unknown. The stress levels in the organization are higher in periods of change as compared to normal circumstances. The reasons for change and the objectives of change should be clearly communicated to all concerned. The involvement and cooperation of employees makes the change process less tedious and stressful.

13.10　LIMITATIONS OF THE STRESS INTERVENTIONS IN EXISTENCE

There is no consistency in the stress management inventions administered by organizations. The gamut of options available poses a problem for the organizations to choose the best solution. Most organizations use more than one technique for stress management (Murphy, 1996; Smith, 2007). There is also no consistency in the outcomes desired from these interventions. The variability of measures used to assess the benefits complicates the process of comparison between different studies. Kirk and Brown (2003) studied the effectiveness of employee assistance programs and found that these are perceived by the employees as an important resource and they do positively impact the employee's mental health. In contrast, Murta et al. (2007) reviewed the 84 stress intervention studies and found link between the process evaluation and the outcome evaluation in only less than half of the studies. The determinants of effective inventions or outcomes are difficult to point out because of incomplete recording of the process evaluation. Richardson and Rothstein (2008) in their meta-analysis of the 55 stress interventions found that organizational interventions were few and relaxation interventions were most commonly used. Also, the outcomes were measured on the basis of psychological variables as opposed to the physiological and organizational measures.

Giga et al. (2003a) reviewed the impact of stress management inter-ventions applied in the United Kingdom to gauge their effectiveness. They found that majority of the interventions had been administered at an indi-vidual level and did have a positive human and/or organizational benefits. They concluded that to have a beneficial long-term gain, the interven-tions should be applied at the organizational level. Giga et al. (2003b) also found that there is tendency in the researcher to evaluate the benefits of stress management interventions over a short time period. Murphy (1996) found that three-fourths of the studies were administered to all the workers and not to the highly stressed ones. None of the interventions was consistently able to produce job/organization-oriented beneficial results, such as improvement in job satisfaction or reduction in absenteeism. To obtain such results, the stress interventions should be modified to tackle the source of stress (Murphy, 1996).

13.11 GUIDELINES FOR EFFECTIVE INTERVENTIONS

13.11.1 STRATEGY

Given the proliferation of stress in all the industries and its widespread consequences spanning over individual and organizations, it is imperative to have a health and productivity strategy integrated with the organiza-tional strategy. The organizational strategies should ameliorate work–life balance (Dewe et al., 2010). The focus should be on achieving the orga-nizational benefits of reduced absenteeism, sick leave, presenteeism, and attrition.

13.11.2 STRESS AUDIT

The company needs to appoint health and safety officer. The assessment of stress level prevalent in the organization should be made through stress audits at regular intervals. The stress audits provide a picture of the phys-ical and mental health of the employees of the organization. The measures to mitigate stress can be taken based on the stress audit.

13.11.3   MIX OF INTERVENTIONS

The sources of stress should be identified and measures taken to eliminate them. This would ensure the well-being of large number of employees. A mix of *primary, secondary,* and *tertiary interventions* aimed at individual and organization should be used to manage the occupational stress (Dollard et al., 2003).

13.11.4   FOCUS ON ORGANIZATIONAL BENEFITS IN THE LONG-TERM BENEFITS

The stress management interventions bring benefits in the long term. The focus of the stress management interventions should be to reduce the levels of the stress in the organization over a long term. There should be caution observed to evaluate the benefits in the short term.

13.11.5   RECORD AND REPORT

Proper record of the intervention administered, duration of intervention and benefits derived should be maintained and reported to all concerned. These can serve as a base for future stress management endeavors. The record also provides the data for measurement of effectiveness of various interventions.

The time-management and assertiveness training should be administered to the employees to make them better prepared to handle stressful situations. The stress prevention measures should be taken to eliminate or minimize the problems. Special care and priority should be given to those employees suffering from acute levels of stress. Administer interventions to stressed employees, use interventions directed organizational benefits, use mix of interventions, assess the benefit in the long term, and record and report the finding for further use.

HEALTH AND PRODUCTIVITY STRATEGY IN INDIA

It is difficult to measure the benefits of these programs as they are intangible. The ICICI's Head of Human Resources points out the importance to continue with these programs as attempts are being made to assess the benefits (Mehra, 2014). The best results are obtained through the combination of two or more stress management techniques (Murphy, 1996, Smith, 2007).

Many companies are aiming to have a differentiated health and productivity strategy. Citi tracks the medical bills, insurance claims, and monthly trends of the employees to arrive at a customized wellness packages for different segments. This is achieved through advanced analytics.

The Citi South Asia's Chief Human Resource Officer feels that employee's health determines the motivation, innovation, and delivery that further impacts the end results of the organization. Despite increased focus on health, 59% of the companies do not have a comprehensive health and productivity strategy (Mehra, 2014). Only 32% of the employers offer stress or resilience management.

Currently only 2% of the employers in India are offering financial incentives for participation in health and productivity programs. The financial incentives for health and productivity programs can boost employee participation.

13.12 CONCLUSION

The abundance of research on stress theory highlights its importance in our psychological, social, and psychological health (Lazarus, 2006). The human capital of an organization is an appreciating asset and should be managed well. This is particularly so for service organizations which are dependent on their employees for service creation and delivery. All out attempt should be made to maintain and retain the precious human resources of the organization. They should be trained and educated to increase their knowledge and skills to handle the increased demands in this turbulent, hypercompetitive world better. Stress management is vital for an organization to succeed as is evident from numerous studies linking stress to absenteeism, leave, health costs, insurance costs, low morale, and low productivity. The companies cannot ignore the management of stress in organizations as there are huge risks involved with that. In order to reduce leave, absence, and attrition rate, the employee health and morale should be monitored, evaluated, and enhanced. The workplace design should be novel and ergonomic in nature and the work should be structured in such a manner that it does not become a cause of stress. The management policies and practices should be aimed at minimizing the stress levels. Instead of following a rigid approach, some control over their work area should be provided. The channels of communication should be kept open at all times. Employees should feel appreciated and respected while working for the organization. Various techniques for managing stress in the literature and used by practitioners have been analyzed. They can be administered to individuals or groups and comprise of physical, mental, social, and spiritual elements. The eastern nations have a history of spiritual journey and so interventions touching mind, body, and soul have a long-lasting effect on employees. The organizations should focus on the group interventions as they inculcate bonding within the group and facilitates team working. There is no consensus in the literature over the best stress management technique. The organizations even in the developing world are aware of the repercussions of high level of stress in the organizations. The education and awareness campaigns are going on but still stress management is not considered the primary factor for increased productivity. There is no one solution for all the organizations. Each organization needs to identify the bouquet of stress management interventions for its use that best suit the profile of its employees.

KEYWORDS

- stress
- workplace stress
- organizational stress
- work–life balance
- life satisfaction
- stress management
- stress interventions

REFERENCES

Ackfeldt, A.; Malhotra, N. Revisiting the Role Stress-commitment Relationship: Can Managerial Interventions Help? *Eur. J. Mark.* **2013**, *47*(3/4), 353–374.

Alters, S.; Schiff, W. *Essential Concepts for Healthy Living* (5th ed.), Jones and Bartlett Publishers, LLC: Sudbury, MA, 2009.

Amarnath, N. Companies Want Employees to Gain Spiritual Quotient. *Econ. Times*, 6 February 2011 (online). Available from <http://articles.economictimes.indiatimes.com/2011-02-06/news/28428543_1_stress-management-courses-apex> (cited: 11 September 2015).

Arthur, A. R. Employee Assistance Programmes: The Emperor's New Clothes of Stress Management? *Br. J. Guidance Counsell.* **2000**, *28*(4), 549–559.

Avey, J. B.; Luthans, F.; Jensen, S. M. Psychological Capital: A Positive Resource for Compating Employee Stress and Turnover. *Hum. Resour. Manage.* **2009**, *48*(5), 677–693.

Barling, J.; Kelloway, E. K.; Frone, M. R., Eds. *Handbook of Work Stress.* Sage: Thousand Oaks, CA, 2005.

Bhatt, S. Erratic Working Hours take a Toll on Health and Life, 29 May, 2013, *Economic Times*, (online). Available from <http://articles.economictimes.indiatimes.com/2013-05-29/news/39580840_1_office-hours-12-hours-late-hours> (cited: 23 September 2015).

Business Line. *Employees's Stress Level Rising: Survey*, 6 September 2012 (online). Available from <http://www.thehindubusinessline.com/news/employees-stress-level-rising-survey/article3866526.ece> (cited: 11 September 2015).

Business Line. *The Deal with Stress.* 4 July 2013 (online). Available from <http://www.thehindu.com/features/metroplus/the-deal-with-stress/article4880837.ece> (cited: 11 September 2015).

Business Line. *Indian Cos Ahead of Asian Peers in Managing Work Related Stress: Survey.* 20 April 2014 (online). Available from <http://www.thehindubusinessline.com/news/indian-cos-ahead-of-asian-peers-in-managing-work-related-stress-survey/article5930700.ece> (cited: 12 September 2015).

Butts, M. M.; Vandenberg, R. J.; DeJoy, D. M.; Schaffer, B. S.; Wilson, M. G. Individual Reactions to High Involvement Work Processes: Investigating the Role of Empowerment and Perceived Organisational Support. *J. Occup. Health Psychol.* **2009**, *14*(2), 122–136.

Clarke, S.; Cooper, C. L. *Managing the Risks of Workplace Stress: Health and Safety Hazards.* Routledge: London, 2004.

Colligan, T. W.; Higgins, E. M. Workplace Stress: Etiology and Consequences. *J. Workplace Behav. Health* **2006**, *21*(2), 89–97.

Cooper, C.; Cartwright, S. An Intervention Strategy for Workplace Stress. *J. Psychosom. Res.* **1997**, *43*(1), 7–16.

Cooper, C. L.; Dewe, P. J.; O'Driscoll, M. P. *Organisational Stress: A Review and Critique of Theory, Research and Applications.* Sage: Thousand Oaks, CA, 2001.

Cotton, D. H. G. *Stress Management: An Integrated Approach to Therapy, Issue 17 of Brunner/Mazel Psychosocial Stress.* Routledge: London, 2013.

Choudhary, K. *Managing Workplace Stress: The Cognitive Behavioural Way.* Springer India: New Delhi, 2013.

Das, J. Holistic Development Purpose of Work–Life Balance. *Econ. Times*, 11 September 2012 (online). Available from <http://articles.economictimes.indiatimes.

com/2012-09-11/news/33763325_1_work-life-imbalance-work-life-balance-young-adults> (cited: 12 September 2015).

Deshpande, R. C. A Healthy Way to Handle Workplace Stress through Yoga, Meditation and Soothing Humor. *Int. J. Environ. Sci.* **2012**, *2*(4), 2143–2154.

Dewe, P.; O'Driscoll, M. P.; Cooper, C. L. *Coping with Work Stress: A Review and Critique.* John Wiley & Sons Ltd.: Chichester, West Sussex, UK, 2010.

Dollard, M. F.; Winefield, A. H.; Winefield, H. R. *Occupational Stress in Service Professionals.* Taylor & Francis: London and New York, 2003.

Economic Times. Women in Male-dominated Offices Undergo High Stress, 25 August, 2015, *Economic Times* (online). Available from <http://articles.economictimes.indiatimes.com/2015-08-25/news/65847266_1_stress-hormone-high-levels-cortisol-levels> (cited: 2 September 2015).

European Commission. Directorate-General for Employment, Industrial Relations and Social Welfare. *Guidance on Work Related Stress: Spice of Life or Kiss of Death.* Office for Official Publications of the European Communities, 1999.

Folkman, S.; Moskowitz, J. Stress, Positive Emotion, and Coping. *Curr. Direct. Psychol. Sci.* **2000**, *8*(4), 115–118.

Galinsky, E.; Kim, S. S.; Bond, J. T. *Feeling Over-Worked: When Work Becomes too Much.* Families and Work Institute: New York, 2001.

Goyal, M. Indian Women Most Stresses in the World: Nielsen Survey. *Econ. Times*, 29 June 2011 (online). Available from <http://articles.economictimes.indiatimes.com/2011-06-29/news/29717262_1_indian-women-stress-workplaces> (cited: 2 September 2015).

Giga, S. I.; Cooper, C. L.; Faragher, B. The Development of a Framework for a Comprehensive Approach to Stress Management Interventions at Work. *Int. J. Stress Manage.* **2003a**, *10*(4), 280–296.

Giga, S. I.; Noblet, A. J.; Faragher, B.; Cooper, C. L. The UK Perspective: A Review of Research on Organisational Stress Management. *Austr. Psychol.* **2003b**, *38*(2), 158–164.

Gruen, R. J.; Folkman, S.; Lazarus, R. S. Centrality and Individual Differences in the Meaning of Daily Hassles. *J. Pers.* **1988**, *56*(4), 743–762.

Hart, P. M. Predicting Employee Life Satisfaction: A Coherent Model of Personality, Work and Non Work Experiences, and Domain Satisfactions. *J. Appl. Psychol.* **1999**, *84*, 564–584.

Hicks, T.; McSherry, C. *A Guide to Managing Workplace Stress.* Universal Publishers: Boca Raton, FL, 2006.

Hindustan Times. Fight Stress Positively: Dalai Lama, 2011, *Hindustan Times* (online). Available from <http://www.hindustantimes.com/india-news/fight-stress-positively-dalai-lama/article1-645882.aspx> (cited: 23 September 2015).

Johnston, K.; Westerfield, W.; Momin, S.; Phillipi, R.; Naidoo, A. The Direct and Indirect Cost of Employee Depression, Anxiety and Emotional Disorders—An Employer Case Study' *J. Occup. Environ. Med.* **2009**, *51*(5), 564–577.

Kirk, A. K.; Brown, D. F. Employee Assistance Programs: A Review of the Management of Stress and Wellbeing through Workplace Counselling and Consulting. *Aust. Psychol.* **2003**, *38*(2), 139–143.

Kompier, M.; Cooper, C., Eds. *Preventing Stress, Improving Productivity: European Case Studies in the Workplace*; Kompier, M., Cooper, C., Eds.; Routledge: London, 1999.

Lazarus. R. S. *Stress and Emotions: A New Synthesis.* Springer: New York, 2006.

Lazarus, R.S.; Folkman, S. *Stress Appraisal and Coping*. Springer: New York, 1984.

Lazarus, R. S.; Folkman, S. Transactional Theory and Research on Emotions and Coping. *Eur. J. Pers.* **1987**, *1*(3), 141–169.

Lehrer, P. M.; David, H. B.; Woolfolk, R. L.; Sime, W. E. *Principles and Practice of Stress Management* (3rd ed.); The Guildford Press: New York, 2007; pp 46–47.

Mace, S. *The Effects of an Organisational Stress Management CreateSpace*. Independent Publishing Platform, 2014.

McCormack, N. Managers, Stress and the Prevention of Burnout in the Library Workplace. In: *Advances in Librarianship*; Woodsworth, A.; Penniman, W. D., Eds.; Emrald Publishing Limited, 2014; Vol 38, pp 211–244.

McEwen, B. S.; Lashly, E. N. *The End of Stress: As We Know It*. The Dana Foundation: New York, 2002.

McGrady, A. Psychopsysiological Mechanisms of Stress. In: *Principles and Practice of Stress Management*, 3rd ed.; Lehrer, P. M., Woolfolk, R. L., Sime, W. E., Eds.; The Guildford Press: New York, 2007; pp 16–37.

Mehra, C. Healthy Employees Are Good for Business. *Bus. Line*, 1 July 2014 (online). Available from <http://www.thehindubusinessline.com/specials/new-manager/healthy-employees-are-good-for-business/article6167236.ece> (cited: 12 September 2015).

Murphy, L. R. Stress Management in Work Settings: A Critical Review of the Health Effects. *Am. J. Health Promot.* **1996**, *11*(2), 112–135.

Murta, S. G.; Sanderson, K.; Oldenburg, B. Process Evaluation in Occupational Stress Management Programs: A Systematic Review. *Am. J. Health Promot.* **2007**, *21*(4), 248–254.

Naidu, S. Stresses? Help's a Mouseclick Away. *Econ. Times*, 19 February 2004 (online). Available from <http://articles.economictimes.indiatimes.com/2004-02-19/news/27407931_1_online-counselling-counsellors-employees> (cited: 4 September 2015).

Richardson, K. M.; Rothstein, H. R. Effects of Occupational Stress Management Programs: A Meta-analysis. *J. Occup. Health Psychol.* **2008**, *13*(1), 69–93.

Romanowska, J.; Larsson, G.; Eriksson, M.; Wikström, B-M.; Westerlund, H.; Theorell, T. Health Effects on Leaders and Co-workers of an Art-based Leadership Development Program. *Psychother. Psychosom.* **2011**, *80*, 78–87.

Roy, T. L. Companies Organising Stress Busting Courses for Employees to Increase Productivity. *Econ. Times* 15 April 2013 (online). Available from <http://articles.economictimes.indiatimes.com/2013-04-15/news/38556060_1_stress-management-corporate-yoga-client-list> (cited: 10 September 2015).

Saiyadain, M. S. *Organisational Behaviour*. Tata McGraw Hill: New Delhi, 2003.

Sarumathi, K. Temporary is the New Permanent. *The Hindu* 26 August 2015; p 18.

Sengupta, D. *You Can Beat Your Stress*. Excel Books: New Delhi, 2007.

Sengupta, D. Anti-burnout Measures Companies Provide for Staff Counselling. *Econ. Times* 12 July 2011 (online). Available from <http://articles.economictimes.indiatimes.com/2011-07-12/news/29765426_1_senior-managers-burnout-night-shifts> (cited: 23 September 2015).

Seyle, H. *Stress without Distress*. Lippincott: New York, 1974.

Smith, J. C. *Stress Management: A Comprehensive Handbook of Techniques and Strategies*. Springer Publishing: New York, 2002.

Smith, J. C. The Psychology of Relaxation. In: *Principles and Practice of Stress Management*, 3rd ed.; Lehrer, P. M., Woolfolk, R. L., Sime, W. E., Eds.; The Guildford Press: New York, 2007; pp 38–52.

Snyder, C. R.; Lefcourt, H. M. *Coping with Stress*; Oxford University: New York, 2001; pp 66–88.

The Hindu. Five Million Jobs Lost During Years of High Growth: Study, 26 August, 2015a, *The Hindu*; p 18.

The Hindu. 97% of Private Firms Lack Knowledge of Workplace Harassment Law, 26 August, 2015b. *The Hindu*; p 18.

The Hindu. Starting Salaries in India amongst lowest in Asia-Pacific Region: Study, 9 September, 2015c, *The Hindu*; p 20.

Theorell, T.; Osika, W.; Leineweber, C.; Magnusson, H. L. L.; Bojner, H. E.; Westerlund, H. Is Cultural Activity at Work Related to Mental Health in Employees? *Int. Arch. Occup. Environ. Health* **2013**, *86*(3), 281–288.

Totterdill, P.; Exton, R. Work and Organisations in 2020: The Future We Want? *Strat. Direct.* **2014**, *30*(9), 4–7.

World Health Organization. *Raising Awareness of Stress at Work in Developing Countries: A Modern Hazard in a Traditional Working Environment. (Protecting Workers' Health Series 6)*, 2007. World health Organization (WHO): Geneva (online). Available from <http://www.who.int/occupational_health/publications/pwh6/en/> (cited: 12 September 2015).

World Health Organization. *Occupational Health: Stress at the Workplace*, 2015 (online) (cited: 12 September 2015).

CHAPTER 14

STRESS MANAGEMENT IN SERVICE ORGANIZATIONS

UMA NAGARAJAN*

Department of Commerce, BML Munjal University, Gurgaon, Haryana, India

E-mail: uma.anita@gmail.com

CONTENTS

ABSTRACT

The beginning of the 21st century has brought a sea of change to the global economy. The digital revolution has turned traditional method of working on its head. The advent of computers and the internet has metamorphosed work processes and work timings and is continuously challenging existing set ups. The increased use of technology and digitalization in the workplace has led to better productivity, improved services, and lesser costs egging companies to further increase their machine power over manpower. This has led to huge job losses and increased job insecurity. It has also blurred the boundaries between home and work with employees often unofficially working round the clock. The consequence has been the creation of one of the most alarming lifestyle risk factors that the global work force suffers from today—Work Stress. WHO defines work stress as the response people may have when presented with work demands and pressures that are not matched to their knowledge and abilities and which challenge their ability to cope. The stress is often compounded by poor work organization, lack of support and control, and ambiguity in job responsibilities.

This chapter explores the major causes of work-related stress across the globe. A comparative study of work stress levels in different continents show the influence of culture, government policies, economy etc. The study gives an insight as to why stress levels are different in different countries for the same or similar types of jobs. An insight approach is analyzed to how employees handle the stress and manage work life balance for making workplace better and satisfying.

14.1 INTRODUCTION

What is this life if, full of care
We have no time to stand and stare
No time to stand beneath the boughs
And stare as long as sheep or cows

—William Henry Davies

To stand and stare is a luxury that employees of today can scarcely afford.

Vikram Bhatia had been working day and night on a new project launch. It was almost complete and he decided that he would postpone his doctor's visit by another couple of weeks. For the last few months, Vikram had been suffering from severe pain in the back and neck, tremendous physical exhaustion, disturbed sleep, and light-headedness. The symptoms came and went for no apparent reason. Just two more weeks, he told himself, and then I will take a long break and also go to the doctor. That evening, when he was returning home from work, Vikram suffered a heart attack. If he had just sat back and thought calmly, he would have realized that it had been going wrong for a very long time. The intense competition, escalating demands at work and late night client calls meant that he was always overworked. In fact, he prided himself on his ability to make do with less than 5 h of sleep. He was always on the run, so skipping breakfast had become quite the norm for him, but of late he had started grabbing colas and chips for lunch too. After all, they saved time and were less messy. He had been feeling uneasy and short of breath on more than one occasion, but he just attributed it to working a little harder than usual. Everything had added up and here, he was now, in the hospital, fighting for his life.

We are witnessing a major technological and digital revolution that has completely overhauled traditional work processes, timings, and work set ups. The benefits of this revolution have been rapid and far reaching, but the progress has come at a price. Computers, smart phones and internet have ensured an unceasing and uninterrupted flow of information round the clock. This has resulted in blurred boundaries between home and work, unrealistic work expectations, longer hours at work, and shortened hours of sleep. This modern madness of living life in fast-forward mode has given birth to one of the biggest lifestyle diseases of recent times—stress. This silent killer has not only quietly sneaked into everybody's lives, but is also making its presence felt in an increasingly alarming manner every year. The American Psychological Association has linked chronic stress to six of the leading causes of death. It has also been found that acute stress is the leading cause of sudden death, especially in young people with no evidence of coronary disease. Sadly, in spite of the mind-boggling and alarming evidence of the detrimental effects of stress, we still see people working themselves to death, literally and figuratively every day and the numbers only seem to be mounting. Why is this so and can anything be done about it?

The term "stress," as understood in the modern context, was first used by Hans Seyle in 1936. He defined it as the "non-specific response of the body to any demand for change." However, work stress and its consequent problems are relatively recent phenomena. NIOSH (The National Institute for Occupational Safety and Health) defines job stress as "the harmful physical and emotional responses that occur when the requirements of the job do not match the capabilities, resources, or needs of the worker" (NIOSH, 1999). Work-related stress is generated by a number of job characteristics including long hours at work and increasingly high workload, loss of control, low support at work, job insecurity, improper work–life balance, unclear work policies, lack of communication, and appreciation, reporting to multiple bosses, etc. Universally, the major causes of stress range from lack of job control to improper work–life balance.

14.1.1 LACK OF CONTROL

Job control is the freedom and ability of employees to use their skills and make workplace decisions. When employees are given limited freedom to make such decisions, they experience lack of control. This problem is compounded by ever increasing work load and higher responsibilities. An employee who has multiple responsibilities but has limited or little control over decisions experiences extremely high stress at the workplace. Karasek's job demands–control model states that when job control is low and demands are high, a job can be termed as an unhealthy or stressful one, which could in turn, increase the risk of illness or injury related to work. A report by EU-OSHA (European Union for Occupational Safety and Health) shows that 29–55% of workers in Europe felt they had no job control over various aspects of work. Further studies have also shown that lack of job control has a direct effect on lower self-reported health (Milczarek, Schneider, & González, 2009).

14.1.2 LONG HOURS AT WORK AND HIGH WORKLOAD

Workloads are increasingly becoming higher and higher in today's world. The average office timing of 9–5 simply exists on paper. An average employee spends almost 12 h a day at work or doing work-related activities. The increased use of technology and internet has been a boon for companies and a bane for employees. Late-night teleconferences, answering mails at wee hours of the morning, answering queries on vacation and weekends are minimum unwritten expectations in an employer employee contract. The unrealistic job expectations and heavy workload is one of the major factors increasing stress levels.

14.1.3 JOB INSECURITY

In today's dynamic and ever-changing market caught in a web of down-sizing, layoffs, outsourcing, hostile takeovers and the omnipresent threat of technology replacing human labor, doubts and uncertainty over keeping one's job is a thought that always lurks round the corner. The global recession has only exacerbated the fear. A layoff can be sudden and quite unexpected. More often than not, employees are clueless about the reason for the layoff and often find themselves out of job for no fault of theirs. Studies have shown that chronic job insecurity is a stronger predictor of poor health than smoking or hypertension. Living under the shadow of continuous job insecurity is even worse than unemployment. Under a constant cloud of uncertainty, employees are faced with the dual stress of looking out for new job openings while having a job at hand, a job that they fear they may lose anytime.

14.1.4 IMPROPER WORK–LIFE BALANCE

Work–life balance is adjusting or scheduling our day-to-day activities in such a manner as to achieve a balance between work and personal life. However, in today's work environment, this delicate balance is getting rattled. Employees often bring their work home and work at odd hours. Though employees may be physically away from their workplace, they are almost never mentally switched off. As a result, they find it increasingly difficult to spend quality time with parents, spouses or children or to participate in day-to-day activities at home and do other personal work. Finding time for relaxation or recreation becomes a luxury that most can only dream of. This disruption in work–life balance leads to mounting frustration and increased stress. Without maintaining a proper work–life balance, an employee will neither be able to live up to his full potential at his workplace nor will he able to do his personal work at peace. After all, no amount of money or success can take the place of time spent with family.

Stress is a universal problem in today's world. However, the extent, nature, and reasons of stress do vary widely across continents.

14.2 ASIA

Recently, a Chinese banking regulator, Li Jianhua died of a "sudden heart attack." He was just 48 years old. The CBRC (China Banking Regulatory Commission) categorized the cause of death as "long-term overwork." He had apparently been working all through the night to get a report done before sunrise, when disaster struck. Li had enough reason to be stressed. He was, after all, the director of CBRC and one of the main drafters of the market-based rules of China's banking system. Even more disconcerting is a statement released by CBRC soon after his death calling him "a model for party members" and someone "who had little time for his family." The report further states that "To learn from Comrade Li Jianhua, one must be like him, always firm in ideals and beliefs, the broader interest, loyal to the cause of the party and the people, unremitting struggle, sacrificing everything" (Oster, 2014). To ask a nation to emulate a person who clearly died from overwork is a dreadful indication that we are living in a world where work takes precedence over everything, including human life.

Asia consists of a number of developing economies like India, China, Thailand, Malaysia, etc. A report by WHO says that work stress is significantly higher in developing countries as compared to developed economies. It further elaborates that countries in transition are subjected to drastic economic and social changes like increased demand for adapting on the part of the workers, poor working conditions, and inadequate healthcare systems, erosion of cultural value systems, exposure to high pollution, poor hygiene, and sanitation conditions, non-availability of basic resources, etc., all of which lead to further aggravation of work stress.

On the one hand, people in developing economies of Asia are benefitting from the rapid development and surge in employment opportunities, while on the other, the constant lookout for greener pastures and the mounting pressure to be one step ahead of others in a highly competitive atmosphere leads to protracted stress. The high employee turnover ratio in these countries means an employee often find himself saddled with the job of two people, while managers find it increasingly difficult to plan tasks ahead or allocate resources in advance. In an ever shrinking world where businesses are conducted across different time zones with uninterrupted access to the internet and computers, late night teleconferences, video conferencing after office hours, and answering emails at odd times of the day is becoming more of an accepted practice than an exception. Since most of the clients and head offices of MNCs are situated in the USA or Europe, it is often expected that the employees in developing countries would make the necessary adjustments to their schedule and adjust to the time zones of the developed countries. This means an employee in Asia might be taking teleconferences in the wee hours of the morning or late at night, in addition to doing regular office work at normal hours. The poor infrastructure, inadequate transport systems, chaotic traffic, and increasing pollution makes commuting a nightmare. Merely reaching the workplace becomes a high cause of stress for employees.

The soaring inflation rate in Asia is also a major factor in compounding the financial stress of people. Increase in salaries is almost never able to keep pace with sky-rocketing prices and people have to work harder and longer to merely maintain the same lifestyle. A culture which lays down minimum requirements of a house and a car for marriage and a social setup that stigmatizes children who are unable to look after their aged parents physically or financially, only adds to the pressure of an already beleaguered workforce. Lack of a proper state security system or pension system to take care of the elderly, escalating costs of education and health care and the pressure to keep up with the Joneses in a consumer-driven market traps people in a vicious money making cycle with no respite. Cultural values add to stress in other subtle ways too. In South-East Asian cultures, people often avoid saying "no" because it is considered impolite and seen as being disagreeable. The inability to refuse to take on a partic- ular task due to cultural pressure can only accentuate stress for an already overloaded employee.

In a highly competitive environment, with many people striving for a few available promotion slots, employees feel a pressure to stay back and work at least for as long as their bosses do." Why can't you stay back" is the question frequently asked instead of "why should you stay back?" Globalization has led to an increase in opportunities for employment that has in turn led to a whopping increase in the number of first generation college-goers and first generation office goers. As a consequence, the supply of skilled and educated labor has increased exponentially with people working for more than 70 h a week for ridiculously low salaries. A huge majority of them are in the bottom part of the pyramid of Maslow's hierarchy of needs and are striving to move up. The incentive to break the glass ceiling of poverty to achieve a standard of living which was considered completely out of reach is an enormous motivating factor for the young first-generation-educated employees. In such a scenario, ideas like better working conditions, shortened working hours, personal time, and leisure get relegated to the background.

Brain drain has become an increasing cause of concern for developing countries in Asia. Brain drain refers to the emigration of skilled workers from developing countries to developed countries in search of better academic and job opportunities and improved economic status. It is one of the major factors for the plummeting levels of productivity and quality of output in developing countries. Rapid expansion and growth in developing countries has led to an escalating demand for skilled labor. This has led to the absorption of a huge number of graduates and post graduates in companies at an increasing rate every year. The process of trying to meet the headcount for skilled labor and the emigration of the best and talented workforce has led to the hiring of a number of mediocre employees in the skilled labor category. This has in turn reduced productivity and quality of output in a number of fields leading to employees putting in longer working hours. Zhu Qingyang, secretary-general of the Shanghai Human Resources Consulting Association, said: "What insiders have found is that many companies can't find qualified talent. Many young students didn't meet the standard for a professional career person" (Xiaomin, 2007). In fact, the pressure to maintain certain basic productivity and quality levels coupled with non-availability of enough good quality skilled employees is one of the reasons that employees in developing countries work longer hours. The longer working hours, in turn, lead to physical and mental exhaustion and become a prime factor in reducing productivity and output quality, thus trapping them in a vicious cycle. Doing a comparative analysis of working hours of different countries from data from OECD and ILO Jon Messenger, an ILO expert on working hours says "Asian countries tend to work the longest, they also have the highest proportion of workers that are working excessively long hours of more than 48 hours a week." In contrast, working hours in the developed world have reduced from 3000 a year in the 1900s to around 1800 hours a year in the 21st century (Stephenson, 2012). An increasing number of workers in Asian economies are suffering from obesity, depression, migraines, insomnia, and anxiety attacks, and broken relationships—most of which can snowball into more serious and sometimes fatal conditions such as heart attacks, diabetes, or cardiac arrests. The mantra for youngsters today is "work hard and party harder." In order to combat the mounting stress, youngsters let themselves loose on weekends indulging in binge drinking, smoking, and late-night partying. The proliferation of stress-related ailments in the workforce and the increasing number of employees lost due to fatalities and other health-related disorders is creating a potential health-care crisis like situation in a number of Asian countries.

Suicidal devotion to work has in recent times touched alarming highs in Asia, especially among the Japanese and Chinese. The Japanese term "karoshi," literally means, death due to overwork and instances of karoshi are on the rise in Asia. China leads the table with the highest number of work exhaustion-related deaths. According to the China Youth Daily, 600,000 people die from work stress and related effects in China every year. Other reports put the numbers at a daunting 1600 deaths a day (Oster, 2014). A significant percentage of these deaths are of white collared workers doing mentally stressful jobs. The primary killer of karoshi victims is heart attack and stroke caused by obesity, insomnia, listlessness, loss of immunity, diabetes and high blood pressure. It is no wonder that the Chinese have their own term for work-related deaths—guolaosi. However, Chinese employees are not the only ones succumbing to this epidemic. Reports of work-related deaths are on the rise in a number of countries in South-East Asia. From the 40-year-old Taiwanese woman who died of a stroke after late night work-related texting to the 30-year-old Japanese employee in Toyota whose body simply caved in after clocking continuous overtime for 80 h a month for the last 6 months, work-related deaths are on the rise (Economist, 2007). Call it Karoshi, guolaosi or by any other name, a killer of epidemic proportions is on the rise in Asia.

In such a high stress environment, it is not surprising, that a number of young employees have started quitting their jobs with no new offers in hand. There has been an unprecedented rise in the number of people quitting to rejuvenate themselves and take a break, to take a sabbatical, to follow their passions, or merely to rethink their options. However, quitting or taking a break is a luxury that many can scarcely afford. The overwhelming majority of employees simply continue to slog away and sometimes literally work themselves to death. Time and again, different surveys on stress levels have yielded the same result. Grant Thornton's 2006 International Business Owners Survey (IBOS) studied the stress levels of more than 7000 business leaders in 30 countries. Unsurprisingly, Asian countries topped the table. Taiwan reported rising stress levels, with almost 90% of business owners saying they felt under increased pressure in 2005 compared to the previous year. China came a close second with 87% reporting high stress levels, the Philippines was third on 76%, Hong Kong, Singapore, and Malaysia filled the fifth, sixth, and seventh spots. The only country among the top six which is not in Asia is Botswana, in fourth place (Business Leaders' Stress Levels Increased by 50 Per Cent in 2005, 2006).

A survey by Regus in 2012 of more than 16,000 professionals across the globe shows that regionally, more Asia-Pacific respondents see their job as a source of stress than their peers in Europe, the USA, or Latin America. Thailand (75%) had the highest number of respondents that indicated work as the primary cause of stress. This was followed by China (73%), Vietnam (71%), Indonesia (73%), Singapore (63%), Hong Kong (62%), Malaysia (57%), and India (51%). Hong Kong and China had the highest number of people who reported an increase in stress from the previous year (Work is top trigger of stress for Asia-Pacific workers|CFO innovation Asia, 2012). A whopping 85.9% of Chinese workers said that their stress levels had increased from the previous years. Most Asians stressed the need for flexible working hours, which they felt would help to reduce stress levels. It is indeed ironical that though 75% of the world's workforce comes from developing countries, enough study has not yet been undertaken to understand the impact of work-related stress on employees of these countries.

14.3 NORTH AMERICA

Sarvshreshth Gupta, just 22 years old, when he was found dead in the parking lot of his building in San Francisco on 16 April 2015. Was it an accident, a case of unexpected death or did the young man consciously decide to end his misery once and for all? A finance graduate from the Wharton Business School, Sarvshreshth was thrilled when he landed a job as an investment banker in the Goldman Sachs Group Inc. (Moore, 2015). Work was intense and schedules were grueling, but initially the young man felt that he could handle it. The New York Times in its report states that "he was universally liked by his colleagues and was in fact so good at his job that he had become one of the go-to analysts" (The New York Times, 2015). In fact, his exceptional competence and strong work ethics probably led to his large workload. Soon work was spilling over into weeknights and occasionally into weekends too. He told his father that he was often working 20 h at a stretch and was terribly sleep deprived. The long-working hours began to take their toll on his mental and physical health and he started complaining to his father. "This job is not for me. Too much work and too little time. I want to come back home." A few months later he submitted his resignation. But under insistence from his parents and his company that he reconsider his resignation, he gave in and rejoined. The pressure, however, continued to be relentless. A month later, he told his parents "It is too much. I have not slept for two days, have a client meeting tomorrow morning, have to complete a presentation, my VP is annoyed and I am working alone in my office" (Gupta, 2015). His misery and agony ended that day, paradoxically, with the end of his life.

"Gluttons for punishment are rewarded on Wall Street, especially at young levels," says Kevin Roose, the author of a book called Young Money which sheds light on the lives and work of young Wall Street bankers (Cohan, 2015). While Asia seemingly tops the list for overworked harried underpaid employees, the rest of the world is not really having it easy either. A 2013/2014 Towers Watson Staying@Work Survey, conducted by global professional services company Towers Watson, and the National Business Group on Health found that stress is the number one work-related risk factor for employees in the USA (78%) and a bigger cause of concern than obesity (75%) and physical inactivity (73%). Unfortunately, only 15% of employers prioritized improvement of the emotional/mental health of employees or stress reduction in their health programs (Watson, 2014). Research by Stanford and Harvard Business School showed that workplace stress contributes to at least 120,000 deaths in the USA each year and accounts for up to $190 billion in health-care costs (Lynch, 2015). Research by the APA (American Psychological Association) shows that though money (64%) was cited as the primary cause of stress, work (60%) comes a close second (Chart 14.1). It also shows that though workplace stress is on a reducing trend from 2007, the USA still has a long way to go before its people can actually relax. Women reported higher levels of stress than men, which is a trend observed in many countries as shown in Chart 14.2 (Paying with Our Health, Stress in America, 2015).

CHART 14.1 Causes of work-related stress in the USA (American Psychological Association, 2014).

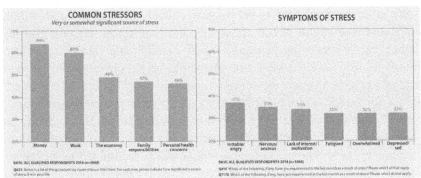

(From Stress in America™: Paying With Our Health. American Psychological Association. http://www.apa.org/news/press/releases/stress/2014/stress-report.pdf, p. 10)

CHART 14.2 Chart showing stress levels by gender (American Psychological Association, 2014).

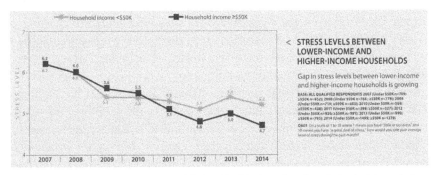

(From Stress in America™: Paying With Our Health. American Psychological Association. http://www.apa.org/news/press/releases/stress/2014/stress-report.pdf, p. 4)

    The burgeoning impact of globalization and the lure of low cost markets have led to developed countries outsourcing their manufacturing and white collar jobs to developing countries. With an increase in outsourcing and the massive influx of skilled employees into developed countries, the threat of losing their jobs always looms large for workers. Those who do manage to retain their jobs are burdened with increasing workloads and longer working hours and of course, the fear, that their own job could be outsourced anytime. To add insult to injury, workers are often expected to cooperate in training of their replacements, a phenomenon referred to as "knowledge transfer." This leads to resentment and loss of morale among the existing workforce. In an uncertain economy just finding its way out of a recession, the perils of not being able to regain employment are very real and palpable.

The continuous migration of skilled intellectual capital from developing to developed countries is undoubtedly a boon for the latter as it helps in increased productivity and improving the quality of output. The vast and increasing brain drain for more than over a decade means that the cream of the intellectual workforce ends up in developed countries. However, there is another side of the coin too. With the migration of best of the workforce from all over the world, competition has intensified tremendously in the workplace leading to higher levels of pressure. There is a constant expectation to deliver the best quality output and improve productivity. With the best resources and the world's most talented skilled labor competing for jobs and promotions, there is hardly any scope for mediocrity. Being part of an economy driven by forces of capitalism and fueled by aggressive cost cutting and profit maximization, employees are continuously asked to meet stiff, and challenging targets. Those who are unable to meet the targets are often ruthlessly laid off with little or no prior intimation. The competition is cut throat and the Damocles sword of job insecurity hangs over everyone's head. A recent survey by the American Psychological Association (2014) found that job insecurity ranked among the top five causes of work stress (Employee distrust is pervasive in U.S. Workforce, 2014).

The Towers Watson's Global Benefits Attitude Survey of 5070 US workers found a huge disconnect between employers and employees on the causal factors of stress. As per the survey, employers felt that lack of work/life balance (86%), inadequate staffing (70%), and technologies that expand employee availability during nonworking hours (63%) were the main reasons for increasing stress in employees (Watson, 2013). Employees, however, felt otherwise. They ranked inadequate staffing as the number one factor for high stress. Low pay increase which was ranked as the second highest factor for high stress by employees was ranked ninth by employers. Employees cited conflicting job expectations as the third highest stress factor. The differences between the employer employee ranking of stress factors show that employers still have a long way to go in understanding the stress drivers of employees and finding ways to combat them. The survey also revealed that most top performing employees stated work-related stress as the main reason for quitting an organization. Says

Brian Kropp, a managing director at the Corporate Executive Board, an advisory firm, "companies in recent years have heaped additional duties on their most promising employees without giving much in the way of support or extra rewards. We tend to put them in the hardest jobs we have, and we tend not to support them. In many ways we go too far" (Frauenheim, 2011). This ends up making the most talented and promising employees prone to the highest stress levels. The continuous pressure to meet unrealistic targets and work in grueling schedules takes a heavy toll on employees leading to deteriorating performance and effort. Employees suffering from high stress levels are not only less productive but also have lower engagement, and higher absentee levels. Wayne Hochwarter, a management professor at Florida State University in Tallahassee probably sums it up best when he says, "The person is left with an empty tank" (Frauenheim, 2011).

The 2010 General Social Survey (GSS) by Statistics Canada showed that 27% of Canadian workers described their lives on most days as "quite a bit" or "extremely" stressful. But more distressing is the fact that 62% of those surveyed said that work was the primary cause of stress (Chart 14.3). The higher stressed employees also were the better educated ones. Among those who identified work as the major stress factor, almost 75% had a college or university education and more than 50% held white-collar jobs in management, professional or technical occupations. Finances and lack of a proper work–life balance were the other top stress factors for Canadians (Crompton, 2011).

CHART 14.3   Chart showing percentage of highly stressed employed population in Canada aged 20–64.

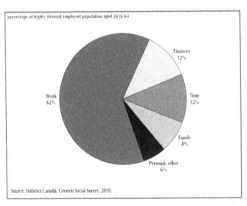

percentage of highly stressed employed population aged 20 to 64

Finances
12%

Work
62%

Time
12%

Family
8%

Personal, other
6%

Source: Statistics Canada, General Social Survey, 2010.

Source: http://www.statcan.gc.ca/pub/11-008-x/2011002/c-g/11562/c-g02-eng.htm.

14.4 EUROPE

Oxford-educated city lawyer, Matthew Courtney plunged to his death from a stairwell at London's Tate Modern. Why did a young and promising lawyer suddenly jump to his death? His heavy workload had undoubtedly been troubling and enervating him. He and other colleagues had also spoken to senior partners at Freshfields Bruckhaus Deringer about their excessive workloads. In fact, he had been doing 14-h shifts in the days before his death. Although the young lawyer had been receiving treatment for depression for some time, the possible adverse impact of his copious working hours on his health cannot be brushed aside (Evening Standard, 2007). More recently in London, an exhausted 21-year-old Merill Lynch intern died of an epileptic fit after pulling off three consecutive "all nighters" (The Guardian, 2013). Moritz Erhardt's epileptic fit could have been brought on by the fatigue and exhaustion of working punishingly long hours in a world where working oneself to the point of death is considered akin to receiving a medal of honor. Though the exact cause of these deaths is not known, what is undeniably clear is noxious impact of the grinding schedules and the death defying work that people put in at their workplaces.

Europe has perhaps the best documented data on stress-related problems at work. Periodical surveys and the advantage of being part of a European Union have proved hugely beneficial in collecting and consolidating data. As per a report by EU-OSHA stress is the second-most commonly reported work-related health problem in Europe. Half of the workers in Europe (51%) believe that work-related stress is a common problem in their workplace. Unsurprisingly, in Greece and Cyprus, more 81% and 88% of workers feel that stress is a common problem and in Slovenia 3/4th of employees cite stress as a major work problem. On a positive note, however, across Europe 45% of employees say that workplace stress is rare (European Opinion Poll on Occupational Safety and Health, 2013).

The deleterious impact of job insecurity can be seen in Europe too with 72% citing it as a major cause of stress. The recession has seemingly compounded the problem and intensified fears of job losses. Heavy workload was also considered to be major stress factor (66%). Facing harassment or bullying at work is another common stress factor for Europeans (59%). More than half the workers felt that lack of clarity on roles and responsibilities was a big stress factor (European Opinion Poll on Occupational Safety and Health, 2013).

Another contrasting trend that showed up in Europe was that while employees were working fewer hours than before, the work intensity was rising and schedules were getting tighter. The positive trend toward shorter working hours is further substantiated by the fifth EWCS (European Working Conditions survey) report which showed a decrease in average working hours from 40.5 h a week in EU 12 in 1991 to 37.5 h a week in the EU 27 in 2010 (36.4 h a week in EU 12). One of the major reasons for the trend toward shorter working hours is the increasing prevalence of part time jobs. The number of people working part time, that is, "short part time" (20 h/week or less) and "substantial part time" (21–34 h/week) has increased from 17% in the EC 12 (1991) to 25% in EU 27 (2010) and the number of people working long hours (working 48 h or more per week) has decreased from 18% in the EC 12 in 1991 to 13% in the EU 27 in 2010 (12% in the EC 12) (Fifth European Working Conditions Survey, 2012). This shows that Europeans are working shorter hours, with an increasing number of people opting for part time jobs. This also explains the impressively high percentages of good work–life balance in Europe. An overwhelming 81% of employees from EU 15 and 73% from EU 10 reported that they were happy with their work–life balance and had sufficient time to balance social commitments with job responsibilities.

Even though more than half the Europeans are stressed out (51%), the numbers there are still reassuringly lower than Asia (60–75%) and the USA (80%). The Europeans further add a feather in their cap by reporting remarkably good percentages of work–life balance (73–81%). So what is it that Europe is doing that helps alleviate the stress of its employees, at least to an extent? An interesting detail thrown up by a Regus survey is that while the USA, Mexico, and India are on or near world average for increase in workplace stress, Canada and France made it to the under average list. Expectedly, the awards for lowest increase in workplace stress went to three European countries United Kingdom, Germany, and Netherlands (Regus, 2009). The irony is that France has made it to the below average list of stressed out countries in spite of the spate of recent employee suicides due to high work stress. In such a scenario, one cannot help wondering what would be the levels of stress in other "above average stressed out countries" across the globe.

The WHO shows that there is reasonable research to conclude that there is an association between working conditions and stress. It also mentions the effects of stress. The Eurofound research on work-related stress in 2010 shows that the primary risk factors which induce stress are a heavy workload, extremely long working hours, lack of control at work, poor relationships with colleagues, poor support at work, and the impact of organizational change. Chart 14.4 shows that impact of stress on men and women in Europe. It should be noted that this is the impact that stress has in a continent whose employees supposedly have relatively lesser stress and better work–life balance as compared to North America or Asia.

CHART 14.4   Health and well-being outcomes by gender (%), in Europe.

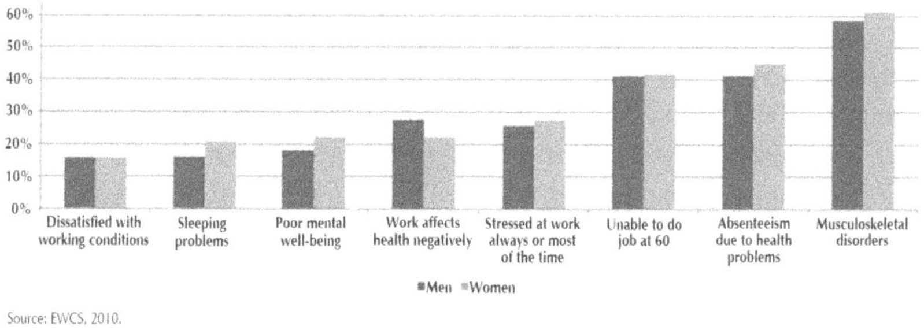

Source: EWCS, 2010.

Source: https://osha.europa.eu/en/tools-and-publications/publications/reports/psychoso-cial-risks-eu-prevalence-strategies-prevention.

Considering that stress is a primary cause of concern across countries and continents, what exactly can be done to combat it? How much of a role does the country, work environment, culture, company, and individual play in elevating or alleviating stress? Studies show that though developing countries do have the highest levels of stress, developed countries are not far behind either. A developed country like South Korea reports notoriously high levels of stress, while some developing economies such as Romania and Bulgaria report far lower levels of stress than their Asian counterparts. Developed countries in Europe report a better life work–life balance as compared to Americans. Though countries like Mexico and Costa Rica are still on the development path, they figure higher on the happiness index than economically developed countries like the USA, United Kingdom, France, Germany, or Japan (Helliwell, Layard, and Sachs, 2015). This shows that though stress is caused by work-related factors like job control, insecurity, and work balance external factors like the state of economy, cultural practices, government, and company policies, etc. also have a significant impact on them.

14.5 CULTURE

When Leonardo Da Vinci said "Every now and then go away, have a little relaxation, for when you come back your judgement will be surer," little would he have expected that a few hundred years later, an entire continent would be following his advice. Europeans have a reputation for having a more relaxed culture than Americans or Asians. While the Americans get a shot in the arm from increased productivity and economic growth and Asians pride themselves on their ability to work long and hard and expand their list of achievements, the average European takes his annual long vacation very seriously, having fun, and relaxing when necessary. While the developing nations in Asia are on the run to own a minimum of one car and a house and the Americans fight it out to own bigger and better cars and houses, the Europeans' self-worth seems less linked to the size of his car or house. In fact, Mauro Guillen, Wharton management and sociology professor says, "It is a sign of social status in Europe to take a long vacation away from home. Money is not everything in Europe; status is not only conferred by money. Having fun, or being able to have fun, also is a sign of success and a source of social esteem" (Reluctant Vacationers: Why Americans Work More, Relax Less, than Europeans—Knowledge@ Wharton, 2006). Europeans also tend to really disengage from work while on vacation unlike Asians and Americans who never really switch off and are still connected through their laptops and blackberries. The European culture encourages people to take vacations, limit overtime at work and maintain a good work–life balance. Since a major reason for stress is the driving urgency to keep up with the Joneses and fit into cultural practices that encourage overtime, the European culture and societal practices deserve a special mention for their role in limiting stress levels, especially in today's world where everyone is out on a rat race.

14.6 GOVERNMENT POLICIES

14.6.1 TAXATION

Different taxation policies in different countries play a major role in influencing work attitudes and work hours. Ed Prescott, a Nobel Prize-winning economist of Arizona State University points out those European countries has much higher tax rates than American countries. This discourages the average European from working too hard because he knows that he will be reaping very small margin of the additional benefit. A lot of European countries top the table when it comes to countries with the highest taxes. As per data from the OECD, the all-in rate, that is, the top personal income tax and social security contributions rate is more than 55% in a number of European countries. It is highest in Slovenia (61.05%), followed by Belgium (59.4%), Finland (57.2%), and Sweden (56.86%). Other than these countries, the all-in rate stands at 55% or more in Denmark, France, and Portugal and is greater than 52% in a number of other European countries. The all in rate of Canada is 49.5%, 48.6% in the USA, and 35% in Mexico.

14.6.2   LEAVE POLICIES

As per a report by the Centre for Economic and Policy Research (CEPR), the United States is the only developed country which has no mandatory paid holidays or paid vacation. Almost one in four Americans has no paid vacation and no paid holidays. No wonder, the United States is being dubbed the "no-vacation nation" (Barber, 2013). The European Union's (EU) Working Time Directive (1993) has set a minimum of four weeks or 28 vacation days per year for all EU member countries. However, several EU members offer more leave to their employees. As per data from the CEPR (Centre for Economic and Policy Research), Austria, and Portugal show the way with 35 days of paid vacation and holidays together, Spain follows a close second with 34. Italy and France give 31 days, while Belgium and Germany give 30 days. United Kingdom follows with 28; Norway 27, Denmark, Finland and Sweden with 25. Canadians get 19 days and even the indefatigable Japanese get 10 days of paid vacation, but the Americans get none (Chart 14.5).

Studies show that Asians and North Americans get far less holidays and vacation time than their European or South American friends. An interestingly titled study called the vacation deprivation study conducted on behalf Expedia.com by Northstar, a global consulting firm gives us an insight in to vacation utilizing habits across continents. While Europeans use almost 100% of their vacation time (28 days average), the Asians and North Americans are quite the workaholics. Europeans also got 12% more vacation time than the global average. The average American got only 15 days of paid leave in 2014 and used only 14. The average worker in the

Asia-Pacific (Hong Kong, India, Malaysia, Japan, Thailand, South Korea, and Singapore) got 16 days, but did not utilize 4 of those days. Thailand topped the list of vacation deprived countries offering a measly 11 days of vacation, but South Korea took the cake with employees using only 7 of the 15 days offered. Japan came in a close second with employees utilizing only 10 of the 20 days offered. The reasons for nonutilization of vacation ranged from the inability to fit in vacation into work schedules to expressing a desire to encash it or carry it over. However, universally 80% of workers stated that vacations make them feel happier which shows that vacations indeed play a role in mitigating stress levels. Sixty-one percent of the British (global average 56%) said that they would switch jobs if they could have more holiday time and 33% said that they would prefer more holidays to a pay rise. Undoubtedly, holidays help people feel happier, more relaxed, and less stressed and worried. Andy Washington, Managing Director of Expedia.co.uk says "It's almost paradoxical. People spend more time away from work, and might just be a better performing employee" (Expedia's 2014 Vacation Deprivation Study, 2014).

CHART 14.5   Paid vacations and paid holidays OECD nations, in working days.

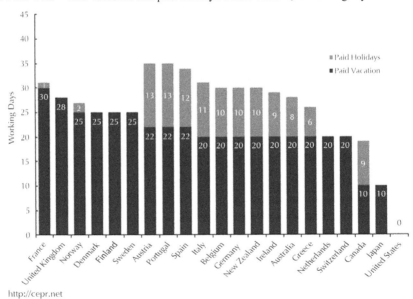

http://cepr.net

Source: http://www.cepr.net/press-center/press-releases/united-states-still-a-no-vacation-nation. (Published under https://creativecommons.org/licenses/by/4.0/)

14.6.3 UNIONIZATION

Europe has more organized levels of unionization with better influence in politics and boardrooms. These unions, therefore, help in lobbying for policies which are more employee friendly. In the United States, the policies are more employer centric rather than employee centric. Europe also has one of the best social security programs in the world offering wide range of benefits ranging from unemployment assistance and pension programs to disability and health care benefits. Few countries in Asia have any unionization which benefits employees.

14.6.4 MATERNITY, PATERNITY, AND PARENTAL LEAVE

Paid maternity leave is at least 14 weeks in most countries in Europe with some countries offering more than 18 weeks. Canada offers around 12 weeks paid leave and most countries in Asia too offer between 12 and 13 weeks of paid maternity leave. The United States is woefully lacking with respect to maternity leave and qualifies as the only developed country that does not offer any paid maternity leave (ILO Database of Work and Employment Laws, Maternity Protection, International Labour Office, 2009).

Paternity leave, though a necessity, has not been given its due importance by a number of countries. Europe stays ahead of the curve here too with most countries in the EU giving around 10 days of paid paternity leave. Some like Iceland and Slovenia generously offer 90 days of paid paternity leave. The concept of paid paternity leave is a rarity in Asia. A handful of countries like Indonesia (2 days) and Philippines (7 paid days) and offer paternity leave (ILO Database of Conditions of Work and Employment Laws on Maternity Protection and ILO NATLEX, 2009). With no policies for paid maternity leave in place, the United States has a long way to go before policies for paid paternity leave can be brought in.

Unfortunately most countries do not even acknowledge or recognize the need for parental leave. The ILO defines parental leave as a relatively long leave available to either parent allowing them to take care of an infant or young child over a period of time following the paternity or maternity leave period. The duration of parental leave is different in different countries and the payment during such leave is minimal and could also be unpaid in a number of cases. Parental leave is quite common in countries in developed countries, Eastern Europe and Central Asia, but is an almost unheard of phenomenon in other countries (Maternity and paternity at work, 2014).

14.6.5 COMPANIES

"All work, no play makes Jack a stressed out employee."

With so much ado about work-related stress, its catastrophic side effects and the appalling increase in "karoshi," it is easy to believe that every human being in the world drags himself out of bed on a Monday morning, curses his, and slogs away unwillingly and resentfully all through the year. Nothing could be actually be farther from the truth. There are a number of people all over the world who really like to go to work every day and actually think their companies are incredibly wonderful places. The company's culture and work environment undoubtedly have a huge role to play in creating such a positive perception, which helps mitigate stress levels and brings out the best from an employee. Glassdoor Career Trends Analyst Scott Dobroski says "Company culture is among the top five factors people consider" when weighing a job offer (Dill, 2014). A good work environment would be one which creates a culture based on trust and nondiscrimination, fosters excellence, and inculcates a feeling of pride and camaraderie in the employees. A survey of the best companies to work for (2015) by Fortune.com and Great places to work, ranks Google. com as the number one company to work for, for the seventh consecutive year (Noyan, 2016). Google also tops the list in a survey by glassdoor. com of the best US companies to work for and comes 2nd insurvey by Economic times ranking the best 50 Indian companies to work for. An employee with Google says, "The company culture truly makes workers feel they're valued and respected as a human being, not as a cog in a machine. The perks are phenomenal." What is it that Google.com is doing right that helps it to top the list of best workplaces globally and individually across continents? Googlers state that they feel a sense of pride in being part of a company or project that helps change our world. They are constantly thrown open to new challenges and love being part of an environment that, they feel, recognizes, and rewards employee potential. Seventy-nine percent of employees in Google feel that promotions go to deserving candidates which speaks volumes about the sense of trust and fairness in the system. An overwhelming 94% of employees felt that they carry a lot of responsibility in the organization and said that they were given the freedom to carry out tasks without micromanagement from their bosses. Their generous 18 weeks paid maternity leave policy goes a long way in ensuring that they retain their women employees. Paternity leave is between 12 and 18 weeks (Otani, 2015). They provide preventive health checkups for spouses of employees and encourage flexible or even part time work options, if needed. Recently, they have started working on system wherein addition to the annual appraisal, employees can give real-time feedback to their colleagues through "GThanks." Their famous nap pods, mouth-watering free food, onsite haircuts, and medical checkups, complimentary laundry, spa, and gym create a workplace that employees simply don't want to leave. Other unique benefits like allowing well behaved canines at work, providing company bicycles for employees to move around their huge campus, adding a healthy dose of practical pranks to double the fun quotient, etc. make this company an employee's dream place to work for (Fortune, 2016).

Another company which has endeavored to create a work environment that benefits both employer and employee with rich payoffs to both is Southwest Airlines. One of the reasons Southwest Airlines flies high is their warm and genial work culture which churns out loyal employees and in turn creates loyal customers. The term "employee recognition" takes on a new meaning here. Employees recognized by customers are featured on newsletters, on the intranet, lauded by the CEO on videos with honorary dinners to boot. Employees are not only made to feel like part of one big family but are also empowered to take decisions on the job that would help customers. A customer recounts the moment when she and her children had gone to the airport to see her husband off for a 6-month deployment to Kuwait. Understanding the emotion of the moment, Southwest employees helped her to go till the gate to see off her husband. Her children were, in fact, allowed to go till the plane to give their father a send-off hug (Gallo, 2014). The empowering of employees to take on the spot decisions to help improve customer experiences not only gives them recognition but also instills in them the feeling of satisfaction and happiness of having put a smile on someone's face. As Kelleher, former CEO of Southwest puts it. "The core of our success, that's the most difficult thing for a competitor to imitate. They can buy all the physical things. The things you can't buy are dedication, devotion, loyalty" (Price, 2009). It is no wonder that the employee turnover here is only 2% and, many of South West's original employees are still with the company (Makovsky, 2013).

The recent spate of deaths has prompted companies to regulate work schedules and redefine work boundaries. In June, Goldman introduced a policy of no "all nighters" for its summer interns. The interns are expected to leave the office before midnight each day. The full time junior bankers have been instructed to stay away from away from office between 9 pm Fridays to 9 am Mondays and violations are strictly monitored. Barclays has brought in policies to ensure that analysts do not work for more than 12 days in a row. Bank of America insists that analysts take leave of at least four weekend days a month (Cohan, 2015). In a speech to a group of young bankers in July, Jamie Dimon, chief executive of JP Morgan Chase said "You've got to take care of your mind, your body, your spirit, your soul, and your health. If you neglect these things, you will destroy your personal relationships. You'll destroy your life. You won't be healthy. You won't enjoy it" (La Roche, 2015).

14.6.6 STRESS AND PEOPLE IN THE FAST LANE

Mahatma Gandhi's words "There is more to life than increasing its speed." are more relevant today than ever before. So how do the people running on the fastest lanes complete the marathon? What helps the top people handle the pressure of deadlines, expectations and managing multiple tasks day in and day out?

The President of the United States, Barrack Obama, who arguably has the most stressful job in the world, says that his steady and calm temperament and his morning exercise routine are two big factors which help him handle stressful situations. "I don't get too high, don't get too low," he says. He talks about the importance of being with family, saying that it helps to give him perspective and keeps him grounded. He further adds that we live in a very fast paced 24 × 7 world "Everything is a crisis, everything is terrible, everything is doomsday. But a year later nobody is talking about it." He feels that in retrospect things are not even half as bad they are purported to be. So if we calm down and simply do our best about handling a situation, things would more often than not, turn out quite fine (The Huffington Post, 2015).

Top sportspersons are often faced with the twin pressure of managing crowd expectations and defending their titles and winning matches each time they step onto the field. Usain Bolt, holder of 11 World championship titles and the fastest man alive talks about how he manages the expectations each time he has to defend his title. "I've learned over the years that if you start thinking about the race, it stresses you out a little bit. I just try to relax and think about video games, what I'm gonna do after the race. Stuff like that to relax a little before the race (H.D., 2011). You've got to do this for yourself first. If you do it well, then people will love you (McRae, 2010). For over a quarter of a century, "God of cricket" Sachin Tendulkar was India's mainstay batsman and the leading "run getter." Each time he came out to bat, a country of a billion held its breath in anticipation and folded its hands in prayer. Work came to a standstill and resumed only after he was out. On being asked about how he coped with the enormous pressure and unrealistic expectations from fans, he says, "Pressure gets to you because your mind is either delving into the past, or is peeping into the future. If your mind is preoccupied with past failures, or is anticipating future ones, the pressure builds. Your mind has to be in the present. Only then will you not feel the pressure." He finds meditation techniques to be a very useful way of handling stress, especially the techniques which require a person to clear the mind of outside matters and simply focus on their breathing. He also adds that achievements matter only when you have people to share it with and having the right people around you helps a person to go on a positive cycle of success, health, and happiness (Lele, 2010).

Bill Gates, founder of Microsoft believes that the mantra to stress handling is to simply "keep it simple." Talking about the best learning from Warren Buffett he says that he was amazed at his ability to boil things down. "To just work on things that really count, to think through the basics. It's a special form of genius" (Crippen, 2009). Richard Branson founder of the Virgin Group puts things in an entirely new light. He says that stress is all about the right prioritization "If I lost the whole Virgin Empire tomorrow, then I'd just go and live somewhere like Bali. Now if there was a problem with my family, health wise ... that's a problem" (Adgey, 2008). While some CEOs like Susan Wojcicki (CEO of YouTube) feel that taking time off can offer new insights and perspectives and be a huge factor in stress reduction (YouTube CEO: How I balance a big family, work, 2014), others like Jeff Bezos, Founder, Chairman and CEO of Amazon.com feel that the major cause of stress is inaction over something that one has control over. Says Jeff Bezos, "I find, as soon as I identify the cause and make the first phone call or send off the first email, it dramatically reduces any stress that might come from it" (How Do You Deal with Stress Jeff Bezos, 2015). Obviously different strategies and plans work for different people. The key is to find the right stress buster. Some CEOs have devised their own unique ways to handle stress. Robert Freedman, CEO of ORC Worldwide, says that he spends two minutes every morning drawing on a napkin while he drinks his coffee. This helps him clear his mind for the day. John Benson, CEO of eFinancialCareers. com seems to have summed it up beautifully when he says, "Remaining disciplined by managing both time and productivity well is the backbone to good stress management. Time management allows me the opportunity to exercise regularly and to switch off the office so I can get colour and perspective in my life. That means putting the BlackBerry literally out of reach, spending time with my family, enjoying the outdoors and trying something new. Adding texture to life isn't indulgence. It's necessary if you're to be creative, centered and fully productive" (Kneale, 2009).

14.7 CONCLUSION

Stress by itself is not completely bad. Studies have shown that an optimum level of stress, also referred to as eustress (good stress) helps us to improve our productivity and performance at work and encourages us to overcome challenges. The problem occurs when the thin line between eustress and distress (bad stress) is breached. When work pressures become too high, deadlines too tight and schedules too grueling, when work–life balance becomes a myth and job insecurity begins to cast its dark shadow on our daily lives, then stress begins to soar to unmanageable levels playing havoc with the minds and body of workers culminating in a host of lifestyles diseases and at times, even in death. To control and prevent the spread of what could possibly be one of the biggest lifestyle killers the world has ever seen, it is imperative that employers and employees work together with the required support from their respective governments. Governments have to bring in necessary laws for minimum wages, leave policies, nondiscrimination at work, and also provide employees with a minimum pension, social security, etc. which would undoubtedly be the first step in relieving stress. Employees have to necessarily take a long-term view of things and understand that not every task is critical and not every mistake is fatal. Employers have to encourage employees to take time off from work, ensure that employees have a good work environment, and promote a good work–life balance. Confucius said "Choose a job you love and you will never have to work a day in your life," but in today's fast paced world, choosing a job that one loves is seemingly not enough. Companies will have to go the extra mile to create a work culture and environment that employees love and if they do, the partnership between employee and employer would undoubtedly be one which would yield rich dividends for both.

KEYWORDS

- **stress**
- **work–life balance**
- **employment**
- **workload**
- **WHO**
- **vacation**
- **karoshi**
- **job control**

REFERENCES

Adgey, B. *Richard Branson on Stress*, 22 April 2008. Retrieved from http://www.tycoon-system.com/qa/question-richard-branson-stress (accessed on 8 September 2015).

American Psychological Association. *Employee Distrust is Pervasive in US Workforce*, American Psychological Association, 23 April 2014. Retrieved from http://www.apa.org/news/press/releases/2014/04/employee-distrust.aspx (accessed on 13 September 2015).

American Psychological Association. *Paying with Our Health, Stress in America*, American Psychological Association, 4 February 2015. Retrieved from http://www.apa.org/news/press/releases/stress/2014/stress-report.pdf.

Barber, A. *United States Still a No-Vacation Nation*, 24 May 2013. Retrieved from http://www.cepr.net/press-center/press-releases/united-states-still-a-no-vacation-nation (accessed on September 5, 2015).

Bolt, H. D. *I Want to Do Wild Things*, 12 December 2011. Retrieved from http://espn.go.com/olympics/story/_/id/7294360/olympics-usain-bolt-being-fastest-man-world-espn-magazine (accessed on 10 September 2015).

Business Leaders' Stress Levels. *Business Leaders' Stress Levels Increased by 50 Per Cent in 2005*, February 2006. Retrieved from http://www.gtrus.com/main.php?year=2006&c hapter=press&page=pr/2006/pr_0602 (accessed on 8 September 2015).

CFO Innovation Asia Staff. *Work is Top Trigger of Stress for Asia Pacific Workers*, 2 October 2012. Retrieved http://www.cfoinnovation.com/story/5609/work-top-trigger-stress-asia-pacific-workers (accessed on 12 September 2015).

Cohan, W.D. *Deaths Draw Attention to Wall Street's Gruelling Pace*, 3 October 2015. Retrieved http://www.nytimes.com/2015/10/04/business/dealbook/tragedies-draw-attention-to-wall-streets-grueling-pace.html?_r=0 (accessed on 6 October 2015).

Crippen, A. *Bill Gates' Best Advice from Warren Buffett: Keep Things Simple*, 22 June 2009. Retrieved from http://www.cnbc.com/id/31487207 (accessed on 7 September 2015).

Crompton, S. *What's Stressing the Stressed?*, 13 October 2011. http://www.statcan.gc.ca/pub/11-008-x/2011002/article/11562-eng.htm.

Dill, K. *The Top Companies for Culture and Values*, 22 August 2014. Retrieved from http://www.forbes.com/sites/kathryndill/2014/08/22/the-top-companies-for-culture-and-values/ (accessed on 4 September 2015).

EU-OSHA. *European Opinion Poll on Occupational Safety and Health*. EU-OSHA: Luxembourg, May 2013. https://osha.europa.eu/en/safety-health-in-figures/eu-poll-press-kit-2013.pdf.

Eurofound. *Fifth European Working Conditions Survey*. Publications Office of the European Union: Luxembourg, 2012. http://www.eurofound.europa.eu/publications/report/2012/working-conditions/fifth-european-working-conditions-survey-overview-report.

Evening Standard. Lawyer Who Fell to Death from Tate Modern was Driven Mad by Cannabis, 15 May 2007. Retrieved from http://www.standard.co.uk/news/lawyer-who-fell-to-his-death-from-tate-modern-was-driven-mad-by-cannabis-6582847.html (accessed on 2 September 2015).

Expedia. *Expedia's 2014 Vacation Deprivation Study*. Retrieved http://inside.expedia.co.uk/node/338 (accessed on 13 September 2015), retrieved http://reviews.greatplace-towork.com/google-inc (accessed on 7 September 2015); http://stats.oecd.org/index.aspx?DataSetCode=TABLE_I7.

Frauenheim, E. *Today's Workforce—Pressed and Stressed*, 16 December 2011. Retrieved from http://www.workforce.com/articles/today-s-workforce-pressed-and-stressed (accessed on 8 September 2015).

Gallo, C. *Southwest Airlines Motivates Its Employees with a Purpose Bigger than a Pay check*, 21 January 2014. Retrieved http://www.forbes.com/sites/carminegallo/2014/01/21/southwest-airlines-motivates-its-employees-with-a-purpose-bigger-than-a-paycheck/ (accessed on 6 September 2015).

Gupta, S. *A Son Never Dies*, 2015. Retrieved from http://www.wallstreetoasis.com/forums/a-son-never-dies-by-sunil-gupta-moving-letter-from-a-father-of-an-ibanking-analyst (12 September 2015).

Helliwell, J. F.; Layard, R.; Sachs, J. *World Happiness Report 2015*, 2015 http://worldhappiness.report/wp-content/uploads/sites/2/2015/04/WHR15.pdf.

ILO. *ILO Database of Conditions of Work and Employment Laws on Maternity Protection and ILO NATLEX*, 2009a. http://www.ilo.org/wcmsp5/groups/public/---ed_protect/---protrav/---travail/documents/presentation/wcms_146268.pdf.

ILO. *ILO Database of Work and Employment Laws, Maternity Protection*, International Labour Office, 2009b. http://www.ilo.org/wcmsp5/groups/public/---ed_protect/---protrav/---travail/documents/image/wcms_146196.pdf.

ILO. Maternity and Paternity at Work. In: *ILO Report*. ILO: Geneva, 2014. http://www.ilo.org/wcmsp5/groups/public/---dgreports/---dcomm/documents/publication/wcms_242617.pdf.

Jobs for Life. *Japanese Employees are Working Themselves to Death*, 19 December 2007. Retrieved http://www.economist.com/node/10329261 (accessed on 6 September 2015).

Kneale, K. *Stress Management for the CEO*, 17 April 2009. Retrieved from http://www.forbes.com/2009/04/16/ceo-network-management-leadership-stress.html (accessed on 2 September 2015).

Knowledge@Wharton. *Reluctant Vacationers: Why Americans Work More, Relax Less, than Europeans*, 26 July 2006. Retrieved http://knowledge.wharton.upenn.edu/article/reluctant-vacationers-why-americans-work-more-relax-less-than-europeans/ (accessed on 9 September 2015).

La Roche, J. *Jamie Dimon on the Single Biggest Challenge a Wall Streeter Will Ever Face in Their Career*, 2015. Available at: http://www.businessinsider.in/Jamie-Dimon-on-the-single-biggest-challenge-a-Wall-Streeter-will-ever-face-in-their-career/articleshow/48025074.cms (accessed on 10 September 2015).

Lele, S. *I Will Cherish this Compliment Forever: Sachin Tendulkar*, 25 February 2010. Retrieved from http://completewellbeing.com/article/i-will-cherish-this-compliment-forever-sachin-full/ (accessed on 8 September 2015).

Lynch, S. *Why Your Workplace Might Be Killing You*, 23 February 2015. Retrieved from https://www.gsb.stanford.edu/insights/why-your-workplace-might-be-killing-you (accessed on 8 September 2015).

Makovsky, K. *Behind the Southwest Airlines Culture*, 2013. Available at: http://www.forbes.com/sites/kenmakovsky/2013/11/21/behind-the-southwest-airlines-culture/#335a954f3fb9 (accessed on 10 September 2015).

McRae, D. *Usain Bolt Warns the World's Sprinters that the Best is Yet to Come*, 30 March 2010. Retrieved http://www.theguardian.com/sport/2010/mar/30/usain-bolt-warns-world-sprinters (accessed on 12 September 2015).

Moore, M. J. *Young Banker Struggled with Quitting Goldman before Death)*, 2 June 2015. Retrieved from http://www.bloomberg.com/news/articles/2015-06-02/young-banker-struggled-with-quitting-goldman-weeks-before-death (accessed on 2 September 2015).

Milczarek, M.; Schneider, E.; González, E. R. *European Risk Observatory Report*, 2009. Available at https://osha.europa.eu/en/tools-and-publications/publications/reports/TE-81-08-478-EN-C_OSH_in_figures_stress_at_work (accessed on 16 September, 2015).

Noyan, B. *Check Out 2016's Best Companies to Work For*, 2016. Available at http://fortune.com/best-companies/ (accessed on 17 October 2016).

OSH. *Stress at Work Facts and Figures*. European agency for Safety and Health at Work: Luxembourg, 2009.

Oster, S. *In China, 1,600 People Die Every Day from Working Too Hard*, 3 July 2014. Retrieved from http://www.bloomberg.com/bw/articles/2014-07-03/in-china-white-collar-workers-are-dying-from-overwork (accessed on 1 September 2015).

Otani, A. *The Ten US Companies with the Best Paternity Leave Benefits*, 30 April 2015. Retrieved from http://www.bloomberg.com/news/articles/2015-04-30/the-10-u-s-companies-with-the-best-paternity-leave-benefits (accessed on 6 September 2015).

Price, S. ISBN: 1626366799, 2009. Available at: https://books.google.co.in/books?isbn=1626366799 (accessed on 5 September 2015).

Regus. *Stress Out? A Study of Trends in Workplace Stress across the Globe*, November 2009. http://www.regus.fr/images/Stress%20full%20report_FINAL_Designed_tcm308-21560.pdf.

Staying@Work™. *Staying@Work™ Survey Report 2013/2014, United States*, June 2014. https://www.towerswatson.com/en/Insights/IC-Types/Survey-Research-Results/2013/12/stayingatwork-survey-report-2013-2014-us.

Stephenson, W. *Who Works the Longest Hours?*, 23 May 2012. Retrieved from http://www.bbc.com/news/magazine-18144319 (accessed on 8 September 2015).

The American Institute of Stress. Retrieved from http://www.stress.org/what-is-stress/ (accessed on 6 October 2015).

The Guardian. *Bank Intern Moritz Erhardt Died from Epileptic Seizure, Inquest Told* (22 November 2013). Retrieved from http://www.theguardian.com/business/2013/nov/22/moritz-erhardt-merrill-lynch-intern-dead-inquest (accessed on 10 September 2015).

The Huffington Post. *Obama Discusses Managing Stress*, The Huffington Post. Retrieved https://www.youtube.com/watch?v=L4OuFaHGB9g (accessed on 8 September 2015).

The National Institute for Occupational Safety and Health. *Publication Number 99-101*, 1999. Retrieved from http://www.cdc.gov/niosh/docs/99-101/ (accessed on 6 October 2015).

TheNewYorkTimes.*ReflectionsonStressandLongHoursonWallStreet*,1June2015.Retrieved http://www.nytimes.com/2015/06/02/business/dealbook/reflections-on-stress-and-long-hours-on-wall-street.html?_r=0 (accessed on 8 September 2015).

Willis Towers Watson. *US Employers Rank Stress as Top Workforce Risk Issue*, 13 November 2013. Retrieved from https://www.towerswatson.com/en/Press/2013/11/us-employers-rank-stress-as-top-workforce-risk-issue (accessed on 5 September 2015).

Xiaomin, X. Chinese Workers Feel More Stress, 20 April 2007. Retrieved from http://www.chinadaily.com.cn/china/2007-04/20/content_854776.htm (accessed on 7 September 2015).

YouTube. *How Do You Deal with Stress Jeff Bezos*. Retrieved from https://www.youtube.com/watch?v=NqVoOC2azZI (accessed on 9 September 2015).

YouTube. *YouTube CEO: How I balance a big family, work*, 2014. Available at http://www.today.com/video/today/56558626 (accessed on 17 October 2016).

INDEX

For Product Safety Concerns and Information please contact our EU
representative GPSR@taylorandfrancis.com
Taylor & Francis Verlag GmbH, Kaufingerstraße 24, 80331 München, Germany

www.ingramcontent.com/pod-product-compliance
Ingram Content Group UK Ltd.
Pitfield, Milton Keynes, MK11 3LW, UK
UKHW021605240425
457818UK00018B/403